Studies and reports in hydrology / Études et rapports d'hydrologie

Recent titles in this series / Titres récents dans cette collection

For details of the complete series please see the list printed at the end of this work

La liste complète des titres de cette collection figure à la fin de cet ouvrage

Mathematical models in hydrology

Proceedings of the Warsaw Symposium
July 1971

Les modèles mathématiques en hydrologie

Actes du colloque de Varsovie
juillet 1971

A contribution to the International Hydrological Decade
Une contribution à la Décennie hydrologique internationale

Volume 3

IAHS–Unesco–WMO
AISH–Unesco–OMM   1974

Published jointly in 1974 by
the International Association of Hydrological Sciences (President: J.-A. Rodier),
19, rue Eugène-Carrière, 75018 Paris,
the United Nations Educational, Scientific and Cultural Organization,
7, Place de Fontenoy, 75700 Paris
and the World Meteorological Organization, 41 Avenue Giuseppe Motta, Geneva
Printed by Adlard & Son Ltd, Bartholomew Press, Dorking, Surrey, England

Publié en 1974 conjointement par
l'Association internationale des sciences hydrologiques (président: J.-A. Rodier),
19, rue Eugène-Carrière, 75018 Paris,
l'Organisation des Nations Unies pour l'éducation, la science et la culture,
7, place de Fontenoy, 75700 Paris
et l'Organisation météorologique mondiale, 41 Avenue Giuseppe Motta, Genève
Imprimerie Adlard & Son Ltd, Bartholomew Press, Dorking, Surrey, England

ISBN 92-3-001031-6 (Unesco)

Table of contents
Table des matières

Volume 3

V

Contents

Opening meeting – Séance d'ouverture

26 July 1971 – 26 juillet 1971

Introductory Speech by Professor Dr. Zdzisław Kaczmarek, Chairman of the Organizing Committee

Ladies and Gentlemen,

I have the honour to welcome among us the Chairman of the All-Poland Committee of the National Unity Front, President of the Polish Academy of Sciences, Professor Janusz Groszkowski. We are pleased that this Symposium has also been honoured by the presence of the Chairman of the Central Office for Water Economy, Minister Zbigniew Januszko, and representatives of the Municipal Authorities. I welcome cordially the President, Professor James Dooge, and officers of the Committee on Mathematical Models of the International Association of Scientific Hydrology*, the representative of UNESCO, Mr. Frederik H. Verhoog, and of the World Meteorological Organization, Dr A. J. Askew. As Chairman of the Organizing Committee, I would like to express my sincere joy at the fact that so many of our colleagues and friends are taking part in this Symposium. According to preliminary estimates, more than 250 delegates from 34 countries are attending. I welcome you cordially to Warsaw, a city always friendly to visitors.

The Symposium is organized by the IASH* Committee on Mathematical Models in Hydrology. I would like to express my appreciation to Professor James Dooge and the Officers of the Committee for their assistance to the Polish Organizing Committee. We are also very grateful for the support and assistance of UNESCO and of the World Meteological Organization.

Although it is the first Symposium organized officially by the Committee on Mathematical Models, it is in fact a continuation of similar discussions held at Fort Collins in 1967 and at Tucson in 1968.

I hope that it will be a very valuable contribution to the development of new methods and techniques in hydrological investigations. Let me express the opinion that the papers and general reports which will be presented at the Symposium are an excellent basis for our discussions. On behalf of the Organizing Committee, I would like to thank the authors of the papers and the general reporters for their very valuable contribution.

Introductory Speech by Professor Dr. Janusz Groszkowski, Chairman of the All-Poland Committee of the National Unity Front and President of the Polish Academy of Sciences

Mr. Chairman, Ladies and Gentlemen,

It is a great pleasure for me to be able to welcome you on the occasion of the Symposium of the International Association of Scientific Hydrology devoted to the application of

---

* The name of the Association was changed from the International Association of Scientific Hydrology (IASH) to the International Association of Hydrological Sciences (IAHS) at the General Assembly held in Moscow, August 1971.

mathematical models in hydrology and water economy. I need not emphasize how much our country is honoured by the presence of so many eminent scientists from all over the world, and I should like to take this opportunity to extend to the Symposium my most sincere and best wishes for fruitful and successful debates. The field of science and technology you represent raises many problems that can be solved only by the joint and unified efforts of numerous scientists, and requires from all of you, a very intense concentration. The mathematical modelling of hydrological processes and phenomena is an extremely useful instrument for analysing and solving many important problems, and has not only cognitive but also a practical significance. We attached a great interest to this Symposium, and I hope that the results of your debates will make an effective contribution to the improvement of our knowledge about hydrospheric processes, as well as about water as an important element in the human environment. In concluding this speech, I should like, on behalf of the Polish Academy of Sciences, to convey my best wishes for a successful conference and to express my hope that your stay in Poland will be useful as well as agreeable, and will contribute to the better tightening of the bonds of friendship among the scientists from different countries.

Introductory Speech by Mr. Zbigniew Januszko, Chairman of the Central Office for Water Economy

Mr. Chairman, Ladies and Gentlemen,

It was with great pleasure that we accepted the proposal to play host to this Symposium of the International Association of Scientific Hydrology. A high level of technology and the development of a number of disciplines of science has been possible only because of the utilization of modern methods, the application of mathematical models and computing techniques. The development of high-speed and reliable electronic computers, initiated in the 1940s and continued at a rapid rate, has expanded more and more the scope of application of the theory of probability, systems analysis, and other mathematical methods in hydrology and water economy. Many problems in this domain have either been already solved by scientists, or will be solved before long. Nevertheless, there still exists a notable array of problems awaiting solution. Governments and economic circles in almost every country are interested in the solution of these problems. The Polish Government also attaches great importance to progress in this field, and has spared no effort and means to further this aim. I am confident that this Symposium, like the previous Symposia held in Leningrad, Fort Collins and Tucson, will mark another step forward in the field of application of mathematical models in water economy in hydrology. We hope also that this Symposium will contribute to better mutual understanding among representatives of the world of science, and to the establishment of friendly contacts. On behalf of myself and the Central Office for Water Economy, I welcome the representatives of Science from many countries gathered here. I wish the participants in this Symposium fruitful debates.

Introductory Speech by Mr. Frederik H. Verhoog, Representative of UNESCO

Mr. Chairman, Ladies and Gentlemen,

UNESCO is glad that this Symposium on mathematical modelling is starting under such perfect conditions. UNESCO sees mathematical modelling as one of the major tools for advancing the science of hydrology. The United Nations system has until now supported the development of hydrology all over the world, especially the establishment of hydrological metworks. Many countries, including the developing ones, now have good hydrological networks. The next step is to give them the possibility of analysing the data which they are acquiring in great quantities. Electronic computers are the only suitable tools available to handle these data, but only, of course, if relaible and suitable computer programmes exist.

Mathematical models are not only excellent management tools, but also point out new problems which, when solved, will make hydrology more scientific. These mathematical models will, furthermore, help to get more effective and cheaper hydrological information systems and engineering works.

I would like to draw your attention to the fact that hydrological mathematical modelling is mostly done by scientists from the developed countries. However, it is the developing parts of the world that are in most need of mathematical models.

The world is looking forward to the moment when mathematical models can be used by everyone and not only by the researcher who has developed them.

Ladies and Gentlemen, I wish you a very fruitful week, thank you.

Introductory Speech by Dr. A. J. Askew, Representative of the World Meteorological Organization

Mr. President, Mr. Chairman, Professor Kaczmarek, Ladies and Gentlemen, Colleagues,

It is a great honour and a pleasure for me to convey, on behalf of Mr. Davies, the Secretary General of the World Meteorological Organization, the most cordial greetings from our Organization to the Organizing Committee and to all the participants at the Symposium.

The organization of such a Symposium is no easy task and the work that has been done by our Polish colleagues in preparing this meeting is to be highly commended. It has been a pleasure for our Organization to provide assistance to the Organizing Committee and to act as co-sponsors of the Symposium. In this regard, I am particularly glad to find that we are once more working together with our colleagues from UNESCO and IASH. This is not the first Symposium that has been so sponsored, and I am confident that it will not be the last. As many of you are no doubt aware, WMO has responsibility within the United Nations family for work in the field of operational hydrology. The recent Sixth World Meteorological Congress, which was held in Geneva this April, took steps to define more clearly and to strengthen WMO's role in hydrology. It appeared a definition of the term "Operational Hydrology" which included, amongst other factors, the development and improvement of relevant methods, procedures and techniques in hydrological forecasting and in the supply of meteorological and hydrological data for design purposes.

WMO is concerned with the question of hydrological forecasting, where mathematical models are becoming increasingly used as an integral part of the forecasting system. In recognition of this, the Organization has initiated a project for an internationally coordinated intercomparison of the use of conceptual models for compiling operational hydrological forecasts. The plan of action has been formulated on the basis of three informal study group meetings, held in Washington D.C., Paris and Tokyo, and it is hoped to start the actual intercomparison phase of the project by the middle of next year.

As can be seen from tbe brief comments I have made concerning the work of WMO, we are primarily concerned with the practical application of mathematical models; this concern stems from the very nature of our responsibilities in hydrology within the international community. However, I consider that such concern should also be shared by everyone working in this field. While I appreciate that many very complex and sophisticated models have been developed mainly as research tools and as a means of studying the land phase of the hydrological cycle, there is an all too frequent tendency to describe as fit for practical application, models which are quite unsuitable because of their complexity, cost of operation and need for copious data.

This is only one of the very important questions which I am sure will be raised and discussed during this Symposium. A glance at the list of participants is sufficient proof that the discussions will be of a very high standard.

Permit me to wish you all, on behalf of the Secretary General of WMO, a very productive and successful meeting.

Last, but not least, I wish to express WMO's deepest gratitude to the Polish National Committee for the IHD, to the Polish National Committee for Geodesy and Geophysics, and to all our Polish colleagues who have helped in the preparatison of this meeting and who have given us such a warm welcome. In particular, I wish to mention the Chairman of the Organizing Committee, Professor Kaczmarek, who has been associated with the work of WMO over many years. We would like to wish him much success in this and in all his other commitments to the advancement of science.

Opening Speech by Professor J. C. I. Dooge, President of the Committee on Mathematical Models of the International Association for Scientific Hydrology

Mr. Chairman, Ladies and Gentlemen,

I speak today not only on behalf of the Committee on Mathematical Models, but on behalf of the International Association of Scientific Hydrology as a whole. I would like to express thanks on behalf of IASH and on behalf of the Committee on Mathematical Models in particular, to the Polish Organizing Committee. They have done a splendid job. It is not only what can be seen here; there is the work which they have been doing over the past two years. I would like to thank them in particular on behalf of the Committee on Mathematical Models for the way they have organized this Symposium so that it fitted into the general programme of that Committee.

I think we have in this Symposium a culmination of the work of the Committee over the past four years, and I hope a good guarantee for the continuation of the work.

It is fitting that we have here today at the opening of this Symposium, the President of the Polish Academy of Sciences representing the scientific side of our subject, and also the Minister in charge of the Central Office for Water Economy reminding us of what is the ultimate objective of all our work. It is fitting that we have here such senior representatives of science and of policy, because these two elements are interconnected in all that we do. As for the Symposium itself, the numbers attending and the number of papers submitted have exceeded our best hopes. The only thing that remains for this Symposium to be an outstanding success is for you, the participants, to make it so by the vigour and incisiveness of your contributions to the discussion.

It gives me great pleasure to declare open this Symposium on Mathematical Models in Hydrology.

Session I   26 July 1971

General Reporter   Dr. Vujica Yevjevich, Professor of Civil Engineering, Colorado State University, Fort Collins, Colorado, USA

Chairman   Dr. Alfred Becker, Institut für Wasserwirtschaft, Berlin, GDR

Structural analysis of hydrological sequences — I   Stochastic hydrology

Vujica Yevjevich

The following papers are reviewed and discussed in this general report. These papers are all published in volume 1 of the Symposium Proceedings.

1. *R. J. G. Bloomer and J. R. Sexton*   Problems encountered in synthetic river flow generation procedures
2. *M. J. Hamlin and N. T. Kottegoda*   Some applications of stochastic flow generation
3. *S. Jovanović, A. R. Dakkak, M. Cabrić and M. Brajković*   Simulation of daily rainfall series using Markov chain models
4. *Henryk Mitosek, Jr.*   Stochastic structure of a process of average monthly flows
5. *Barry M. J. Barton*   The generation of synthetic monthly runoff records for ungauged British catchments.

This group of papers is mainly concerned with the generation of new samples by using the properties of historic samples and experimental methods (experimental statistical, Monte Carlo, data generation, synthetic simulation methods or whichever term is being used by individual hydrologists). Questions for discussion are posed at the end of the review of each paper.

Problems encountered in synthetic river flow generation: by Bloomer and Sexton
This paper presents the authors' work related to the application of the experimental statistical (Monte Carlo) method in generating new samples of monthly and daily river flows for some rivers in England. The time series are analysed structurally and described mathematically by separating the periodic components in some parameters from the stochastic component, by fitting an autoregressive linear model to the dependence of the stochastic component, and by considering the remaining stochastic component of residuals as an approximate second-order independent stationary process.

Seven steps are outlined for the procedures of which five relate to the analysis, testing and description of series structure. The other two steps refer to the generated samples. Special attention is given to the analysis of the range and to the properties of the residuals in the form of the approximate independent stochastic components.

Range analysis
After the range is defined, the discussion in the paper turns to the approximation of the expected range, given by Hurst in the form of a simple power function (Hurst, 1951)

$$R_s = K s^H \tag{1}$$

in which R_s is the expected range (the population mean), s the subsample size, K a constant, and H an exponent, asymptotically with a value of $H = 0.50$ and usually greater than 0.50. The work by other authors is also reviewed and the statement by Kottegoda that maximum H is 0.63 for autoregressive linear models is questioned. It is easy to show, as follows, that this statement is not correct.

In the case of an independent standard normal process, the exact expression for the expected range, given by Anis and Lloyd (1953), is

$$ER_s = (2/\pi)^{1/2} \sum_{i=1}^{s} i^{-1/2} \tag{2}$$

which gives $ER_1 = (2/\pi)^{1/2}$ and $ER_2 = 1.71(2/\pi)^{1/2}$ for $s = 1$ and $s = 2$, respectively.

For these values of s and ER_s, equation (1) gives

$$H_{max} = \frac{\log ER_2 - \log ER_1}{\log s_2 - \log s_1} \tag{3}$$

which for $s = 2$ and $s = 1$ gives $H_{max} = 0.775$. Even for an independent standard gaussian process, the greatest slope of the expected range function in a log-log coordinate system and for $s = 1$ and $s = 2$ exceeds the claim of the maximum value of H being 0.63 for the linear autoregressive models.

For the expected range of the first-order linear autoregressive models see Yevjevich (1967):

$$ER_s = (2/\pi)^{1/2}(1 - \rho^2)^{-1/2} \sum_{i=1}^{s} i^{-1/2} \left[\frac{1+\rho}{1-\rho} - \frac{2(1-\rho^i)}{i(1-\rho)^2} \right]^{1/2} \tag{4}$$

For $s = 1$ and $s = 2$, equation (4) gives

$$ER_1 = \left[\frac{2}{\pi(1-\rho^2)} \right]^{1/2} \tag{5}$$

and

$$ER_2 = \left[\frac{2}{\pi(1-\rho^2)} \right]^{1/2} \left[1 + \left(\frac{1+\rho}{2} \right)^{1/2} \right] \tag{6}$$

The slope through the points $(ER_1, s = 1)$ and $(ER_2, s = 2)$ is given by equation (3) in the following form:

$$H_{max} = \frac{\log \left[1 + \left(\frac{1+\rho}{2} \right)^{1/2} \right]}{\log 2} \tag{7}$$

For $\rho \to 1$, $H_{max} \to 1.00$, as a limit of the maximum slope for the first-order linear autoregressive model and the two smallest values of s (1 and 2).

As already shown by the above brief analysis, the slope H in equation (1) can be as close to $H = 1$ as desired, provided that the first serial correlation coefficient in the first-order autoregressive model approaches the value of 1 and the subsample size is very small.

The authors of this paper show that the sampling variation of the generated independent standard normal random numbers may produce a value of H much smaller than 0.50 and the example cited had $H = 0.44$, though the asymptotic value is $H = 0.50$, as an expected minimum average value. This example shows a very large sampling variation of the exponent H. The authors also confirm the above theoretical analysis that H can be large (between $H = 0.50$ and $H = 1.00$) for the autoregressive linear models. The position taken by the authors, that the feasibility of using H as a model parameter in hydrology should be questioned, is correct.

Some difficulties arise in assessing the largest values of H for various types of linear models and various sample sizes s, because the slope H in log–log coordinates is simply a fit of a straight line through points $(\log ER_1, \log s = 0)$, $(\log ER_2, \log s = \log 2)$, ..., $(\log ER_s, \log s)$, while the functions $ER_s = f(s)$ are not of the type of equation (1), and are not straight lines in log–log coordinates. Fitting a straight line to a population curve which is not a straight line in a given set of transformed coordinates is, at least, a dubious proposition except for preliminary investigations like those carried out by Hurst (1951).

Fitting probability distribution functions
The fitting of probability distribution functions to residuals or independent stochastic components of historic series concerns these authors as it does everybody who has studied the application of the Monte Carlo method in hydrology. As the residuals can also have negative values, the authors used the bounded three-parameter distribution functions. Particularly, the three-parameter lognormal distribution is studied with parameters estimated either by the method of quantiles or by the method of moments. Although the authors are correct that the use of the maximum likelihood method to estimate the lower boundary of this function requires an iterative procedure, this procedure is very simple to implement on a digital computer. It should be noted that the lower boundary of the residuals may often be computed by using the observed values of zeros or the recession curve of river flows with values close to zero.

Preservation of sample properties
The authors raise an interesting problem, namely that the preservation of daily flow statistics in generating the new samples might not be sufficient to preserve the comparable monthly flow statistics, except in the case of the mean. A test performed in that regard with days summed up to make monthly average flows showed the monthly means and variances to be comparable with the historical monthly means and variances. However, the monthly flows produced from generated daily flows showed on average a much smaller autocorrelation than the historical sample. The authors' explanation is correct, namely that the autoregressive models of lower order do not necessarily preserve the higher lag autocorrelation coefficients and, therefore, cannot reproduce exactly the autocorrelation of monthly, seasonal or annual flows.

Similarly, as for the autocorrelation, the generated samples did not reproduce well the lowest flows of the non-overlapping intervals of the historic sample, with droughts of generated samples being, on average, much greater than the historic drought. However, by generating samples of monthly flows directly, the droughts of generated samples and those of the historic sample were close.

The unsatisfactory performance of monthly flows as obtained from generated daily flows was assigned to the problem of the independence of residuals, with the test of independence not being acceptable by the χ^2-test. This leads to the conclusion that the problem of the independence of residuals needs much greater attention.

Questions for discussion
The following questions may be posed for discussion:

(1) How good is the range analysis by the power function, used by Hurst as an approximation, for the study of hydrological stochastic processes?

(2) Why and how did it come about that some investigations of stochastic hydrological processes claimed that the expected range of autoregressive linear models was not to be able to preserve the functions of observed mean ranges, when the opposite is true?

(3) What probability distribution functions are best suited for fitting the independent second-order stationary stochastic components (residuals) of hydrological daily series in general, and daily flows in particular?

(4) What method should be best followed in order that the generated samples of small time interval values (say daily values) may preserve also the general properties of historic samples of large time interval values (say 15-day, monthly, seasonal and annual values)?

(5) What conditions should a method for generating new samples fulfil in order that droughts of generated samples belong, in statistical sampling terms, to the populations inferred from the available historic samples?

Some applications of stochastic flow generation: by Hamlin and Kottegoda
This paper refers to the generation of samples of monthly and 5-day flows for some rivers in England. The generation of these samples is based on the concept that seasonal variations are in the mean and the standard deviation, while no definite seasonal patterns could be detected in the skewness coefficient.

Testing the mathematical dependence models
Four dependence models for the stationary stochastic components were investigated on the flow series. They are the linear autoregressive model (given here in the notation adapted for all four models)

$$y_i = \sum_{j=1}^{m} a_j y_{i-j} + x_i \tag{8}$$

in which y_i is the dependent variable, x_i the independent variable, a_j the autoregressive coefficients, and m the order of the linear model; the moving average schemes of the type

$$y_i = \sum_{j=1}^{m} b_j x_{i-j} + x_i \tag{9}$$

in which b_j are the coefficients of the moving average scheme; the cross monthly correlation type of regression of the form

$$y_i = B_j y_{i-1} + (1 - C_j)^{1/2} x_{i-1} \tag{10}$$

and the fractional noise model given in the paper in the form

$$y_i = (H - 0.5) \sum_{j=1}^{M} j^{H-1.5} x_{i+j-1} \tag{11}$$

in which H is the Hurst slope of the expected range in the log–log coordinate system, and M is the memory or the length of the dependence model.

The probability distribution functions used for the independent stochastic components in the paper were gamma distribution and beta II distribution, as well as the three-parameter lognormal distribution. The Smirnov–Kolmogorov tests were used for accepting or rejecting a given fit to the frequency distributions of residuals of monthly river flows. The split-sample technique was applied, with one half of the historic sample being used for generating the new samples, and the other half being used for testing the generated samples. Various parameters of the generated samples and the other half of the historic sample were compared, by the selected types of tests.

In summary, the reproductions of properties of historic samples was best by the autoregressive linear model of equation (8) (and in these cases, by the first-order linear model), the second best model being the moving average scheme, the third best by the cross monthly correlation type of regression of equation (10), and the least satisfactory was the application of the fractional noise model of equation (11). These results should have been expected, because some of the three latter models have no physical background in the processes of water storage, evaporation and outflow of a river basin.

The use of fractional noise models needs some more discussion. Being a type of linear moving average the equation for the expected values of the range of all linearly dependent models is known to be (Yevjevich, 1967)

$$ER_s = (2/\pi)^{1/2} \left(\sum_{j=0}^{m} b_j^2 \right)^{1/2} \sum_{i=1}^{s} i^{-1/2} \left[1 + \frac{2}{i} \sum_{k=1}^{i-1} (i-k) \frac{\sum_{j=0}^{m-k} b_j b_{j+k}}{\sum_{j=0}^{m} b_j^2} \right]^{1/2} \tag{12}$$

Equation (12) can then be applied to equation (11) by substituting m for M and the b_j coefficients are now

$$b_j = (H - 0.5)j^{H-1.5} \tag{13}$$

It is easy to show that the resulting equation for the expected range of equation (12) with the b_j coefficients of equation (13) is not a power function of the type of equation (1), and that it is not so simple to reproduce the H slope by the moving average scheme called the fractional noise. One further wonders why should the generation of samples by a computer be used in order to obtain the approximate values of the expected range while there is an expression for it? As there is an infinite number of sets of monotonically decreasing coefficients of moving average schemes, one can develop as many new models as one wishes. What is really at stake is the physical background of the hydrological processes which supports the mathematical modelling as well as the statistical inferences performed on thousands of time series to prove that a model is applicable. The authors have thus contributed something to this problem of selecting the realistic dependence models for the stationary hydrological stochastic processes.

Mean and variance of the 5-day flows
The authors state that the means and standard deviations of each time interval in an annual cycle have variances much greater in the case of the 5-day series than for the monthly series. Because the 5-day flows have to be combined as sums of six values

and averaged in order to obtain the average monthly flows, the above statement by the authors must be correct. It can be proven by invoking the central limit theorem.

Negative values of fitted periodic functions
By fitting a set of harmonics, with amplitudes significantly different from the amplitudes in the case of non-periodic series, to the periodic parameters of the mean and the standard deviation, it was shown by the authors that a portion of these fitted functions may be negative. It is also the experience of others that the limited number of harmonics may result in negative values. The authors have designed a particular procedure to avoid this difficulty. However, as in the case of truncated probability distributions, the periodic components may be conceived also as a truncated function at a boundary greater than zero. The author's method of computing or postulating a value c, as the realistic minimum value of periodic components, requires a systematic analysis and tests on a large number of cases.

Significant harmonics of periodic components
The authors used only the main 12-month harmonic in describing the periodicities in parameters, and considered the contributions by other harmonics small because of the small explained variance by each of them. It is, therefore, of practical importance to develop and use the objective methods of inference of significant harmonics.

Autocorrelation of stochastic components
The authors did not find the autocorrelation coefficients of stochastic components of the river flows investigated to be dependent on seasons, or to be peridoic. This is to be expected. The autocorrelation is mainly a function of the river basin response to a given water input, and not so much of the amount and the number of inputs, provided the river basin always responds in the same way. Therefore, if the rainfall contribution to runoff predominates in a river basin in comparison with the small snowmelt contribution to runoff, the autocorrelation of the runoff stochastic component of that river basin should not be periodic.

Questions for discussion
The following questions may be posed for discussion.

(1) What experience is obtained elsewhere on a large number of time series in fitting various linear models of dependence to the hydrological stochastic time series, or to the stochastic components of the periodic-stochastic hydrological time series?

(2) What is the real value of the moving average scheme in the form of fractional noises in hydrology?

(3) What experience is available elsewhere in treating the periodic components when a part of the periodic function exhibits negative values?

(4) What is the best method in selecting significant harmonics in the mathematical description of periodic parameters by Fourier analysis?

(5) Are there any physical reasons for the periodic autocorrelation coefficients of the stochastic component of periodic-stochastic hydrological time series in a pure rainfall producing runoff regime, and how would the nonlinear river basin response hydrographs affect the autocorrelation?

Simulation of daily rainfall series using Markov chain models: by Jovanović et al.
The authors present a method of generating samples of daily rainfall data simultaneously for a group of precipitation stations by using the Markov chain approach in two steps. The first step consists of generating the wet and dry spells of the precipitation process, while the second step generates the daily precipitation amounts inside the already generated wet spells.

Generating wet and dry spells
The time lengths of uninterrupted wet and dry spells are distributed as two geometric
distributions. The non-rainy day is defined for precipitation less than 0.5 mm. In this
case, the Markov chain has only two states with four transition probabilities, estimated
by the sample relative frequencies, thus serving as four parameters of the two geometric distributions. The distribution of the length of rainy spells, k, is then

$$f(k) = p(1 - p)^{k-1} \tag{14}$$

in which p is the transition probability of a rainy day occurring provided that it rained
the day before, and $1 - p$ is the probability of a non-rainy day occurring provided that
it rained the day before. A similar equation is given for the distribution of the length
of the non-rainy days [however, the paper gives the opposite definitions of the transition probability in equation (4) and in Table 1 (vol.1, p.111 and p.112, respectively)].

As it is known that the lengths of spells represented as continuous variables are
gamma distributed, the geometric distribution is a good approximation in this case.
The tests show that equation (14) fits well the observed frequency distributions.

Generating the daily rainfall
The daily precipitation as a continuous variable is divided into seven discrete states in
order to apply the Markov chain concept. Seven probability distributions of conditional discrete variables are developed, as well as an eighth distribution for the transition from the last non-rainy day to the first rainy day, by fitting the two-parameter
gamma distribution functions to the empirical frequency curves. The shape and scale
parameters were estimated by the method of moments. Because shape parameters
were, on average, smaller than ten, it is well known that the maximum likelihood estimation method would give much more efficient estimates than the method of moments.
However, this does not affect the correctness of the method but can make it more
accurate.

It was found that the shape parameter changes both with the state (called the interval in the paper) of the Markov chain, as well from station to station and, in general,
it was increasing with an increase in the rainfall value of the state. The data of the transition frequencies were not accurate enough to draw definite conclusions on both how
the shape parameters change from state to state or from station to station.

It was also found that the scale parameter was almost independent of the state of
the chain (or of the precipitation amount), and was used as a constant. It also looks
as if this parameter changes little from station to station, and may be eventually
inferred to be a regional constant. However, this test was not performed in this paper.

Uniform independent numbers are generated on a computer, and the two geometric
probability distributions of wet and dry spells give an alternating sequence of these
spells. Similarly, for each wet spell, the generated sequence of uniform independent
random numbers and the use of eight conditional probability distributions produce the
set of states of rainfall data inside each of the wet spells. Thus, only seven values of
rainfall corresponding to seven selected states, are produced for each station. This
point requires a more detailed discussion.

In general, when variables and series are discrete, it is appropriate to use the Markov
chain method. When both the variable and the series are continuous, the use of Markov
processes should be by proper selection. However, when the variable is continuous
and the series is discrete, the use of Markov or autoregressive models seems more appropriate. Therefore, the use of Markov chains for the two states of wet and dry spells
can be well justified. However, the replacement of a continuous variable by only seven
states has two problems: (a) the entire range of precipitation values is not well covered,
and (b) there is a need to estimate 14 parameters for seven states. It may be that the

use of Markov models may require much fewer parameters to estimate.

For a given day inside a wet spell of a 122-day flood occurrence period, the same uniform number is used for all 22 precipitation stations. This represents a very high correlation between the rainfall data of all stations. The correlation should practically be unity and only the linear correlation imposed on the data would decrease the correlation somewhat. The problem of preserving the daily rainfall correlation between stations represents a problem that is not solved by this method of generating the daily rainfall series.

Questions for discussion

The following questions may be posed for discussion.

(1) What is the experience of other investigators in the application of geometric probability distributions to the time series of wet and dry uninterrupted spells, and whether the use of the two-parameter gamma distribution would be more appropriate?

(2) In fitting the frequency distributions of conditional discrete variables in simple Markov chains by the two-parameter gamma distribution, why, from the physical point of view, should the scale parameter be so stable a parameter both with the change of the state (rainfall amount) and from station to station? Similarly, why should the shape parameter increase with an increase of the precipitation values, except that the tail conditions impose somewhat different distributions and parameters (distribution of exceedences)?

(3) Is it better and more reliable to use Markov chains or the Markov models in producing the generated samples of continuous variables with the discrete series?

(4) What difficulties and what loss of information and accuracy are produced when a continuous random variable is approximated by a limited number of states?

(5) How the increase in the number of states in approximating a continuous variable by a discrete variable, with the intent to use the theory of Markov chains, increases the number of parameters to be estimated and thus decreases their reliability?

(6) What method should be used for simulating the daily rainfall data at several stations in order not to have a much higher correlation between the generated series or pairs of stations than the correlation between the observed series?

Stochastic structure of a process of average monthly flows: by Mitosek, Jr.

It is difficult for me to agree with some statements of a general nature made by the author. However, these statements are not relevant for the presentation of autocorrelation analysis made for the monthly flows by the author.

This paper is a typical example of forcing statistical techniques to the reality of hydrological processes. In order to apply the well-known theory of Markov chains to the monthly river flows, it was necessary to use the autocorrelation matrix of N gaussian variables. To do that, the monthly river flows are not only standardized by removing the periodicity in the mean and in the variance, but they are also normalized by a logarithmic transformation. After that, the classical multiple regression (correlation and regression) analysis is applied, and the standard Fisher tests on the z-variable are performed. The author comes to the conclusion that such transformed monthly random variables are highly autocorrelated, and for Polish rivers they follow simple Markov chains, or first-order autoregressive linear models. It is not surprising that all monthly flows studied did not show the 'perfect' fit of this model. The experience of many studies shows that the more complex Markov chains or the more complex linear models (say, the higher-order autoregressive models) may be applicable.

This paper points to a possibility of using the partial autocorrelation coefficients to measure, not only the significance of the first autocorrelation coefficient in the stochastic second-order stationary components of hydrological time series, but also whether the higher-order autocorrelation coefficients may be significant, thus showing the higher-order linear models to be the best fits for hydrological time dependences.

The author in his work has assumed that the removal of periodicity in the mean and the standard deviation has produced the second-order stationary normal process and, in this way, the stationary process in the strongest sense. The tests were not performed to ascertain that the covariances (in this case the autocorrelation coefficients) are those of a second-order stationary process. It is quite likely that the Polish rivers, fed by approximately equal quantities of water from rainfall and snowmelt, would show that the autocorrelation functions are also periodic.

Questions for discussion
The following questions may be posed for discussion.

(1) How often has the first-order autoregressive model, or its approximation in the form of simple Markov chains, approximated the stochastic components of monthly river flows?

(2) What advantages can be obtained in using partial autoregression coefficients in studying the order of the linear dependence in hydrology?

(3) Are the conclusions about the time dependence more reliable by using the Markov chains or by using the autoregressive or Markov models?

The generation of synthetic monthly runoff records for ungauged British catchments: by Barton
This paper contains three parts: (a) the introductory statements with the author's justifications of the approach selected, (b) the deterministic catchment model, as the basic contribution of the paper, and (c) the generation of the stochastic rainfall input and the stochastic evaporation output of a catchment. This general report on topic 1 will discuss briefly some statements given in the introduction, and the generation of rainfall and evaporation data, leaving the major part on the deterministic catchment model to be covered by the general report on topic 4.

Positions on generation of river runoff data
The author takes the position that the practical hydrological techniques in generating the new samples of river flows directly from the observed river flow series for the purpose of solving the long-term yield problems have been largely unsuccessful for the three following reasons: (1) too many parameters are necessary and must be estimated, thus making the regional correlation with catchment parameters unsatisfactory, (2) there is a large sampling variability of runoff parameters because they are not only reflecting the physical nature of the catchment but also the prevailing regime, and (3) there is very often a relative shortage of runoff records. Furthermore, the author has a philosophical objection to the generation procedure of runoff samples from the runoff data because of the fact that each river basin is unique, so that the use of regionalized catchment parameters involves a negation of this uniqueness. These statements deserve some brief comments.

The author had in mind a particular method of generating the monthly river flows from the observed runoff data, which requires a large number of parameters for each of 12 months. The present trend in developing versatile methods for generating river flow sequences at any time unit is in the use of structural analysis of time series and the statistical inferences of models with a minimum of statistical parameters to be estimated.

There is no proof that the sampling variability of river flow parameters is higher than the sampling variability of the corresponding parameters of rainfall or evaporation data. This argument is not necessary in supporting the author's proposal to simulate the rainfall and evaporation data and then to synthesize the runoff data by using catchment response models.

The author is correct that runoff records are often too short for a reliable estimate of the necessary models and parameters as the basis for generating new samples of river flows. This fact will continue to limit the application of the generation of runoff samples from runoff data. One does question the statement that the uniqueness of each river basin precludes the use of runoff data of that river basin in simulating new runoff samples.

The first two papers presented in review form in this general report, and written by four authors from Great Britain, are in variance with this author's contention that the last of the three objections has made the approach of using observed runoff data in generating new samples of river flows untenable.

Generation of stochastic climatic variables

The author processes a method of generating samples of basic climatic variables, namely the rainfall input in a catchment, the potential transpiration as the stochastic output measure of basin evaporation, and the 'snowlag' as a quantified estimation of the effect of snow accumulation and melt. In assessing the various sources of error, which are inevitable in this approach, the author has opened a pandora box of problems. As the paper by Jovanović *et al.* has shown, the generation of rainfall data is not as simple as the author's presentation may imply. There is no discussion whether only the average areal precipitation over a catchment is generated or whether data from several stations are produced. The difference between the real total evaporation over a river basin and the potential transpiration at a point in the river basin may be relatively high, involving additional randomness. The stochastic character of snow accumulation and snowmelt is very much simplified by the author's approach. There is no discussion how the groundwater storage and the storage of water in the surface are taken into account.

Questions for discussion

The following questions may be posed for discussion:

(1) How valid are the three major objections (advanced by the author) to the generation of new river flow samples by using available river flow data?

(2) How limiting is the uniqueness of each river basin so that the major information becomes non-transferrable from one catchment to another?

(3) How accurate is the reproduction of catchment rainfall properties in simulating new rainfall samples?

(4) How significant may be the difference between the potential transpiration from a river basin and the real evaporation and evapotranspiration from it, and are these differences subject to a large stochastic variation?

(5) How realistic is the author's concept of the 'snowlag' in reproducing the stochastic or the periodic-stochastic character of snow accumulation and snowmelt?

Concluding remarks

The general reporter acknowledges the contributions by all of the authors to the topic of structural analysis of hydrologic sequences, in general, and to the problems of developing new techniques in generating samples of hydrological periodic-stochastic time series, in particular. It is his hope that the essences of the authors' contributions have been properly assessed and adequately summarized for the purpose of stimulating discussion at the symposium, as well as for consideration by future readers of the proceedings of this symposium. Apart from posing many questions for discussion after reviewing each of the five papers, these general questions may also be advanced for discussion:

(1) What is the basic value of generating techniques to hydrology and water resources?

(2) In which direction should the research for better generation techniques be pointed, so that the best result can be expected?

(3) Assuming that all available methods of structural analysis of hydrological time series, as periodic-stochastic processes, may be classified either as the stationarity approach or the non-stationarity approach, which of the two approaches promises a more reliable reproduction of time series properties inferred from the samples available?

REFERENCES

Anis, A. A. and Lloyd, E. H. (1953) On the range of partial sums of a finite number of independent normal variates. *Biometrika* 40, 35–42.

Hurst, H. E. (1951) Long-term storage capacities of reservoirs. *Trans. Amer. Soc. civ. Engrs* 116, No.776.

Yevjevich, Vujica (1967) Mean range of linearly dependent normal variables with applications to storage problems. *Wat. Resour. Res.* 3, No.3.

DISCUSSION

R. J. G. Bloomer:

I would like to thank Professor Yevjevich for his review of the paper and for putting so much effort into his general report. However, I wish to make three points in order that there should be no misunderstanding.

(1) The results given for the H values from autoregressive models were based only on *very short samples*, though care was taken to ensure that, in this portion of the work, the synthetic and historic samples were of the same length.

The results, therefore, confirm the instability of H estimates from small samples. It is too early to be dogmatic on the relative merits of autoregressive and self-similar models; both have a place in certain situations. All that could be said was that, in our results, autoregressive models were suitable in some cases.

(2) The independence of daily residuals is a most important requirement before generation can proceed. The runs test is a very sensitive measure of this and appears a more reliable index than serial correlation coefficients. The fact that the residuals were not independent in our results indicated that another model should be tried, though time did not permit this.

(3) In many cases, where low flow sequences are important, the preservation of longer term properties of data (obtained from daily generators) is a necessary criterion of a good model.

I would like to make two points concerning the paper by Hall and Johnston:

(1) It is clearly very important to extend the above work, and to examine whether short-term periodicities in *discharge* existed and whether these affected gauging techniques.

(2) Professor Cavadias has stressed the need for new gauging methods. The Water Resources Board is very conscious of this and work is currently directly towards ultra-sonic and electromagnetic methods. It is too early to say whether either technique was viable or accurate, but preliminary results from the ultra-sonic study were encouraging.

M. J. Hamlin:

Dr. Kottegoda, my co-author, has something to add about distributions. In order to comply with the reviewer's request, I will limit my remarks to two points and then allow Kottegoda to present the additional information which we would like to put to the meeting. I would first like to emphasize something which Professor Yevjevich said about negative flows. In the models we have studied, we have sometimes had negative flows in the fitted deterministic component. This however applies only to the 5-day model and not to the monthly model. I would also like at this stage to say something about Bloomer's paper. We have also had trouble with the non-preservation of long-term parameters in short-term data.

I believe that the most satisfactory approach is to simulate or generate 5-day data, which we feel can be done successfully, and then to break this down randomly into smaller time units, rather than to generate the smaller time units in the first place. Of course, it is necessary to have the proper correlation structure when breaking down the 5-day units but I believe that this is possible.

N. T. Kottegoda:

In this graph (Fig.1) I have shown the kurtosis and the skewness of several series of British river flow data which have been analysed using time intervals of 1 day, 5 days and a month. The monthly data present only a small problem with regard to distribution because these are confined quite close to the point of normality, i.e. point C, and the distribution of monthly data could be either lognormal or a simple two or three parametric gamma distribution. When we decrease the time unit it is found that, as expected, the skewness increases and so does the kurtosis and a type III or gamma function may not be appropriate for some time series.

This graph is based on the work by the English statistician, K. Pearson. For the

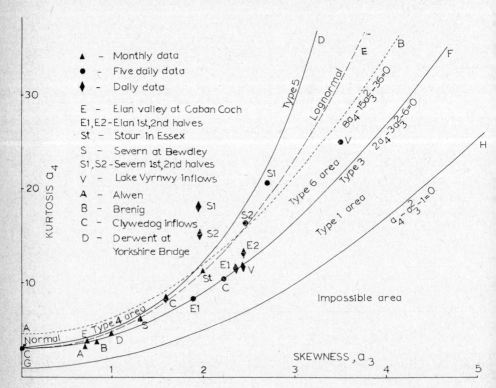

FIGURE 1. Residuals of Markov models.

representation of the distribution of hydrological data, the type III function is not very flexible because its use implies a fixed relationship between the kurtosis and skewness as represented by curve *CF* and the appropriate equation. The type V function is based on a similar relationship. In contrast, the type I and type VI functions have wider applicability. In addition to the Pearson type functions, the lognormal function which N. L. Johnson has recently shown in a paper in *Biometrika,* can be represented by the curved line *CE*.

The points representing the time series are located in the type VI zone and, in many cases, there seem to be significant differences between the points and the criteria which justify the use of the lognormal or type III functions. Therefore, my contention is that instead of indiscriminately using the lognormal or the type III functions, a closer approximation may be achieved in some time series by the type VI function which is more flexible.

S. Jovanović:
As stated in our paper, a first-order Markov chain model was applied for generating daily rainfall depths. In order to obtain the transition probabilities, the relative frequencies are not used to estimate the probabilities p_{ij}, as done in the case of the previous model which described the probability of the appearance of wet and dry spells and where the system has but two possible states. The characteristic of the latter model is the theoretically infinite number of states. In this study, a limited number of 20 states (and not 7 as mentioned in the general report) has been taken, the transition probabilities being determined according to the two-parameter gamma distribution function. This method limits the maximum daily rain depth which can be obtained by simulation to $r_i = r_j = 68$ mm; it represents the rain depth corresponding to state $i, j = 19$. It is true that at certain raingauging stations more important depths of daily rain had been observed, but those were local storms which have little influence on the flood wave appearance in the main water course. It should be pointed out that in this study, rains covering the largest part of the Sapadna Morava basin and lasting two or more days are of interest. Moreover, only those depths of daily rainfall that represent the appropriate states can be obtained by simulation, namely 20 different depths only. Although this could be considered as a drawback of the method, it did not represent a serious problem, as a great number of combinations had been obtained by simulation, both as to the length of rainy spells and the distribution of daily rainfall depths in the spell.

As to the shape parameter of the gamma distribution, its value increases with an increase in the rainfall value of the state, and it varies within rather wide limits for individual raingauging stations. It is the consequence of the pluviometric regime of the area and it is in accordance with the nature of the investigated phenomenon. The scale parameter varies much less from station to station. On the other hand, a precise tendency concerning the value of this parameter for particular states could not be determined; therefore, a constant value has been adopted for each station (i.e. for each transition probabilities matrix). Thus, for the transition probabilities matrix (22 matrices for as many raingauging stations), we had to determine one value of the shape parameter for each state and one value for the scale parameter per matrix, which means 21 parameters per matrix. Such a number of parameters did not represent insurmountable difficulties while computing the probabilities of transition. I wish to stress that analyses are being undertaken to prove the existence of a relationship between the mentioned parameters of the gamma distribution and the topographical, climatical and other characteristics of the site.

I. Rodríguez-Iturbe:
I will be very short. I just want some clarification in the discussion of the paper by Hamlin and Kottegoda. In the comparison of the models used in the paper, there are

claims of best fitting of one model over another, and I personally believe that fit depends on the purpose of the simulation. If the best fit claim is made on the absolute basis of the difference between historical and generated values at each time, then I believe that the claim is not correct, because, due to the same stochastic nature of the models, each simulated trace is just one of the many equally likely possible sequences that can occur. I think it is better to perform the comparison between the models on the basis of what one is interested in preserving, and, of course, minor variants we all can preserve with the different models. But there are problems like extremes to consider, and what I at this moment do not know, is how the authors of some of the papers claim the so-called best fitting.

P. Johnson:

I wish to highlight one or two of the more important features of the research carried out and summarily described by Barton in his paper.

Firstly, it is noted that a number of the techniques used by Dr. Barton were in part abstracted from the results of other research workers. The way in which these techniques have been adopted and modified to enable the generation of monthly stream discharge data is seen to be one of the most important aspects of the research. The combining of a stochastic data generating technique with a deterministic catchment model is also an important feature.

It is noted that the entire catchment model consists of three distinct phases. The first of these is hydrometeorological, being concerned chiefly with the development of effective input to the succeeding two phases. In the first phase, three streams of data, precipitation, potential evaporation, and 'snowlag', are combined. These three streams may either be recorded data, if the model is being used for simulation only, or stochastically generated data. Data generation is not necessarily limited to a Thomas and Fiering method as used by Barton.

The second phase of the model is a soil moisture accounting procedure which, by its formulation, attempts to allow for special variation of catchment soil water storage characteristics. The third phase seeks to represent the general persistence of catchment drainage.

Stochastic properties of precipitation, and hence generation of monthly precipitation, are based on Thiessen means for a catchment. Stochastic properties of both evaporation and snowlag were developed using simplifying routines. The relative deviations of a month's evaporation about the monthly mean was almost the same for each record, implying a consistent deviation for all catchments. Use was made of this feature in developing a master matrix of statistical coefficients of potential evaporation for use in generating a set of potential evaporation data for any catchment in the country.

The quantified concept of 'snowlag' is considered to be a new one. It represents the ratio of the water equivalent of a snowpack remaining at the end of the month, to the water equivalent of the snowpack existing at the beginning of the month plus the total precipitation occurring during the month. Statistical properties of this parameter enable the influence of snowpack persistence to be compounded with generated precipitation and evaporation data. In a manner somewhat similar to that adopted for evaporation data generation, the snowlag statistical coefficients are represented in two master matrices, the one used being decided by the annual average number of days on which snow on the catchment is observed to lie. In a country in which precipitation data are generally more readily available than streamflow data, the concept of routing stochastically generated precipitation data through a deterministic model to generate streamflow data would seem to be a useful one. Further development of the work started by Barton to synthesize weekly or daily streamflow data would therefore seem to be warranted.

M. J. Hamlin:

Clearly the question concerning the best model is fundamental to our discussion this morning. The best fitting model depends of course on what one wishes to preserve. We have looked at a number of parameters. The preservation of means and variances is straightforward. We have included range, runs, run lengths, runs above the mean and runs below the mean. It is unfortunate that our paper is not yet available to you since much of this information is given in the paper.

In addition, we have studied the correlation structure of both the real and the synthetic data. I have two slides which represent figures from the paper and, since you do not have these in front of you, perhaps I may be permitted to show them. The first slide is Fig.4 of our paper (vol.1, p.22) and shows the comparison of the correlograms between real and synthesized data. One of the techniques we used with the longer periods of record was to split the original data into two halves. One half was used to formulate the model from which the data were generated and these generated data were then compared with the second half of the original data. The three correlograms show, in turn:

(1) Data synthesized from the second half of the record compared with the first half.

(2) Data synthesized from the first half of the record compared with the second half.

(3) A comparison of the two halves of the record.

The second slide which is Fig.1 of the paper (vol.1, p.18) shows the variation obtained between the means, variances and skewnesses for the two halves of the record when compared with points obtained from the synthesized data. The full line represents the line joining the harmonic fitted values and the dotted lines represent the 95 per cent confidence limits.

The final two slides show sets of historical data and synthesized data plotted one above the other. The first is for the River Severn at Bewdley and the second is the River Stour which drains the chalk catchment to which we refer.

In both cases the generated data give a critical sequence more severe than the historical data. This brings me back to Rodriguez-Iturbe's question concerning the use to which generated data are to be put. In our studies we have assumed the data are required for reservoir design and we have paid particular attention to range, run sequence and minimum flows.

D. R. Dawdy:

I wish to compliment Professor Yevjevich on his discussion. Further, I agree whole-heartedly with the point which he made concerning the fact that the Hurst phenomenon is the most important problem in stochastic hydrology today. Most of the researchers in stochastic hydrology are concerned with that problem. In answer to one of the questions which Professor Yevjevich raised, at least indirectly, stationary autoregressive models cannot pressure the Hurst phenomenon except by either distorting the auto-correlation function, as was pointed out by Matalas, or by having a rather excessively high order autoregressive model, as was pointed out by Fiering. In the generation of drought and flood data, which are the extreme values with which we are most concerned in stochastic simulation, it seems that Markovian models are inadequate. This was pointed out by Bloomer in discussing the run test as a test of adequacy for Markovian models.

Equation (2) in the general report of Professor Yevjevich is an asymptotic result for autocorrelated sequences and is exact only for independent standard normal processes. Similarly, equation (4) is not proven to be exact either, but rather was postulated based on analogy with equation (2). It was then checked by Monte Carlo simulation, but not proved. [Although Professor Yevjevich references a paper by himself (Yevjevich, 1967) the derivation and checking of equation (4) was the major point of the dissertation of Sutabutra (1967).] For a non-stationary process, the expected value of the Hurst phenomenon has been presented quite ably in the papers of Mandelbrot and Wallis.

I would like to make one point concerning the discussion of the paper on daily rainfall simulation. The assumptions about the generating process for a model determine the resulting probability distribution function. We cannot test the adequacy of a model by looking at the data which were used to determine the parameters for the model. If the parameters are derived from the data, then, in the limit, the stochastic similation should reproduce data with those parameters.

A further point which I would like to make is that fractional gaussian noise (FGN) is *not* a form of moving average. The parameter, M (for memory), in the Mandelbrot and Wallis formulation is an operational parameter, whose size is limited only by the capability of a digital computer. It is *not* a distributional parameter. Varying M has no influence upon the results of FGN generation, for M is chosen 'large' in relation to the study period of interest.

In closing, I have some slides which I wish to discuss. The question was raised concerning the best method for selecting, by Fourier analysis, the significant harmonics of the periodic parameters for stochastic models. My viewpoint is that the best method for selecting harmonic periodic parameters is not to do so. The Harvard model formulation which preserves the means month by month (or any other period) is, at this time, much superior to harmonic analysis. If the method of harmonics is to be used as a smoothing function, then it should preserve certain statistical characteristics more efficiently than alternative methods. To date, harmonic analysis has not been shown to do this.

As an example of the problems introduced by harmonic analysis, the first slide (Fig.2) is from a thesis at Colorado State University (Mejia, 1971). The fitting of the monthly means by harmonics for the Chixoy River in Guatemala is shown by the dotted line, whereas the solid line connects the recorded monthly means. Standard statistical tests indicated two points.

Firstly, a clear-order autoregressive model was adequate. Secondly, the method of harmonics fitted the monthly means adequately in terms of explained variance. However, as can be seen, the harmonics introduce spurious oscillations into the monthly values, particularly for months of lower mean flow.

The second slide (Table 1) shows that the first-order serial correlation coefficient is 0.623, and at lag ten the serial correlation is still 0.4. In fact, all of the first ten serial correlation coefficients are significantly different from zero at the 95 per cent level. A stepwise regression was used to determine the structure of the autoregressive model, and, as shown in the slide, the first, fourth, sixth, eighth, ninth and tenth-order coefficients were significant. All of these results were based on the use of the Harvard model to determine the random component. However, when the method of harmonics was used to derive the residuals, the first-order serial correlation for this streamflow series was negative, and there was no significant serial correlation coefficient at any lag!

The third slide (Fig.3) shows the peculiar correlation structure for the Chixoy River. It results from the fact that the river basin contains a karst region. The heavy solid line is a correlogram fitted by a so-called broken-line process (Mejia, 1971). This is a stochastic generation model which can model complex correlation structures with few parameters. It may well be a good tool for use in some cases where Markovian models are not adequate.

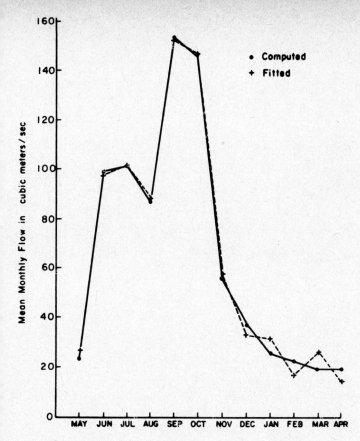

FIGURE 2. Monthly means for the Chixoy River computed from the historical record (1962–1963 to 1969–1970) and fitted by a sum of harmonics.

REFERENCES

Mejia, J. M. (1971) On the generation of multivariate sequences exhibiting the Hurst phenomenon and some flood frequency analysis. Doctoral Dissertation, Colorado State University, November 1971.

Sutabutra, Prathet (1967) Reservoir storage capacity required when water inflow has a periodic and a stochastic component. Doctoral Dissertation, Colorado State University, June 1967.

Yevjevich, V. M. (1967) Mean range of linearly dependent normal variables with applications to storage problems. *Wat. Resour. Res.* 3, No.3

R. G. Quimpo:

I would not like to be drawn into this controversy but after hearing the remarks and questions I have to have my say. Before I make any comment I would like to add some results to those of Kottegoda and Hamlin which I think corroborate some of their findings.

Using a self-similar model as a basis to obtain an equivalent autoregressive model (a least square criterion on the kernel function was used to determine equivalence), we generated 2400 years of annual flow values. We found that the R/S versus N plot from data generated by the 'equivalent model' had less scatter than that for data generated by the basic self-similar model.

As regards the comparison of models, one common property of equations (8), (9) and (11) which may be used as a basis for comparison is the fact that they are equa-

TABLE 1. Autocorrelation and autoregression coefficients for the residuals $\epsilon_t = (x_{p,\tau} - \mu_\tau)/\sigma_\tau$ of Chixoy River monthly flows

Lag k (months)	1	2	3	4	5	6	7	8	9	10
Autocorrelation coeff. r_k	0.623	0.496	0.464	0.531	0.456	0.530	0.513	0.549	0.347	0.411
95 per cent confidence limits for r_k	0.157	0.158	0.159	0.160	0.161	0.161	0.162	0.163	0.164	0.165
	−0.178	−0.179	−0.180	−0.181	−0.182	−0.184	−0.185	−0.186	−0.187	−0.188
Autoregression coeff. ϕ_k	0.399	(NS)*	(NS)	0.178	(NS)	0.227	(NS)	0.272	−0.252	0.122

* NS is entered for 'not significant', meaning that its inclusion in the regression equation will not make any significant reduction in the error sum of squares because its F ratio is too small (49).

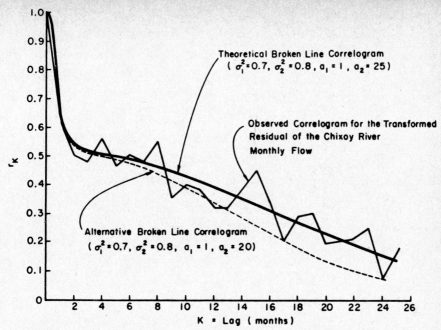

FIGURE 3. Correlograms of the Chixoy River monthly flows.

tions of linear systems. Since these systems are characterized by their kernel functions, a comparison of these kernel functions would give some insight into the differences and similarities among them.

Finally, a comment on the self-similar model being also a moving-average model, at least in the form of type I and II approximations to fractional gaussian noise. Equation (11) describes a scheme wherein a set of basic variables is weighted and added to obtain another variable. I think it does not matter whether N is finite or not. Regardless of what value N takes, the model weights (in a sense, averages) are the values of several variables. Since the equation is recursive, one seems to be justified in calling it a moving average model. Of course, this is a matter of semantics.

J. A. Cole:

I would like to mention the work I have undertaken with my colleague, J. D. F. Sherriff, on a stochastic model of daily rainfall for the Dee basin in North Wales. This used a two-stage procedure:

(a) Wet spells are defined as a run of days, all having more than 0.13 mm of rainfall. From their empirical distributions of run durations, samples of wet and dry spells are taken alternately to build up a composite sequence of wet, dry, wet, etc. (Spell durations were shown to be uncorrelated.)

(b) Within a wet spell, a first-order Markov chain is used to create the daily rainfall amounts.

The model was tested for compliance with its source data via the following statistics:

(1) daily rainfall means,
(2) daily rainfall variances,
(3) daily rainfall first order autocorrelations,

(4) monthly rainfall means,
(5) monthly rainfall variances,
(6) 4-day run frequencies, based on all 625 combinations of 5 rainfall categories.

Tests (1), (2) and (3) were satisfactory, which was hardly surprising as they dealt with statistics incorporated explicitly within the model. Tests (4) and (5) showed some small but significant departures, such as to indicate the need for higher than first-order persistence to be built in. The run test (6) was a useful confirmation of realistic daily patterns, but it was subjective and no assessment of significance is as yet possible. I would like to urge mathematicians to devote research towards improving the power of such distribution-free tests.

Finally, although areal rainfall patterns have been sampled in the Dee study and used to relate a single time series to events at other sites, this is a poor solution to the multi-variate problem. A better approach lies in either (i) 'magic carpet' models with continuous deterministic weather trends within which storms or storm cells have stochastic development, (cf. Amorocho's general report in session X), or (ii) strictly multivariate models, incorporating cross and lag correlations within a discretized time and space framework.

P. E. O'Connell:

I would like to comment on the problem of the independence of residuals posed in Bloomer's paper. The inability of the daily model used to reproduce the autocorrelation structure of the monthly flows can be attributed to the nature of the model used for the daily flows. Low order Markovian models fitted to daily data, having already taken account of periodicity in the means and standard deviations of the daily values, will fail to preserve the long-term persistence that is frequently observed in monthly and yearly data. In fact, monthly totals accumulated from data generated by daily low order Markov models will, in general, be almost independent on a yearly basis. The inadequacy of low Markovian models in representing daily streamflow has already been encountered by Quimpo (1967) at Colorado State University and, in testing his residuals, he found that the first serial correlation coefficient of the residuals was significant, according to Anderson's test. However, these residuals could have been fitted by a first-order moving average process which is tantamount to using a mixed autoregresive moving average model overall. Such models can possess long memories and preserve the long-term persistence observed in daily, monthly, and yearly data (O'Connell, 1974). A further point worthy of note is the danger of distorting the low frequency or long-term effects in the data by taking out means and standard deviations.

REFERENCES

Quimpo, R. G. (1967) Stochastic model of daily river flow sequences. HYD. Paper 18, Colorado State University, Fort Collins, Colorado.
O'Connell, P. E. (1974) A simple stochastic model of Hurst's law. *Mathematical Models in Hydrology* (Proceedings of Warsaw Symposium 1971), vol.1, pp.169–187; IAHS Publ. No.100.

T. Kaijser:

1.   In this first paragraph I want to clarify some remarks I made at the general discussion of the Symposium.

Let T_n, $n = 1, 2, \ldots$, denote the time of the beginning of the nth rainfall in some fixed area and let t_o be a fixed 'starting time'. Let

$$X_1 = T_1 - t_o$$

$$X_n = T_n - T_{n-1}, \qquad n + 2, 3, \ldots$$

and define

$$N(t) = \max(n|T_n \leqslant t_o + t)$$

The random variables X_n, $n = 1, 2, \ldots$, are thus the waiting times between the start of two consecutive rainfalls, and the random variable $N(t)$ is equal to the number of rainfalls in the time period $(t_o, t_o + t)$.

Now suppose that the random variables $X_1, X_2, \ldots$, are independent and equally distributed with a common density function defined by

$$f(t) = \lambda e^{-\lambda t} \tag{1}$$

It is then well known [see for example Feller (1966, p.11)]

$$Pr[N(t) = n] = e^{-\lambda t} \frac{(\lambda t)^n}{n!} \tag{2}$$

from which we see that $N(t)$ is Poisson distributed with parameter λt. Thus, if it is reasonable to assume that the time lengths between two consecutive rainfalls are independent, equally distributed random variables with a common density function given by (1), then it is no surprise if the number of observed rainfalls is approximately Poisson distributed, since we know from (2) that the theoretical distribution of the number of rainfalls in *exactly* Poisson distributed.

2.   In this paragraph I want to make some remarks on the simulation method described in the paper by Jovanović *et al.*

Let me quote the following lines in their paper (the last 2 lines of the second section). 'The simulated sequences were analysed and graphed in Figs.2 and 3. The χ^2-test proved again that the fit was quite good at a significance level of 5 per cent.' However this is not true. The performed χ^2-test does not prove anything new about the fit. It is not possible to prove anything about the accuracy of a model using simulated series. Simulation methods are useful for obtaining distribution functions which are too cumbersome to calculate by ordinary analytical methods. Simulations cannot be used to enlarge the list of observation data.

Finally, in order to avoid any misunderstanding, it seems to me, looking at the first performed χ^2-test (vol.1, p.113), that the model describes the hydrometeorological system quite well.

REFERENCE

Feller, W. (1966) *An Introduction to Probability Theory*, vol.2. Wiley, New York.

Closing remarks by the general reporter

The preceding discussion has shown that the organization of Symposia by using the system of general reporters can be very successful. Progress in science occurs when controversial subjects are openly discussed, allowing the participants to ascertain on what they agree or disagree. I will comment briefly on a couple of points only.

Equation (2) of the general report is not an asymptotic, but an exact equation. By generating a large number of samples, one can convince oneself that it is the exact equation. The equation related to fractional noise, as given in the paper by Hamlin and Kottegoda, is a moving average scheme. Once the parameters h or n of that equation are selected, they give monotonically decreasing coefficients, typical of moving average schemes. As many models of that type can be designed as one wishes, provided that the condition of monotonically decreasing weights of the moving average scheme is satisfied. This position may be a controversial subject.

As it concerns the remarks by Dawdy on the harmonics of the mean monthly values, one finds that they can often be fitted by less than six harmonics. Sometimes all the six harmonics give significant results, though this depends on the technique used for the significance test. The spurious (sampling variation) harmonics are usually the harmonics of the highest frequency.

General Reporter   Dr. Karol Krajewski, Assistant Professor at the Institute of Environmental Engineering, Warsaw Polytechnic University, Warsaw, Poland

Chairman   Dr. V. I. Korzoun, Hydrometeorological Service of the USSR, Moscow, USSR

Structural analysis of hydrological sequences — II

Karol Krajewski

As a result of the illness of Professor Kartvelishvili, I have been entrusted by the Organizing Committee with the gratifying task of discussing some of the reports delivered during the debate on the first day. This is probably because I was the least busy member of the Organizing Committee, but I had only a few days for preparation, and I lacked the knowledge and experience of Professor Kartvelishvili. I think therefore that I undergo this honourable duty best by only presenting a short summary and asking the participants here, especially the authors of the papers, to be good enough and expand their contributions by discussing these subjects which, unfortunately, I shall be able only to broach. I shall discuss the following papers (all published in volume 1 of the Symposium Proceedings):

1. *R. T. Clarke*   The representation of a short period of experimental catchment data by a linear stochastic difference equation
2. *S. Dyck and Ch. Kluge*   Investigations on the structure of frequency distributions of floods
3. *C. Cunnane and J. E. Nash*   Bayesian estimation of hydrological events
4. *G. P. Kalinin, N. V. Nikolskaya, V. M. Evstigneyev and V. A. Zhuk*   Spatial and time analysis and estimation of river flow
5. *Rafael G. Quimpo*   Structural relation between parametric and stochastic hydrology models
6. *Vujica Yevjevich*   Significance tests of periodicity in hydrological time series

Having acquainted myself with these papers, I came to the conclusion that they can be discussed within the rough framework of two groups. The first group — this of course is an optional division — would include the papers by Dyck and Kluge, Cunnane and Nash, Kalinin *et al.* and Yevjevich. The interest of the authors of these papers was concentrated on methodological problems and research of observation material aimed at gaining a better knowledge of the statistical properties of river flow. As regards papers of the second group, I think that Clarke and Quimpo dealt in more detail with the problem of the formal description of hydrological processes which make possible a direct construction of mathematical models.

The paper by Dyck and Kluge discusses a number of variants in the utilization of all accessible information about floods to gain a more accurate determination of their probability, and for better modelling. Account has been taken of maximum flows, for example, on the basis of annual maximum flows, or on the basis of maximum flows exceeding a certain assumed level. The paper describes an original method of flood analysis based on the modelling of the maximum flood flows. The method, utilizing information about precipitation and retention, makes it possible to estimate the distribution of maximum flows.

The paper by Cunnane and Nash describes a different method of evaluating the distribution of annual maximum flows, namely one which utilizes Bayes law. The paper stresses that by using Bayes law it is possible to obtain an evaluation of *a posteriori* distribution quantiles taking advantage, as usual, of the knowledge of the probability distributions of local floods.

The paper by Kalinin *et al.* contains an interesting discussion on the possibility of utilizing certain generalizations of the relationships (obtained from an analysis of ample observation material) between such frequently used parameters as coefficients of skewness, coefficient of variance and coefficient of excess, C_s, C_v and E. The paper comprises a number of equations allowing the determination of these coefficients: the coefficients of skewness C_s and excess E, as a function of the variance coefficient C_v for different characteristic flows and for different periods.

Yevjevich's paper discusses the problem of the detection of periodicity in observation sequences of precipitation and streamflow.

Clarke's paper deals with the construction of a mathematical model comprising a description of the interdependence of the observation records in the form of a certain stochastic differential equation. The problem is considered from the viewpoint of the acquisition of linear and nonlinear equations for use in short-term forecasting of hydrological phenomena. The paper includes a comparison of that model with a model based on multidimensional distribution of probability.

Results obtained by the authors indicate that the proposed new type of models may be successfully used, given — as the authors emphasize — an appropriate dense observation network ensuring adequately accurate data required for the forecasts.

The paper by Quimpo contains a theoretical discussion on the relationships between parametric models and stochastic models in hydrology. The paper examines the possibilities of the reconciliation of the techniques used for their construction. Particular emphasis is placed on the dependence between a unit hydrograph and hydrological series, known as the implementation of the stochastic process. I think that the author raised one of the basic problems involved in modelling, namely, he tried to link the static investigation of the phenomena defined by a certain system of parameters and the examination of these phenomena as processes variable in time and space. Obviously, these processes can be described only within probabilistic categories.

I apologise for not having formulated problems for discussion on these six papers and hope that I may partly be excused by the circumstances. However, three issues related to the papers are worth discussing.

The *first problem* is the choice of the best way of using information on the properties of phenomena and processes, having at our disposal very short observation sequences. The point is that we can often discuss very sophisticated questions connected with the construction of a model, whereas life and practice itself provides us only with very limited observation material on which we can base our hypotheses. This is a difficulty all of us have no doubt encountered. In many countries this is a crucial problem as observations have not long been carried out. The *second problem* is how to utilize possible generalizations in the given situations. The problem arises how to utilize for the mathematical modelling of phenomena not only information directly related to that region and a given phenomenon but also information acquired from the elaboration of some, often voluminous, material for many countries, many rivers, and many climatic areas. Probably it would be extremely useful to utilize such information, but it is obvious that various hazards are involved.

The *third and final problem* is rather loosely connected with the papers, but, in my opinion, it is interesting enough. What techniques could be suggested for the verification of the constructed mathematical model, above all, from the point of view of its physical justification? All who are engaged in modelling, especially the modelling of processes, have encountered the problem that it is hard to judge whether a model and the obtained results, can be regarded as good or bad. When the problem is carried to a certain extreme, one may say that the best thing to do is to obtain, by means of computation (using, for example, a digital computer), an identical process to that which served for the determination of the parameters of the model. But we also know that the acquisition of an identical process would know nothing more about that process. We would prefer that the process we have generated is similar enough to the prototype to satisfy us, and differs from it so much that it can be of some use to us. I shall be glad if the little I have said will help to clarify at least some of the problems raised by the authors and the issues which preoccupy us all.

DISCUSSION

C. Cunnane:
I would like to add to the general reporter's remarks on the paper by Cunnane and Nash which discusses the type of inference which, given different kinds of information, can be made about a population and presents a method whereby two kinds of information may be combined into a single estimate. The estimation of a quantile of a population of annual maximum floods is used to demonstrate the method. A Gumbel (type 1) distribution is assumed although this is not vital to the points being demonstrated. Given an annual maximum sample, the quantile can be estimated by any standard method and a confidence interval for this estimate can be computed. If a large number of random samples is available and such an interval computed from each of them, a stated proportion of these intervals would contain or cover the population value. Such intervals are called sampling confidence intervals. However, they are not the kind required by the practising engineer because there is only one and not a large number of samples available. He wants an interval which allows him to say that the population value lies in the interval with a stated probability.

If there is information on the population available in addition to the sample data, we can derive a Bayesian estimate and a Bayesian confidence interval which allows the required kind of statement to be made. Such additional information is usually available. For instance, the examination of many records of annual maxima floods in the British Isles shows that the coefficient of variation C_v lies, in nearly all cases, between 20 and 60 per cent. If a single sample gives a C_v value which is only 5 per cent, then we know that it is an unusual sample and that a confidence interval computed from it will be unrealistically small. It is little consolation to know that a realistic confidence interval would be obtained from almost any sample but the one available. Bayesian estimation allows the extra or prior information to be combined with the sample data to provide realistic looking answers.

The method is one in which the likelihood function of the parameters, given the sample data, is multiplied by the equivalent of weighting factors which express the prior belief or probability that the parameters have particular values. The prior probability is obtained by use of empirical relations between population parameters and catchment and climatic indices. In certain extreme cases it is almost obligatory to use this prior information and, as a general strategy, we recommend using it with the sample information in as many cases as possible.

V. Vevjevich:
The general reporter has stressed that, if one wishes to use the techniques of stationary stochastic processes, the hydrological stochastic part must be made stationary. As stated in my general report (session I), the methods used by hydrologists in analysing hydrological time processes are two types: by stationary and non-stationary approaches. Many colleagues who use some existing methods of generating new samples do not

stress that they belong to the non-stationary approach. These methods do not test whether parameters are periodic or not. Mathematical techniques of treating the non-stationary stochastic processes are in infancy in comparison with existing mathematical techniques for the analysis of stationary processes. The general reporter's question is whether one should go for the stationary approach or should stay with the non-stationary approach? I tried to emphasize in my paper the importance of discriminating between the deterministic periodic parameters and the stochastic stationary part, separating them in order to apply the methods of stationary processes to the stochastic part. My position is that the stationary process approach has more promise in hydrology than the non-stationary approach.

The comments on particular questions raised by the general reporter are as follows. For the mathematical modelling of stochastic processes in hydrology and water resources, it seems that only the combination of three parallel or combined lines of of attack can produce the most reliable models. One line is the conceptual formulation of the model, based on both the probability theory and stochastic processes as well as experience. A significant trend exists at present for developing new conceptual stochastic models. However, without a physical support, mainly from the responses of hydrological environments to inputs, one can never be sure whether these new models are good or not. Another line, sometimes downgraded by theoreticians in probability theory, is by using inferential statistics on a large number of samples. Some opinions exist that the statistical inference of stochastic processes is a relatively undeveloped field. My position is that only the combination of the conceptual approach to stochastic models, the analysis of physical environmental responses and other physical information, and the stastical inferential technique on a large number of series of a variable, would produce the most reliable mathematical models.

The general reporter asked the question how one extracts the maximum information from a given quantity of data? Most practitioners of hydrology, especially advisors to developing countries, often state that the smaller amount of available data the simpler the statistical methods of information extraction should be used. The approach should be the opposite, namely, the lesser the amount of data available, the more powerful techniques should be used to extract the full information. For 10 years of hydrological data one cannot afford to lose, say, three or four years by an incorrect, approximate or inefficient estimation technique. For 40, 50 or 70 years of hydrological data one can afford to lose, say, an equivalent of five or six years of data by an approximate estimation technique. The irony in the world is that its humid regions have most data, while the arid regions have relatively little data. In humid regions the variance of a random variable is small, its skewness coefficient is also rather small. The more arid a region, the larger is the variance of a random variable, the larger are its skewness coefficients. This means that for the same information the larger should be the sample of historical data. Where one needs most data for reliable information, one has the least.

The point well underlined by the general reporter is the problem of regional information. Many hydrological and meteorological services in the world have concepts leading to a diffusion of information rather than to a concentration of information. The plotting of isolines of a parameter may often represent a diffusion of information available at a set of points, whilst, in practice, one needs to concentrate the information at points of interest.

The design of regional networks and the processing of data should be based on the concepts of concentrating information at points of maximum interest.

The verification of models requires some kind of honesty with ourselves. If one uses series of a variable for a hundred stations and finds that five series are not fitted by the model whilst 95 per cent are fitted well, the tail part of the distribution of the testing parameter would imply those proportions. Often, papers are presented in which the models are checked on a very limited number of series, however, if the same models are applied to a very large number of series, say hundreds, the bulk test may reject that model. Tools are available at present — mainly computers, numerical methods and mathematical techniques — in order to check models on a bulk of data, from many regions, from different climates, and under other conditions. The potential exists to check and accept or reject models, say, under the assumption that 95 per cent of cases give a good fit, and reject these models when much more than 5 per cent of cases do not produce a good fit.

R. G. Quimpo:

First of all, I would like to thank the general reporter for his kind comments. I would like to address myself to two of the problems which he posed.

My paper was concerned mainly with the relation between the parameters of stochastic models and those of parametric (deterministic) models. The present work has been motivated by what some refer to as a dichotomy in present-day approaches to hydrological modelling. I hope that this attempt at relating parameters of one of those of the other would contribute to the eventual rejoining of these two dichotomous paths.

The general reporter has suggested that we discuss the physical justification behind the stochastic models and relate these to those underlying parametric (deterministic) concepts.

From the paper, I have shown that the use of an autoregressive scheme automatically implies that, conceptually, a storage model is assumed. Of course this is understood and well known to many investigators, but the point here is, that by using this approach, the technique could be extended to more complicated models.

The next thing that I would like to point out deals with the use of short-term data to develop streamflow generating schemes. My approach allowed me to use unit hydrograph parameters to estimate parameters of Markov-type models. If the theory can be verified more extensively than we have done so far, short-term information from unit hydrographs would allow us to estimate parameters of stochastic models.

The present result is very meagre and I have barely nibbled at a very big problem. Perhaps I could highlight some questions that need to be answered before significant progress is made. In my development, I started with a fairly general approach but, in order to arrive at a closed form solution, I assumed that the input to the basin could be described by a correlation structure which is negatively exponential. Even then I came up with a rather complicated equation so that I had to further assume that the input was 'white noise'. This resulted in a very much simplified correlation structure which allowed me to integrate the resulting equation. It is obvious then that questions may be raised regarding the justifiability of the assumption regarding the structure of the input. Along this line, help is needed from other investigators. We know that many people are studying the problem of rainfall modelling, and here I just took a model for rainfall which somebody else proposed. I purposely picked that particular model because it simplified the mathematics for me. The problem then is to find a more realistic rainfall model which one could insert in the present formulation and still render it amenable to solution. I realise the difficulty that this involves but I hope that this will start something which we could all join together in attempting to solve.

G. P. Kalinin:

The general reporter raised today very interesting problems which, in fact, are key problems in hydrology. The continuation of our common efforts for further development and progress in hydrology depends, to a large extent, on the proper solution of these problems.

I would just like to comment on some aspects of the presentation made by our general reporter — time is too short for discussing all of the problems in detail.

The first question concerns the possibilities of the theoretical development of a probabilistic model of the river flow formation, leading to the calculation of this flow. I must say that in hydrology the situation is unsatisfactory in this respect. It can be seen that in different countries as well as in different regions of the same country, the choice of model used is very subjective. The next step is to check how the selected model corresponds with the relatively limited empirical material. More significant progress and the formulation of the proper model can be achieved only by the combination of parametric hydrology and some theoretical calculations leading to the development of mathematical models with well defined limited of their applicability. This problem is, however, very difficult, and I do not think that we will be able to solve it completely in the next 10 to 15 years. But the problem should be raised. At present it would be very important to determine, even from a theoretical viewpoint, the amplitude of the oscillations of the hydrological characteristics. I think that, for some of the hydrological characteristics, we should be able to define these amplitudes.

My second comment pertains to stochastic hydrology, and I think that with some modification we should take more advantage of the hypothesis which proved its value in statistical physics: I mean the hypothesis of ergodicity. As we know, using the hypothesis of ergodicity, we can substitute regional observation by time observations. Unfortunately, this hypothesis cannot be transferred in its complete sense to hydrology or to other natural sciences, since all the statistical parameters, at least, are regionally variable. The problem emerges, therefore, what use can be made of this hypothesis? The basic concept is that we can consider hydrological processes as subject to some defined distribution laws. In such a case, if we normalize our parameters and take into consideration their variability, we can empirically define the distribution laws. But, in this case, we rely on bulk historical material and the problem is reversed: instead of taking the ready-made mathematical receipts and transferring them to hydrology, we analyse the hydrological data in order to determine the distribution laws which can then be used in the analysis.

At present, a large group of my associates are working on this problem. We have generalized the unusually large amount of data from all over the world and have obtained a most interesting picture which still needs further verification before being stated precisely. But, on the basis of the first phase of our work, we can say that, in principle, our two first parameters, namely mean flow and coefficient of variability, define, to a considerable degree, the flow oscillations, since other parameters are correlated with these two. For us it is extremely important finding an answer to the question of which moments shall be used for a sufficient accuracy of the analysis.

Considering various possibilities of solving this problem, I would like to indicate the historic hydrology or, as some call it, the paleohydrology. Owing to the present achievements of geologists and geochemists, we have certain possibilities of a more precise determination of the time periods when some processes occurred in the past, and this way we can reconstruct the hydrological data. I doubt, however, if this approach could lead us to any positive results in structuring the flow distribution curves valid for present conditions. The reason is that historically both the climatic factors and the climatic conditions were changing to a considerable degree. But it is worth mentioning here that all of us — hydrologists from various countries — in most cases work under the assumption that the correlation coefficient between flow series, which is always used in all kinds of hydrological studies, is not a function of time. I am deeply convinced that this is our common mistake. If we take any factor and analyse it from the paleohydrological and paleogeographical viewpoint, we come to the conclusion that in long time periods, thousands of years or some tens of thousands of years, some factors are quite significantly changing in time. This way we can distinguish certain trends which explain the fact that we could note significant correlation. It should be stressed that the correlation coefficient in flow series is changing in time: the longer the period of record the higher is the correlation coefficient. Since our flow series are usually not very long this tendency is not so apparent. In the temperature series the variability is still higher but if we take such phenomena as the oscillations of water levels in the Caspian Sea and other types of lakes, we shall note quite clearly that the correlation coefficient is changing in time.

I hope that these historical comments are worth mentioning since they throw some light on some problems of statistical hydrology.

O. Pfafstetter:

According to our experience in stochastic models, we have found that the simulation of low flows during drought periods is poor. We sometimes even get negative values. I would say that any defects in the stochastic models are more easily seen in drought than in flood seasons. To improve the model we tried to separate the baseflow from the total flow and treated only the surface runoff by stochastic methods. The separation of the baseflow imposes complicated problems. Thus, we think that the most promising solution would be to generate rainfall by stochastic methods and subsequently compute runoff with parametric models. I think this method would improve stochastic generation.

W. G. Strupczewski:

There appears in hydrological investigations a strong tendency to utilize, as much as possible, the information contained in observational data. This topic was discussed in the very interesting paper by Dyck and Kluge.

Two approaches were developed with the purpose of obtaining additional information concerning the characteristics of hydrological phenomena: (1) with relation to events which occur several times in a year (rise, decrease), the probability of an event results from an analysis of all occurrences during one year; (2) in determining the probability distribution, one utilizes information contained in the observational data for related events which were recorded for a longer time period. It is possible to use both sources of information simultaneously. The utilization of this information gains importance particularly when observational series are short, since then the estimation of characteristics determined by traditional methods is evidently inaccurate. However, one should not forget that in the search for a source of additional information we have to construct more complicated mathematical models. These models are founded on a large number of assumptions which can be verified only on statistical grounds, while the observational data necessary for such verification are scarce.

For example, in my paper motivated by the first approach (Strupczewski, 1967, a, b), the number of occurrences per year of an event under study was assumed to have Poisson probability distribution. Also the annual probability distribution was assumed to be Fisher-Tippet (type I). It turned out that with the rise frequency higher than 2, it is more effective to apply the latter model instead of the traditional one.

Nevertheless, if I had to decide which model was to be taken as a basis for a project of a hydrotechnical object, I would be in some trouble. I would start to consider the reasons for these additional assumptions.

Since the effectiveness of the 'all events' model increases with rise frequency, one is inclined to work with such a definition of a rise event which results in a high annual frequency. Yet if a too small value is selected for the definition of a rise, the identification of such an event may become more difficult. In such a case, in spite of the high theoretical exactness of the model, I doubt whether it is exact in reality.

The second-mentioned approach brings into analysis a multi-dimensional variable. The move from a single variable analysis to multi-dimensional analysis requires a knowledge of a relatively much larger random sample. In addition to the traditional model, one has to assume the distribution type of the multi-dimensional variable.

I would like to draw attention to one more problem in the area of statistical hydrology forecasting which is of similar character. The predictors accounted for in the forecast form a sub-vector chosen from the vector of potential predictors $\{X_1, \ldots, X_k\}$ based on the procedure aimed at minimizing the forecast error (Kaczmarek, 1970 a, b). Let us assume here, for simplicity, that the number of predictors, h, which enter the forecast is given beforehand. Then out of (k/h) different sub-vectors, one has to chose the one which minimizes the mentioned error. The sample correlation coefficient between this sub-vector and X_0 predicted has the minimal value

$$\hat{R}_{X_0, Z_{opt}} = \max_{Z_i} \hat{R}_{X_0, Z_i} \qquad i = 1, 2, \ldots, \frac{k}{h}$$

In estimating the exactness of the forecast, the sample value

$$\hat{R}_{X_0, Z_{opt}}$$

is taken for the real value of the correlation coefficient. If the exactness of the forecast is not satisfactory one extends the vector of potential predictors and repeats the calculation.

It seems that the estimation of the exactness of the forecast is too optimistic here. A procedure for the choice of the optimal predictor system baffles the estimator

$$\hat{R}_{X_0, Z_{opt}}$$

The relevant estimation will usually exceed the real value, more frequently so if the difference $(k - h)$ gets larger. This would mean that the effects from new sources of information gained by an increase in the dimension of the $\{X_k\}$ vector, are smaller than is present admitted. Empirically, this fact is most easily checked by the Monte-Carlo technique. Let us assume that the given correlation matrix for the vector $\{X_0, X_1, \ldots, X_k\}$ contains correlation coefficients for the general population

$$\Pi_{pot} = (r_{ij}) \qquad i, j = 0, 1, \ldots, k$$

Let us generate M random samples of N elements. On the base of each sample let us estimate the correlation matrix $\Pi_{pot}^{(m)}$, where $m = 1, 2, \ldots, M$. Assume $h = 1$ and consider k probability distributions for the absolute values of the correlation coefficients from the N-dimensional sample

$$f(|\hat{r}_{0l}|) \qquad l = 1, \ldots, k$$

where f stands for the density function.

Construct next k conditional distributions

$$g(|\hat{r}_{0l}| \, | \, |\hat{r}_{0l}| = \max_{l=1,\ldots,k} |\hat{r}_{0i}|)$$

Comparing the average values for unconditional and conditional distribution with $l = 1, \ldots, k$, we note that the average value is smaller in the first case. Similarly, we can prove that

$$|E(\hat{r}_{0l}^{(M)})| = |\frac{1}{M} \sum_{m=1}^{M} \hat{r}_{0l}^{(m)}| \leqslant |E(\hat{r}_{0l}^{(S_l)})| = |\frac{1}{S_l} \sum_{S_l=1}^{S_l} \hat{r}_{0l}^{(S_l)}|$$

where the index S_l denotes random samples for which

$$|\hat{r}_{0l}| = \max_{i=1,\ldots,k}|\hat{r}_{0i}|$$

Of course

$$\sum_{l=1}^{k} S_l = M$$

and

$$\lim_{M \to \infty}|E(\hat{r}_{0l}^{(M)})| = |\hat{r}_{0l}| < \lim_{M \to \infty}|E(r_{0l}^{(S_l)})|$$

REFERENCES

Kaczmarek, Z. (1970a) *Metody Statystyczne w Hydrologii i Meteorologii:* Warszawa.
Kaczmarek, Z. (1970b) Bledy w modelach prognostycznych (Errors in forecasting models). *Wiad. Sl. H. i M.* **VI** (1).
Strupczewski, W. G. (1967a) Wyznaczanie prawdopodobienstwa powtarzajacych sie zjawisk (Determination of the probability of repeating phenomena). *Acta Geophys. Pol.* **XV** (2).
Strupczewski, W. G. (1967b) Wyznaczanie rozkladu prawdopodobienstwa rocznego niektorych zjawisk na podstawie wszystkich ich wystapien (Determination of the probability of annual distribution of some events on the basis of all their occurrences). *Acta Geophys. Pol.* **XV** (3).
Strupczewski, W. G. (1969). Ocena charakterystyk statystycznych pewnego zjawiska przy wykorzystaniu informacji zawartych w obserwacjach innych pokrewnych mu zjawisk (Estimate of the statistical characteristics of some events on the grounds of the information gained from observations of other correlated phenomena). *Acta Geophys. Pol.* **XVII** (2 and 3).

Closing remarks by the general reporter

I shall take advantage of these four minutes left and try to make a brief summary of today's discussion, although I think that four hours would not be enough for that purpose. In my opinion, the discussion concerned very important problems on which depend our successes in mathematical modelling. It is an undisputed fact that we always have to pay for the lack of data, provided we agree that observation material is this most important and precarious treasure on which we base all further hypotheses. The previous discusser emphasized the fact that the more complicated the mathematical model, the more assumptions it contains and, in certain situations, there is the risk that the number of these assumptions may lead us astray, which means that we are likely to obtain a result which may prove totally useless in our further practice.

This is also the problem of what should be maximized when choosing and verifying a model. The previous discusser pointed out that the minimization of the sum of square deviations of the residuals is, in his opinion, incorrect. Probably it will become necessary to introduce new criteria, apart from the classic criterion which originates from the least squares method, and make ample use of the methods of optimization for the verification of the constructed models which are often multi-parametrical. I think that our discussion was fruitful, but, in common with every fruitful discussion, it produced more questions that answers. Perhaps, satisfactory answers to some of these questions will come in the course of the next few days of this Symposium.

General Reporter   Professor G. S. Cavadias, McGill University, Montreal, Canada

Chairman   Professor R. G. Quimpo, Department of Civil Engineering, University of Pittsburgh, Pittsburgh, Pennsylvania, USA

Structural analysis of hydrological sequences — III

G. S. Cavadias

The following papers are reviewed and discussed in this general report. All are published in volume 1 of the Symposium Proceedings.

1. *M. J. Hall and P. M. Johnston*   Stochastic analysis of velocity fluctuations in a natural stream channel
2. *A. V. Rozhdestvensky*   Methods for the mathematical description of the structure of river runoff variations
3. *A. I. Davydova*   Spectral analysis of the structure of hydrological series
4. *I. O. Sarmanov*   New forms of correlation relationships between positive quantities applied in hydrology
5. *G. A. Grinevich, A. G. Grinevich and A. F. Solovjova*   Multidimensional statistic modelling of hydrological processes on the basis of structural analysis of the empirical hydrographs of runoff

The papers belonging to topic 1 deal with the modelling of flow phenomena using methods in the time and frequency domains and, in the case of the paper by Sarmanov, using the correlation theory of the multivariate gamma distribution.

We shall examine the papers of this section approximately in the order of increasing time scale. This will enable us to follow better the change of emphasis in the models proposed by the authors.

The paper by Hall and Johnston deals with the nature of short-term fluctuations of the velicity of natural stream channels. The authors describe the preliminary phase of the analysis of an experiment on velocity measurements. Their analysis deals mainly with velocity fluctuations having an average period of the same order of magnitude, or larger, than the current meter exposure time (macro-scale turbulence).

Using an interesting measuring technique, they produce velocity time series which they analyse firstly by fitting a second degree polynomial to eliminate the trend and then by computing the spectra of the de-trended series.

The tentative conclusions of this paper are that:

(1) There are no apparent dominant periodic components in the series examined.

(2) The shape of the spectrum indicates that the series are generated by a low order autoregressive process, and, consequently, velocity fluctuations are not random in time.

During further stages of this study, it may be advantageous, due to the non-stationarity of the velocity series, to use the time-domain approach described by Box and Jenkins (1970).

The paper by Grinevich *et al.* proposes a model for the structure of monthly flows. This is essentially an additive model consisting of a Casic trend, a homogeneous

Poisson process that generates groupings of wet and dry years, an autoregressive process for yearly flows, a seasonal term and a random term. This type of additive decomposition has a long history, especially in the analysis of economic time series. The early work by Shiskin and Eisenpress (1957) was based on a computer-oriented adaptation of the traditional ratio-to-moving-average approach and is still in use by economists. Later, Rosenblatt (1963), proposed a combination of spectral and regression methods to fit an additive model of the form

$$y(t) = c(t) + s(t) + u(t)$$

to non-stationary series where the trend-cycle term $c(t)$ and the seasonal term $s(t)$ are considered to be deterministic functions of time, whereas $u(t)$, the residual or irregular component, is stochastic.

In contrast, the model proposed in this paper, considers more than one sotchastic term in the decomposition and consequently, the properties of the individual components cannot be estimated analytically. For this reason the authors advocate the use of a Monte-Carlo simulation to generate artificial sequences based on the properties of the observed flows. It is assumed that the irregular component follows a distribution of the Pearson family. As long as the decomposition of hydrological time series by analytical means remains intractable and the investigation has to resort to Monte Carlo methods, the success of the resulting model, such as the one presented in this paper, can be judged only by comparison of the statistical properties of the generated sequences and the observed flows, both in the time and the frequency domains.

The papers by Davydova and Rozhdestvensky address themselves mainly to the controversial problems of the existence of periodicity in the long-term fluctuations in river runoff (Vitha and Soucek, 1967; Rodríguez and Yevjevich, 1967). Since the conclusions of the authors are opposed in this controversy, it may be useful to discuss these papers together.

The paper by Davydova reports on the estimation of the spectra of more than 100 rivers in the Northern Hemisphere with periods of record ranging from 60 to 100 years. By examining the individual spectra and using standard tests of significance, the author concludes that cycles in annual runoff of 2, 5, 7 and 11 years have been demonstrated. The next part of the paper is a spectral analysis of atmospheric circulation variables between 1891 and 1968, followed by a cross-spectoral analysis of the above variables and the flow records. According to the author, this analysis corroborates the conclusions of the first part of the study, concerning the existence of long-term cycles.

The paper by Rozhdestvensky also uses spectral analysis for modelling annual flow series. The flows at ten stations in the Dnieper basin are analysed by first computing the empirical autocorrelation functions for different origins and lengths of record, and then generating artificial first-order autoregressive sequences of the same lengths and autocorrelation coefficients. The mean of the autocorrelation functions of the flows indicates that the assumption of a first-order autoregressive process is not unreasonable. On the other hand, the results show that both the autocorrelation function and the spectrum depend on the origin and the length of record, so that according to the author they cannot be considered as typical of the generating process but only as descriptive of the observed flows.

For the decomposition of the annual flow series, Rozhdestvensky uses a binomial filter (Tick and Woodbury, 1963; Jenkins and Watts, 1969). Examination of the components obtained in this way (not shown in the paper) indicates that their properties also depend on the origin and the length of record used. The study includes a cross-spectral analysis of the annual streamflow series and the Wolf numbers which arrive at the same conclusion.

Returning now to the paper by Davydova, we may formulate a number of questions concerning the methodology used. In particular, the fact that 'basins located close to each other exhibit periods of different duration' needs an explanation if we assume that the long-term periods found in the study are not due to sampling fluctuations. In addition, a physical explanation is also needed for the long-term periods in atmospheric circulation which, presumably, are the causative factors of the corresponding cycles in runoff.

Based on the conclusions of this study, Davydova proposes an additive model for annual flows, consisting of the harmonic components at the periods mentioned above and a random term. This model is then used to simulate the flows of a number of rivers. It is to be noted that for a valid test, the model must be applied to years that are not part of the record used in its derivation.

The paper by Sarmanov gives some results from the theory of the symmetrical and non-symmetrical multivariate gamma distribution developed by the author in previous papers. The gamma distribution has been shown to fit well many hydrological series, and consequently, in studying the correlation between two such series, we must take this fact into account. Since it is not possible to find an exact transformation from the gamma to the normal distribution, this problem has been approached in the past with the use of an equi-probability transformation (Thom, 1957; Cavadias, 1958). This is a univariate, numerical procedure and does not permit study of the general properties of the multivariate gamma distribution. The paper by Sarmanov is therefore an interest-addition to the literature, especially when it is supplemented by applications that permit definition of the areas of validity and the limitations of the approach.

It is to be noted that, in most areas of the application of statistics, special methods have been developed and adapted to the nature of the variables encountered. It is hoped that hydrologists will not only take over useful elements from existing statistical theory, but they will continue developing specific methods, adapted to the nature of hydrological variables.

REFERENCES

Box, G. E. P. and Jenkins, G. M. (1970) *Time Series Analysis, Forecasting and Control:* Holden-Day, San Francisco.

Cavadias, G. S. (1958) An approach to forecasting the spring runoff in Quebec. *Proceedings of the Western Snow Conference.*

Jenkins, G. M. and Watts, D. G. (1969) *Spectral Analysis and its Applications:* Holden-Day, San Francisco.

Rodriguez, I. and Yevjevich, V. (1963) *Proceedings of the Business and Economics Section:* pp.94–131. Amer. Stat. Assoc.

Rodriguez, I. and Yevjevich, V. (1967) Sunspots and hydrologic time series: *Proceedings of the International Hydrology Symposium,* September 6–8, Fort Collins, Colorado.

Rosenblatt, H. M. (1963) Spectral analysis and parametric methods for season adjustment of economic time series: *Proceedings of the International Hydrology Symposium,* September 6–8, Fort Collins, Colorado.

Shiskin, J. and Eisenpress, H. (1957) Seasonal adjustments by electronic computer methods. *J. Amer. Stat. Assoc.* 52, No.280, 415–449.

Thom, H. C. S. (1957) A statistical method of evaluating augmentation of precipitation by cloud seeding: *Final Report of the Advisory Committee on Weather Control:* Washington.

Tick, Leo J. and Woodbury, Max A. (1963) Some time series techniques useful in life sciences. *Proceedings of the IBM Scientific Computing Symposium.* Statistics, pp.265–275.

Vitha, O. and Soucek, U. (1967) Variability in hydrologic phenomena. *Proceedings of the International Hydrology Symposium,* September 6–8, Fort Collins, Colorado

DISCUSSION

M. J. Hall:

My co-author and I wish to thank Professor Cavadias for his useful comments, both written and verbal, on our paper. He has, however, raised two points which we feel require some further elaboration.

Firstly, we wish to point out that the measuring technique described in our paper was specifically devised with the object of quantifying the errors associated with the widely-used standard methods of velocity–area discharge estimation using current meters. Our procedures are unsuitable for routine gauging operations owing to the time and effort required to transform the raw data into a computer-compatible form. Even with the aid of a chart reader connected directly to a card punch, we estimate that each hour of the velocity time series absorbed three man-hours of effort. While the need for improved techniques of discharge estimation in rivers is recognized, the current meter is still one of the most widely-used field instruments in hydrometric practice (and probably will remain so for some years to come in many areas of the world), yet there is relatively little information available on the errors associated with discharge estimation by the velocity–area method.

Secondly, although we refer to the form of the spectra shown in Figs.5–7 of our paper (vol.1, pp.35–37) resembling that of a low-order autoregressive process, our intention at this stage of the work was not to fit such models to the data. The point raised by the general reporter in his verbal presentation concerning the use of partial autocorrelation coefficients is well taken; however, our initial objective was to establish that there exists an obvious correlation structure with the time series of measurements in a river. The presence of such a correlation structure throws considerable doubt on the assumptions of randomness in such series made by previous investigators (e.g. Kalinske, 1945; Dickinson, 1967).

Although the velocity time series which we analysed did not appear to possess dominant periodic components, other investigators, who have studied the behaviour of rivers in the same region of the British Isles, have reached a different conclusion. For example, Johnson (1971) reported the presence of several harmonics with wavelengths in excess of 25 min at three gauging stations maintained by the Yorkshire Ouse and Hull River Authority. The amplitudes of the oscillations quoted by Johnson are surprisingly large. At Skelton on the River Ouse, maximum amplitudes equivalent to 25 per cent of the average velocity at 75 per cent of the depth and 12 per cent at 25 per cent of the depth were recorded. However, the average velocities of the time series obtained by ourselves at Kilgram on the River Ure were approximately three times as large as those quoted by Johnson (1971) for the River Ouse at Skelton. This difference tends to indicate that well-defined periodic components may only be observable in deep and slowly-moving rivers.

Since submitting our paper, we have had the opportunity to extend our analysis to the velocity time series recorded with the Hilger Watts meters. These data provided filtered spectra which were virtually identical to those obtained with the Ott meters at the same points of depth. In addition, studies of the cross-correlation between time series recorded at different points in the vertical have tended to indicate the presence of a complex resonance phenomenon in the suspension of the (lighter-weight) Ott meters. These results will be discussed more fully in a separate report.

REFERENCES

Dickinson, W. T. (1967) Accuracy of discharge determinations. *Hydrology Paper No.20*, Colorado State University, Fort Collins, Colorado.
Johnson, D. (1971) Velocity oscillations and current meter measurements: report by E. M. Wilson on a lecture to the Pennines Hydrological Group. *Proc. Instn civ. Engrs.* 49, 405–406.
Kalinske, A. A. (1945) Application of statistical theory to measurements of velocity and suspended sediment in rivers, *Trans. Am. geophys. Un.* 26, 261–264.

R. J. G. Bloomer:
I should like to follow Hall's remarks and explain the concept of some further work that is being done in Great Britain on this subject. Quite obviously, one of the most

vital aspects concerning the fluctuations in velocity is its effect on the short-term fluctuation in river discharge. Some experiments have recently been carried out by the Yorkshire River Authority on behalf of the Water Resources Board in which 40 current meters were placed in a river section simultaneously. Readings from these meters were taken every 30s for a period of many hours. It is intended, once the vast quantity of data obtained have been reduced to a computer compatible form, to analyse these data and to see whether the results that Hall has already described, so far as velocity at a single point is concerned, are applicable to the integrated velocity, that is the discharge in the river.

I would like to refer secondly to Professor Cavadias's comments on advanced methods of gauging. I think many of us are well aware that the usual forms of gauging adopted at present are in some respects rather crude. An experiment currently in progress in the United Kingdom is, among other things, studying the feasibility of using ultrasonic methods of average velocity determination. Preliminary study is also under way at present on the electromagnetic technique for flow measurement, wherein the magnetic flux created by the mass flow of the river is calibrated and used to obtain an instantaneous measurement of discharge.

E. Hansen:

I would like to make a few comments on the paper by Hall and Johnston. The authors consider the problem of determining the mean velocity at a point in a turbulent stream based on a velocity record of limited length. From the measurements, they conclude that the velocity fluctuations are not completely random (white noise). This is a well-known fact from the theory of turbulence.

In their conclusion the authors write 'Owing to the presence of a correlation structure, the effective number of independent observations available in computing the standard error of the mean velocity will be less than the total number of data'. This is a correct but, I would say, rather vague statement. Should it not be possible to say something more specific about how the standard error of the sample mean depends on the length of the record?

For this purpose let us consider a stochastic process $\{x(t)\}$, having a mean value μ_x and an autocovariance function $C_x(\tau) = \sigma_x^2 e^{-T/T_x}$, where σ_x^2 and T_x are the variance and the time scale, respectively. This is the continuous analog to the first-order autoregressive model for discrete data. Based on a sample record of length T, we may obtain the following estimate, $\bar{\mu}_x$, of the population mean μ_x:

$$\bar{\mu}_x = \frac{1}{T} \int_0^T x(t)\mathrm{d}t \tag{1}$$

$\bar{\mu}_x$ is a stochastic variable and it is easy to show that it is an unbiased estimate of μ_x. Furthermore, it is easy to show that the variance of $\bar{\mu}_x$ is obtained from the expression

$$\mathrm{var}\ \bar{\mu}_x = \frac{2}{T} \int_0^T \left(1 - \frac{\tau}{T}\right) C_x(\tau)\mathrm{d}\tau \tag{2}$$

or by the introduction of the exponential covariance function we find

$$\mathrm{var}\ \bar{\mu}_x = \sigma_x^2 \frac{T_x}{T} \left\{1 + \frac{T_x}{T}(e^{-T/T_x} - 1 - 1)\right\} \tag{3}$$

This equation shows that the standard error of $\bar{\mu}_x$ depends on the ratio of the length of the record T to the time scale T_x of the process $\{x(t)\}$.

Interpreting $\{x(t)\}$ as the turbulent velocity process, it becomes obvious from (3) that, for larger values of T/T_x, the sample record will only contain information about the true mean value corresponding to a number of $\frac{1}{2}T/T_x$ independent observations.

Thus, from this analysis, we have learned that the time scale T_x of the underlying process plays a central role in evaluating the sampling properties of the sample mean.

I. P. Druzhinin:

I would like to draw your attention to one important feature of the paper presented by Grinevich and his associates.

Ten years ago at Irkutsk, we initiated some studies on the models of the multi-annual oscillations of river runoff. The studies were undertaken because of the discrepancies between the models and the real properties of the above-mentioned process. Professor Yevjevich in his works is particularly concerned with one of these discrepancies.

We tried to evaluate the amount of information upon which the models were relying at that time. We came to the conclusion that the result was just 25–30 per cent of all necessary information.

As was underlined in the general report, the important feature of the paper presented by Grinevich *et al.* is not only the development of a set of models describing a complex process, but also complementary analysis of some unknown properties of this process. These properties pertain to the grouping of wet and dry years. This phenomenon was almost disregarded in other models. However, we attach considerable importance to it since it indicates other sources of information which so far are unused. There is a large amount of unused information in the hydrometric observations recorded in the past.

These facts justify our hopes for considerable improvement in the mathematical models describing the oscillations of the river runoff process. The work of Grinevich and his colleagues is one attempt leading in this direction.

M. J. Hall:

We should like to thank Eggert-Hansen for his supplementary remarks on the need for records which are long compared with the time scale of the process when computing unbiased estimates of the mean for realizations from a stochastic process with an exponential covariance function. If he consults the references cited in our paper, he will find that the assumption of such a covariance function, although well established in turbulence theory, is by no means universally accepted for velocity measurements by a current meter in a river. Our results indicate that the assumption of randomness in the time series of velocities made by previous investigators would not be correct at the Kilgram gauging station, and that a stochastic model having an exponential covariance function would indeed provide an adequate representation of the time series. The fitting of such a model would be a prerequisite to the determination of optimum exposure times for current meters in river gauging.

N. Kottegoda:

The chairman, Dr. Quimpo, and, earlier, Professor Cavadias, made some comments about the number of parameters to be used in an autoregressive process. I would like to give a few points from my own experience. There seems to be three basic methods of determining the number of parameters which one could estimate from a short series of data. All these are based on the serial correlogram or the spectrum. Partial autocorrelation coefficients have been dealt with by Jenkins and Watts (1969, chapter 5), and these could give a fairly good idea of how far we should go in adding parameters to a process. But that alone may not solve the problem nor give the best model for the system. As Dr. Quimpo added, we might also look at the variance explained by the auto-correlation coefficients.

My experience of this is that more than about 90 per cent of the variance which one could account for by an autoregressive process can be explained by the first term. However, rivers with a high groundwater component may be different. This is due to basic carryover effects. In such cases, one should include additional terms. In commonly found types of monthly series, a first-order Markov process explains up to about 95 to 99 per cent of this variance, and practically nothing more is achieved by the additional terms.

There is a third and perhaps more important technique for determining the number of parameters, and this involves the examination of the serial correlogram of the residuals, for independence. We have been led to believe that the 95 per cent confidence limits will give us a satisfactory answer to this. If we accept this hypothesis, its limitations notwithstanding, we will find that, in the case of most series, including daily series, about four or five terms will suffice for an autoregressive process.

I think that through the judicious use of all three methods one could arrive at an optimum number of parameters for a particular time series.

REFERENCE

Jenkins, G. M. and Watts, D. G. (1969) *Spectral Analysis and its Applications:* Holden-Day, San Francisco.

D . R. Dawdy:

The problem presented by Kottegoda and Hall is an important one is stochastic hydrology. The problem is how to preserve long persistence with a small number of parameters. The parsimony of parameters is important operationally. As Fiering (1967) pointed out, Markovian models cannot preserve persistence without the use of a model of order of about one-third the length of the persistence. The fractional gaussian noise models (Mandelbrot and Wallis, 1968) and broken-line models (Mejia, 1971) are able to preserve persistence with a parsimony of parameters. Probably other alternative models will be developed. Another problem is how to identify the need to preserve long-term persistence. This need must be identified before models which preserve persistence in any form will have other than academic interest.

Perhaps the answer to our Russian colleague's question of how to model long-term drought (sequences with long runs) is not to be found in 11-year sun spot cycles or other such deterministic cycles of global phenomena. The fact that these global phenomena have different effects on adjacent basins is a contradiction of the theory. Droughts and floods are always occurring somewhere, sunspots or not. Certainly droughts cannot be modelled by lower order autoregressive models. Rather, they must be modelled by stochastic models which include long-term persistence in their structures. Operationally these should be models with few parameters. The parameters then should be determined in such a way as to preserve statistical characteristics of interest to the modeller. That approach is partly the answer to many of the questions being raised in many of the papers on stochastic hydrology in this Symposium.

REFERENCES

Fiering, M. B. (1967) *Streamflow Synthesis:* Harvard University Press.
Mandelbrot, B. B. and Wallis, J. R. (1968) Noah, Joseph and operational hydrology. *Wat. Resour. Res.* 4, No.5.
Mejia, J. M. (1971) On the generation of multivariate sequences exhibiting the Hurst phenomenon and some flood frequency analyses. Doctoral Dissertation, Colorado State University, November 1971.

V. Yevjevich:

The problem of selecting the number of parameters in mathematical models is not simple to answer. In general, the more parameters used in a model, the more data are needed to estimate these parameters reliably. The problem is one of optimization between the number of parameters and their reliability. The question has been posed by Dawdy whether the currently used stochastic models are good or not. The point beyond that question is what the physics of a phenomenon implies. What is in nature which makes the dependence to be of a Markovian or any other type? Basically, one should ask oneself, what are the physical factors and/or processes which makes a given type of dependence? The solar radiation input to the earth's surface, as the prime mover of the water cycle, has a high stochasticity but relatively small dependence in time. What makes the high persistence in many hydrological processes, such as runoff and evaporation, are the physical processes of water storage and outflow. The Markovian models should be defended because the physical and the statistical analyses show that the autoregressive models are often the best and simplest models to use in practice. From all the linear models available, the Markovian models (as moving average schemes with an infinite number of terms) have been shown to fit dependences of most hydrological stationary stochastic processes well. One can produce a multitude of linear models of moving average type, but which one of those models is predetermined by natural physical processes? As was pointed out in the opening remarks, a bridge is necessary between the deterministic and stochastic approaches to hydrology. A physical understanding, like Professor Kalinin showed for groundwater, is necessary to determine what in nature makes for a type of stochastic model to be valid. The differences in the concepts of persistence in hydrology will not be resolved by stating one or the other position, but by an analysis of the physics of geophysical and hydrological processes.

In the non-stationary approach of generating hydrological series, the autoregression technique is rarely used for time units less than a month. The autoregression approach to daily flows would require 365 autoregression equations, with several thousands of parameters. It is not feasible in practice. The rapid increase in the number of parameters required to be estimated with a decrease of time interval of discrete series limits the autoregression technique of the non-stationary approach to the description and simulation of hydrological processes. The question of the optimum of the number of parameters in hydrological models will surely be posed very often in the future, and it will not be settled without a better understanding of the physics of hydrological processes.

E. Plate:

I would like to address myself to the problem of how best to represent data by a Markov model. It is customary to use the lag-one correlation coefficient obtained in the usual manner from adjacent data points. However, this coefficient depends on the spacing of the data and is therefore of no particular analytical significance. One should keep in mind that a first-order Markov process implies a certain form for the complete autocorrelation function, $R(\tau)$, which is

$$R(\Delta\tau) = R(0)e^{-\beta\tau} \tag{1}$$

This funciton is determined by the coefficient β which should be selected to give the best fit of equation (1) to the whole autocorrelation function rather than just to the one value

$$R(\tau) = R(0)e^{-\beta\Delta\tau} \tag{2}$$

which is the lag-one correlation. One should therefore replace the present technique of lag-one calculation by the method of fitting equation (1) to the data by a suitable curve fitting technique. I do not wish to create the impression that I advocate the representation of hydrological data using a Markov model, but I do think that if one finds it expedient to use this model, one should determine β in a more sensible manner than by the use of the lag-one calculation.

I. Rodríguez-Iturbe:

In the last comment do you say that the second order autoregressive scheme can be different in some manner from the first-order Markov model in the framework of the paper? I do not think so. No autoregressive scheme will have a second derivative theory in the continuous case (of course this is no argument for discrete cases), and all that Plate implied comes true for any order autoregressive scheme. In fact, many of the crossing properties of all autoregressive schemes are the same if the first autocorrelation coefficient is the same. Therefore, I think that it is doubtful what can be gained by just increasing the number of parameters of the autoregressive schemes.

C. C. Kisiel and L. Duckstein:

Structural analyses of hydrological sequences can be pursued within the following frameworks: scientific, conceptual, statistical inferences, and decision. The classification is arbitrary and there may be some overlap. The theme of this discussion is that structural analyses must clearly define the objective function. Are they to develop fundamental scientific understanding or are they intended to solve an immediate design problem? The former is, generally, a long-term continuing activity that at times leads to interminable and indecisive dialogue. The latter goal requires acceptance of the situation as it is at the time of decision: uncertainty in models, model parameters and socio-economic losses to be incurred as a consequence of errors of estimation. Imbedded in the latter goal is the relative importance of variables entering the decision: physical (chemical, biological), hydrological (models), economic and social. If, in a given management problem, the hydrological models are not important, why should time be invested in their improvement? And if this still holds true across a large class of management problems, why persist in the effort except perhaps for the personal satisfaction of having reached some measure of scientific understanding? But even for scientific investigation we have a limited budget. Hence, a need exists for operations research studies of models and of data. When a conflict exists, is it not better to investigate areas which influence a decision? One may argue that data taken today may influence decisions to be made in the future using new models, new developments and new technologies. It is thus important not to jeopardize the probability of future generations using long-term records at locations which are of no importance at present. We suggest that prospective studies using perhaps the Delphi method, be included in the operations research studies of models to be developed and data to be taken.

The structural analyses have been largely pursued in a conceptual modelling framework. The scientific approach involving mathematical physics (e.g. in terms of the partial differential equations; mass, momentum) is largely neglected. Why not synthesize very detailed hydrographs based on the entire set of basic conservation equations for the basin, so as to judge what is lost in models of monthly and annual time series? So much of the orientation is in terms of *a priori* Markov, self-similar and other stochastic processes. The data analysis and subsequent parameter estimation is forced into the mould of those processes. But who has demonstrated that these processes are consistent with the hydrodynamics of the hydrological cycle? Or does it make no difference? If so, why?

D. R. Dawdy:

My viewpoint, perhaps, has not been stated clearly in my remarks so far. I do accept linear approximations in hydrology. Markovian models do solve certain problems. However, there seems to be some who, either think that Markovian models represent the real world, or that Markovian models can solve *all* problems. Some discussers seem not aware that alternative models are becoming available to solve some of the problems discussed here. Plate pointed out that one of the problems in Markovian modelling is that the parameters have meaning only in the continuous case. Crossing properties for Markovian models are determined by the first-order serial correlation. Therefore, many of the statistical properties which we wish to preserve are determined by an arbitrary partitioning of the data. The choice of a model should be made in a decision framework. If a Markovian model will serve for the job at hand, by all means use it. If it does not, one should not be forced to use it. However, alternative models are often misrepresented and, therefore, misunderstood. Models should be investigated concerning their properties, the stochastic properties which each model can preserve, and the need for determining what properties should be preserved. Hopefully, we would chose a model with as few parameters as possible which will serve our purpose. The playing around with statistical tests to determine the order of a Markovian process to model water resources problems is, to a great extent I think, a fruitless endeavour, much like a dog chasing its own tail. Alternative models with few parameters may preserve the extreme values, the long runs, the droughts. These alternative models to the Markovian are a valid field for research. Anyone who says that such alternative models should not, or are not, worth studying is only fooling himself. Let us hope that he does not fool others.

A. Reznikovszky

Yevjevich was discussing the possibilities of using some physical criteria for the evaluation of the quality of the stochastic models which we use in hydrology. I think it is very difficult to make such physical assessments of our models, since the formation of hydrological phenomena follows very complex rules.

I doubt, therefore, if any significant progress can be made in this respect. But, consequently, it should be stressed that the stochastic models in hydrology are not the targets in themselves. They are just the tools for the solution of various design or operational problems of the water resources system which will be discussed at the end of this Symposium. I think — and I have sufficient reasons for such a conclusion — that if we compare solutions of water management problems achieved by the application of different hydrological models, in very many cases the resulting designs or operational rules for water resources systems will not be so very much different.

I think that in such a case many controversies, including the present discussion, will be inessential. Just in case the solutions are different to a significant degree, we shall concentrate on a more thorough physical analysis of the nature of these processes which are modelled by means of the stochastic models. In this case, most probably, the application of the methods and criteria of the statistical decisions theory would be fully justified. We should be more cautious with the decisions in the design of the top-importance systems and we can afford less caution when the system in question is less important.

M. Roche:

Je ne sais pas si ce que je vais dire se rapporte bien au sujet de la séance de ce matin, mais comme Cavadias a lui-même évoqué ce point, je me permets de faire quelques commentaires à ce sujet. Il s'agit de l'influence des facteurs physiques sur les séries climatologiques et hydrologiques et, en particulier, de ce fameux problème des prétendus cycles ou de la périodicité des éléments climatologiques ou hydrologiques.

J'ai toujours été très étonné qu'on se contente de poursuivre cette analyse directement sur les séries elles-mêmes. Quels sont en effet les facteurs extra-terrestres qui peuvent agir sur ces chroniques? Ce sont soit les quantités d'énergie qui proviennent de l'extérieur du système terre-atmosphère, soit les influences que pourraient avoir les attractions des différents corps célestes sur la répartition des masses d'eau et des masses d'air dans l'atmosphère.

Passons sur les variations cycliques que la terre engendre elle-même par sa rotation autour du soleil et qui constituent les variations saisonnières bien connues. Si l'on examine ces différents facteurs, on s'aperçoit qu'il n'en reste guère plus qu'un qui puisse être pris en considération: la quantité d'énergie solaire qui fait marcher la marmite d'où sortent les précipitations et les débits des rivières qui en sont la conséquence. Donc, pour mettre en évidence une variation cyclique quelconque due à une influence extérieure au système terre-atmosphère, on doit normalement la rechercher dans la variation des quantités d'énergie solaire disponible aux confins de l'atmosphère. J'avais toujours cru que cette quantité d'énergie était constante, puisque aussi bien on l'appelle constante solaire, et je n'ai pas eu connaissance jusqu'à présent de résultats tendant à prouver qu'elle varie et que, par exemple, l'existence des taches solaires a sur elle une influence notable. Je pense donc que, s'il y a quelque chose à chercher, c'est dans ce sens qu'on doit le faire et je demande si quelqu'un dans l'assemblée est au courant d'études de variations de cette constante solaire, faites de manière scientifique et statistique.

Closing remarks by the general reporter

In my closing remarks I would like to comment specifically on some of the fundamental questions raised in the papers of this session.

Two of the papers were devoted to the controversial question of long-term cycles in hydrological variables. Although most hydrologists at the present time view the existence of such cycles with some scepticism, it is also true that in many recent symposia a number of papers have been presented supporting their existence.

Consequently, since the question cannot be considered as settled, I would like to invite more comments from the audience on this topic. Several authors and discussers have dealt with the question of the properties of historical records which should form the basis for simulation models of hydrological variables In this connection, there is a fundamental question that merits consideration: should we adopt completely the operational approach discussed by Kisiel, according to which we must give preference to the preservation of the properties relevant for the operation of a particular project? We must not forget that the development of hydrology as a branch of geophysics is relatively recent, and it is very important when building hydrological models to keep physical considerations in mind and not to subordinate them completely to operational efficiency. The exclusive use of operational criteria in judging the performance of simulation models may result in a situation parallel to that of the early, purely empirical days of hydrology, thirty years ago.

To be more specific, assume that we have a relatively short hydrological record, which is just one realization of the unknown underlying process. According to the present state of the art of simulation, we can preserve several important properties. such as mean variance, autocorrelation coefficients, zero-crossing statistics, of the sample records. However, given that we can only reproduce the properties of the sample and not those of the underlying process, it is necessary to keep physical consideration, in addition to operational characteristics, always in mind, in order to have reasonable confidence that the simulation model represents the underlying process. This is a question on which additional comments from the audience would be helpful.

Session IV   27 July 1971

General Reporter   Dr. Ignacio Rodríguez-Iturbe, Associate Researcher, Instituto Venezolano de Investigaciones Cientificas, Caracas, Venezuela

Chairman   Mr. Leo R. Beard, Director, The Hydrologic Engineering Center, US Corps of Engineers, Sacramento, California, USA

Structural analysis of hydrological sequences — IV

Ignacio Rodríguez-Iturbe

The papers discussed in this general report are as follows. All are published in volume 1 of the Symposium Proceedings.

1. *I. V. Khomeriki*   Elements of runoff periodicity in tasks of modelling
2. *S. G. Agarkov, I. P. Druzhinin and Z. P. Konovalenko*   Cyclical structure of the hydrological series and the nature of individual components
3. *N. S. Korganova and N. A. Kartvelishvili*   On deterministic components in natural streamflow
4. *Ramachandra A. Rao and R. L. Kashyap*   Stochastic difference equation modelling of hydrological processes
5. *P. E. O'Connell*   A simple stochastic modelling of Hurst's law
6. *P. Torodovic and D. R. Dawdy*   Stochastic point-rainfall simulation
7. *J. R. Wallis and N. C. Matalas*   In hydrology *h* is a household word

Introduction

A very general classification of mathematical models for use in hydrology can be made when looking at the purpose of the modelling process. Thus, we may have models for short range analysis and models for long range studies. Short range analysis arises mostly in the input—output type of studies, where the deterministic approach has been most useful because of the purpose of the model: one is looking for the output which the model predicts when it is subjected to a certain input. Long range studies deal mostly with problems where hydrological simulation is necessary, in this case a stochastic model is used to preserve certain estimates of the statistical properties of a measured record, and to use those estimates to generate many equally likely sequences.

The equally likely sequence can each be used in the place of the historical record, and a spectrum of equally likely optimal designs may be developed (Rodríguez-Iturbe *et al.*, 1971a).

Six out of the seven papers reviewed in this general report can be seen to be dealing with the long range type of modelling, the only exception is the paper by Todorovic and Dawdy which may fit in any of the two categories mentioned above. We will not discuss the papers according to their numeration but rather as they seem to fit into the general discussion.

Rainfall simulation

Rainfall simulation presents special characteristics which makes it quite different from runoff simulation. When working with runoff data, we can fix the sample size according to the purpose of the analysis. On the other hand, with rainfall processes, we are dealing with a random number of random quantities because, not only is the amount of rainfall a random variable, but also the number of storms in any given period has to be considered as a random variable. The model presented in the paper by Todorovic and Dawdy can be used to study the distribution of the number of storms and the amount of precipitation during storms in a given period of time. A stochastic model of this kind will be very useful for stochastic streamflow generation when used as the input in a structure-imitating deterministic rainfall-runoff model. Thus the model will build-in the proper autocorrelation structure for the streamflow record, and the independent stochastic component will be supplied by the rainfall input generated by a scheme like the one proposed by Todorovic and Dawdy.

Among the random numbers of occurrences, the maximum value or values may be of most interest and it is precisely the distribution of the maximum storm rainfall in time that constitutes the main result of this paper. To determine the previous distribution, two other distributions must be specified. The first is $H_v(x)$, the distribution of the size of the maximum storm, given that v storms have occurred. The second is $P(E_k^t)$, the probability that k storms will occur in a time period of length t.

Todorovic and Dawdy assume that the times between rainfall events are independent identically (exponentially) distributed random variables. From there it follows that $P(E_k^t)$ is a Poisson distribution. For $H_v(x)$, they assume an exponential distribution and then proceed to simplify their general expression for the distribution of maximum storm rainfall in time.

The paper by Todorovic and Dawdy is an original and interesting work which should be the source of further research to relax the assumptions made with respect to the distributions, $H_v(x)$ and $P(E_k^t)$.

Streamflow simulation

The class of models most frequently used for stochastic streamflow simulation has been the Markovian, in which the present value of the process is a function of its own past, and a random component independent of the past,

$$x_t = a_1 x_{t-1} + a_2 x_{t-2} + \ldots + a_n x_{t-n} + \epsilon_t \tag{1}$$

The parameter n represents the memory of the process or the span of time during which an event meaningfully affects the following events. These models are supposed to preserve the lower order moments (mean, variance and sometimes, skewness) and the first, or higher, autocorrelation coefficients. Rodríguez-Iturbe (1969) showed that, unless sequences are quite long (by hydrological standards) the reliability of the estimates of the lag correlation coefficients will be unacceptable and, because of that, only those cases where $n = 1, 2$ have practical hydrological importance.

The paper by Rao and Kashyap presents an algorithm for the parameter identification of a general class of models which include, as special cases, the autoregressive, the moving average, and the mixed type of models. The basic model under consideration is written as

$$y(i) = a_0 + \sum_{j=1}^{n} a_j y(i-j) + \sum_{j=1}^{m} a_{n+j} w(i-j) + w(i) \tag{2}$$

where $y(i)$ are the values assumed by the process, also referred to as the measurements, and $w(i)$ is an independent random process uncorrelated with past values of $y(i)$.

The algorithm is recursive and the minimization of the prediction error in the least squares sense is the criterion used in its development. At each time step are computed the estimates of the parameters in the model as well as the predicted estimate of the variate at a future instant. Personally, I have no objection to the development of the algorithm but I do not agree with Rao and Kashyap on several important points.

Rao and Kashyap maintain that the main reason why much of the work so far has been carried out using models with a small number of parameters is the amount of computation involved in the identification of parameters for models like equation (2).

They go further and say that 'whereas the system identification problem has received considerable attention in hydrology, the statistical prediction of hydrological events has not been studied in depth'.

I believe that the reasons for using a small number of parameters in models like equation (2) are not mainly computational but arise from two other aspects:

(1) In most hydrological time series, no advantage is gained with the increase of the number of parameters. In other words, when a model is behaving poorly what is wrong is probably the model itself and not the number of parameters.

(2) The statistical reliability of the estimators is extremely poor for high-order models and thus their use is not guaranteed.

I also believe that there is some confusion in this paper in respect to 'prediction' and 'simulation'. The authors state that 'perhaps more important is the use of prediction in the operation of water resources systems' and that 'the well-known prediction methods based on autoregressive models can easily be used for prediction, if their parameters are known. However, the use of autoregressive or similar models becomes cumbersome if the parameters are updated at every time step and the updated version of the model is used for prediction'.

We believe that in hydrology, models like equation (2) are mostly used for simulation with the purpose of generating sets of equally likely traces and it is the spectrum of these equally likely sequences which is going to be used in the design and operation of water resources systems. Models like equation (2) are not used to predict what is going to happen in the future. If this were the case, then there would not exist a stochastic philosophy behind them. Because simulation and not prediction is the main purpose of these models, we do not worry about updating their parameters at every time step.

Rao and Kashyap use three criteria for the comparison of the performance of the models:

(1) The mean square error

$$\frac{1}{k} \sum_{j=1}^{k} e^2(j)$$

(2) The mean of the absolute values of the error

$$\frac{1}{k} \sum_{j=1}^{k} |e(j)|$$

(3) The mean value of the error

$$\frac{1}{k} \sum e(j)$$

where $e(j)$ represents the prediction error or difference between the real value and the model value.

We believe that these criteria could be used as a first step in the comparison of several models and for the rejection of those models which give much higher values than others, but they should not be used to choose one model over another when there is no great difference in the values obtained. A more meaningful evaluation of a stochastic model can be made by studying whether it is able to reproduce certain parameters which are not built into its structure and which have important hydrological significance (e.g. extreme values). Besides, a stochastic model by its very nature is not supposed to give a perfect fit to the historical trace because this is only one of the many which occur according to the probability distribution built into the model (Rodríguez-Iturbe et al., 1971a).

The authors also perform a logarithmic transformation on the model and it would be interesting if they would comment on the following two points:

(1) How do they deal with zero flows?
(2) How do they preserve the statistics of the original series?

The paper by O'Connell also deals with a model which is a mixed moving average-autoregressive process, but now the emphasis is on the modelling of the so-called Hurst's law. This states that

$$R_n = Kn^H \qquad H > 0.50 \tag{3}$$

where R_n represents the adjusted range of a geophysical time series, n is the length of the series and H is a constant for each series with a 'typical' value of 0.72. Hurst (1961) also showed that, asymptotically, for a normal independent series

$$E(R) = 1.25\, n^{0.50} \tag{4}$$

This result was also proved using a different approach by Feller (1961). Mandelbrot and Van Ness (1968) showed that the theoretical asymptotic proportionality to $n^{0.50}$ holds for any finite memory process, but this type of process may exhibit the n^H behaviour initially as a transient state. Nevertheless, extensive experiments by Wallis and Matalas (1970) tend to discredit the transient hypothesis in respect to hydrological time series. The investigation of Mandelbrot and Van Ness (1968) had a direct impact upon hydrological modelling where Markovian models had been mostly used. This was pointed out by Mandelbrot and Wallis in a series of papers (1968, 1969, a, b, c, d) where they proposed discrete approximations to the process known as fractional gaussian noise (FGN) which, in turn, can be proven to behave according to the n^H law. Thus, the FGN type of models can preserve the Hurst phenomenon and this is an advantage over the traditional autoregressive models. Nevertheless, they also have their disadvantages which are studied in some detail by Rodríguez-Iturbe et al. (1971 b, c) and commented on by Scheidegger (1970) from a different point of view.

O'Connell uses a model called an autoregressive integrated moving average process (ARIMA) developed by Box and Jenkins (1970). The ARIMA model is quite flexible for the description of stationary and nonstationary behaviour in time series. The particular ARIMA model used by O'Connell is just a mixed autoregressive moving average process where the order is 1 for both the moving average and the autoregressive schemes:

$$Z_t = \psi Z_{t-1} + A_t - \theta A_{t-1} \tag{5}$$

where ψ and θ are constants and the A's represent uncorrelated random deviates (white noise). O'Connell shows extensive experiments where the model given by equation (5) behaves according to the n^H law up to large values of n. This result does not seem unexpected because Wallis and Matalas (1970) have shown that 'simple Markov models can reproduce the observed high values of K and H at the expense of greatly distorting the correlation structures of the series', which seems to be the case in the model given by equation (5). For the case of the Mississippi River near St. Louis, USA, where $r_1 = 0.294$ and $H = 0.79$, O'Connell suggests the use of the ARIMA model with a value of $\psi = 0.95$ and $\theta = 0.80$, this will mean that the following correlation structure is preserved:

$$\rho_K = 0.95\rho_{K-1}$$

and thus we have

$$\rho = 0.294, \rho_2 = 0.279, \rho_3 = 0.265, \rho_4 = 0.252, \rho_5 = 0.240, \ldots \rho_{10} = 0.186$$

This seems unrealistic. All the existing evidence is that there is a strong decreasing trend with lag in the autocorrelation structure of hydrological series. This does not mean that the autocorrelation function dies after a few lags because there is indeed a great difference in the long run behaviour of two models, one with an autocorrelation which dies shortly and the other with an autocorrelation which, although undistinguishable from zero by statistical tests, does not become effectively zero until a large amount of time lags. This last effect is clearly shown by Mejia *et al.* (1971).

A typical value of r_1 for annual runoff series can be considered to be 0.175, taken as the average of 140 river gauging stations, presented by Yevjevich (1963). In order to reproduce a first autocorrelation coefficient in the range $0.1-0.2$ jointly with a value for the exponent H of 0.7, Table 1 of O'Connell's paper (vol.1, p.177) shows that we will need to use values of ψ in the range $0.95-0.99$. This, I believe, will greatly distort the correlation structures of the hydrological series. It seems that, although a model like equation (5) may have a long memory fixed with a simple parameter ψ, it does so with an autocorrelation function which is uncommon in hydrological reality because it will have to be close to unity and the correlation structure is given by

$$\rho_K = \psi\rho_{K-1} \qquad K \geqslant 2 \tag{6}$$

There are also some questions about 'persistence' and the 'Noah' and 'Joseph' effects which I would like to address to O'Connell and also to Wallis and Matalas.

Wallis and Matalas start their paper with an interesting discussion of the Hurst phenomenon and hydrological persistence: the tendency for high flows to follow high flows and for low flows to follow low flows.

The authors analyse two measures of persistence: the serial correlation function and the Hurst exponent H. Both have their particular shortcomings which are very well discussed in the paper and which will be commented on later; more important is the fact that although both the correlation function and the Hurst exponent are closely related to persistence, neither of them is a *direct* measure of the phenomenon and thus do not permit a clear attack on the Joseph effect. We believe that direct measures of persistence are found in the so-called crossing theory which gives the tools to study the properties of run lengths, run sums and crossings. Moreover, the same parameters which control those characteristics are those which control the behaviour of the extreme values making it feasible to study the Noah effect. These ideas are presented in a series of papers by Rodríguez-Iturbe *et al.* (1971 b, c), but what we wish to emphasize at this point is that neither the study of the Hurst phenomenon nor the study of the correlation function will allow the engineer to make quantitative statements about the magnitude of the peak values and the length of the runs that a particular model is going to generate. 'It is in answer to this kind of question that we think crossing theory, and a model adapted to the conditions which the theory requires for its application, can shed the most light at the present moment' (Rodríguez-Iturbe *et al.*, 1971 b).

The comments of O'Connell and Wallis and Matalas with respect to the previous philosophy will be very welcome.

Wallis and Matalas present an interesting discussion on the estimation of H and ρ_ν. From a sequence of N events $x(1), \ldots, x(N)$, they form two sequences $x(1), x(2), \ldots, x(N - \nu)$ and $x(\nu + 1), x(\nu + 2), \ldots, x(N)$. If the two sequences are denoted $x_1(\nu)$ and $x_2(\nu)$ then they define r_ν the estimate of ρ_ν as

$$r_\nu = \frac{C[x_1(\nu)x_2(\nu)]}{S[x_1(\nu)]\, S[x_2(\nu)]} \tag{7}$$

where $C[x_1(\nu)x_2(\nu)]$ is the covariance of $x_1(\nu)$ and $x_2(\nu)$ and $S[x_1(\nu)]$ and $S[x_2(\nu)]$ are the standard deviations of $x_1(\nu)$ and $x_2(\nu)$. The authors present graphs which show that r tends to underestimate ρ, the magnitude of the bias depending upon N, ρ, and the sequence generating process. For given values of N and ρ, the bias is larger for filtered type 2 FGN processes than for Markovian processes, with the bias increasing as H increases. For given values of H and ρ, the bias decreases as N increases.

Wallis and Matalas do not mention in their paper which estimator they used to obtain the covariance function in equation (7). Generally two types of estimators are used when the observations $x_1, x_2, \ldots, x_n$, come from a discrete time series:

$$(a) \quad C_{xx}(k) = \frac{1}{N} \sum_{t=1}^{N-k} (x_t - \bar{x})(x_{t+k} - \bar{x}) \tag{8}$$

$$\left.\begin{array}{c} \\ \\ \end{array}\right\} \quad k = 0, 1, \ldots, N-1$$

$$(b) \quad C_{xx}^{*}(k) = \frac{1}{N-k} \sum_{t=1}^{N-k} (x_t - \bar{x})(x_{t+k} - \bar{x}) \tag{9}$$

The estimator $C_{xx}^{*}(k)$ is an unbiased estimator of the autocovariance $\gamma_{xx}(k)$, whereas $C_{xx}(k)$ is only asymptotically unbiased as the record length N tends to infinity,

$$EC_{xx}(k) = \begin{cases} \gamma_{xx}(k)\left(1 - \dfrac{|k|}{N}\right), & |k| \leqslant N \\[2mm] 0, & |k| > N \end{cases} \tag{10}$$

(assuming $\bar{x} = 0$)

However the biased estimator $C_{xx}(k)$ has a smaller mean square error than the unbiased one $C_{xx}^{*}(k)$.

The manner in which the authors write equation (7) suggests that they used a different estimator in which both the means and the standard deviations are computed separately for the series $x_1(\nu)$ and $x_2(\nu)$,

$$C_\nu^{**} = \frac{1}{N-\nu}\left[\sum_{t=1}^{N-\nu} x_t x_{t+\nu} - \frac{1}{N-\nu}\left(\sum_{t=\nu+1}^{N} x_t\right)\left(\sum_{t=1}^{N-\nu} x_t\right)\right] \tag{11}$$

This estimator may be biased or unbiased depending whether one divides by N or by $N - \nu$. The main disadvantage of the estimator given by equation (11) is that both mean values and both standard deviations change with the lag order.

We would appreciate clarification of this point by the authors because, if they used the biased estimator then this could be the reason for the underestimation of ρ, an underestimation that, as shown by equation (10), will decrease with the sample size N and will increase with the lag k.

The reason why the bias changes with the type of process could also be explained if we modify equation (10) for the case when $\bar{x}$ is not zero, and want to correct for the mean as given in any one of the formulae (8), (9) and (11).

The bias of the covariance estimator, $C_{xx}(k)$, is then given by

$$E[C_{xx}(k)] = \gamma_{xx}(k) \left(1 - \frac{|k|}{N}\right) - \left(1 - \frac{|k|}{N}\right) \text{var}\,(\bar{x}) \tag{12}$$

Thus, the effect of correcting for the mean is to increase the bias still further. The term var $(\bar{x})$, is, in the case of a first-order autoregressive process given by (Rodríguez-Iturbe, 1969):

$$\text{var}\,(\bar{x}) = \frac{\sigma^2[1 + \rho(1)]}{N[1 - \rho(1)]} \tag{13}$$

and although we do not know it for the discrete type 2 FGN processes, it is known that, for a continuous stationary process (Jenkins and Watts, 1969)

$$\text{var}\,(\bar{x}) = \frac{1}{T} \int\limits_{-T}^{T} \left(1 - \frac{Y}{T}\right) \gamma_{xx}(y)\,dy \tag{14}$$

Intuitively, because of the form of $\gamma_{xx}(y)$, var $(\bar{x})$ will be higher for the type 2 FGN process than for the Markov model, and this would explain why the bias in the figures presented by Wallis and Matalas is larger for type II than for a Markov model.

The type II error analysis presented by Wallis and Matalas is extremely interesting and I agree wholeheartedly with their conclusion that the inherent low power of the tests is one of the causes why the hydrologist is not able to detect long-term persistence in real life correlograms. For values of ρ_1 and N commonly encountered in flow sequences, the type II error of the most powerful of the tests was greater than 70 per cent! This clearly shows how high the probability is of accepting a false hypothesis when significance tests are dogmatically applied on hydrological correlograms, and the numbers 0.05 and 0.95 are treated with the same superstitious awe that is usually reserved for the number 13. This way of thinking has been very common in the past and the writer congratulates Wallis and Matalas for bringing out a point which is in very much need of research in hydrology.

Wallis and Matalas end their paper with the conclusion that, in order to make more reasonable the resemblance between synthetic and observed sequences of comparable length, it is necessary to use estimates of the characteristics in the synthetic generating processes that are different from those extracted from the historical sequences. This has large practical importance in synthetic hydrology and is justified because the synthetic traces will produce serial correlation coefficients and Hurst's H's which will be smaller than those estimated from the historical sequences which, in turn, are smaller than the true ones.

The remaining three papers have something in common because all of them deal with the question of periodicities in annual hydrological series.

The paper by Korganova and Kartvelishvili uses maximum likelihood techniques for investigating the presence of harmonics in annual runoff series from different parts of the world. I agree with their conclusion that the unstability of the harmonics and the insignificant values of the amplitudes 'provides all with grounds to believe that the harmonics can be explained by illusory effects due to inadequate accuracy and insufficient length of hydrological series'. Agarkov and his co-authors arrive at a completely different conclusion from that of Korganova and Kartvelishvili. They state that 'the cyclic structure of the annual river runoff of the USSR is confirmed by means of more strict methods'. I do not agree with the previous conclusion.

The method of analysis used by Agarkov *et al.* is indeed the most appropriate for the type of problems investigated, but some of their statements are not clear. They say that the coherence index 'makes it possible to estimate the course coordination of two oscillating processes having the same frequencies, but whose amplitudes and phases change irregularly over time. If the phase difference between oscillations is constant, then the oscillations are called coherent while the coherence index is 1'.

The following comments seem in order. From Cramer's spectral representation, any stationary series can be considered as a sum of components or frequency bands, each component being statistically independent of the others (Cramer, 1940). The theory of stationary processes tells us that, not only is the component centred at frequency w_j independent of all the other components of the variable, but it is independent of all components of another variable except the component centred on w_j (Granger and Hatanaka, 1964).

A direct measure of the square of the amplitude correlation at frequency w is given by the coherence function. Even if the amplitudes are fully correlated, it is possible that the corresponding frequency components will have different phases. Thus, the statement of the authors that 'if the phase differences between oscillations is constant, then the oscillations are called coherent while the coherence index is 1', does not seem completely clear.

The advantage of using cross-spectral techniques lies in the fact that, when the cross-correlation function is used, one searches for correlations for different lags in time considering each process as a whole. On the other hand when the coherence function is used, one investigates if there is any correlation between the processes in a very small range of frequencies. Because of this, the purpose of the authors in pursuing with the preliminary filtering of the series is not clear. Moreover, although linear filters do not change the coherence function, nonlinear filters do change the coherence between series (Rodríguez-Iturbe, 1967). We will appreciate the authors' comments on this aspect.

With data which go back for between 30 and 50 years, any results concerning harmonics cannot be guaranteed with periods of 9—11 years or, even worse, 16—20 years. The coherence indices shown in Table 2 (vol.1, p.88) for 10—11 year components of solar activity, atmospheric circulation and river runoff are very much effected by the shortness of the data used, their reliability seems unacceptable to me and their statistical significance does not mean very much because of the enormous type 2 errors involved in the tests.

The paper by Khomeriki is an attempt to correlate model construction and cyclicity calculations in tasks of regulation. I object to the model used by the author which gives annual runoff in the form of an additive function of a stochastic process and a periodic function where the periods are of 10, 15 and 20 years. I do not believe that there exists serious grounds for the previous assumption and, moreover, if low frequency effects are going to be taken into account for storage regulations of annual runoff — and they should be — then the model should be of a completely different kind as was previously discussed when reviewing the papers by O'Connell and Wallis and Matalas.

The writer has made an attempt to review the different papers and raise questions for reply by the authors. We look forward to the general discussion by participants of the Symposium.

REFERENCES

Box, G. E. P. and Jenkins, G. M. (1970) *Time Series Analysis, Forecasting and Control,* 553 pp.: Holden-Day, San Francisco.
Cramer, H. (1940) On the theory of stationary random processes. *Ann. math. Statist.* **41**, 215–230.
Feller, W. (1961) The asymptotic distribution of the range of sums of independent random variables. *Ann. math. Statist.* **22**, 427–432.
Granger, C. W. J. and Hatanaka, M. (1964) *Spectral Analysis of Economic Time Series,* 299 pp.: Princeton University Press, Princeton, New Jersey.
Hurst, H. E. (1961) Long-term storage capacity of reservoirs. *Trans. Amer. Soc. civ. Engrs* **116**, 770–808.
Jenkins, G. M. and Watts, D. G. (1968) *Spectral Analysis and its Applications,* 525 pp.: Holden-Day, San Francisco.
Mandelbrot, B. B. and Van Ness, J. W. (1968) Fractional Brownian motions, fractional noises and applications. *Siam Rev. Appl. Math.* **10**, No.4, 422–437.
Mandelbrot, B. B. and Wallis, J. R. (1968) Noah, Joseph and operational hydrology. *Wat. Resour. Res.* **4** (4) 909–918.
Mandelbrot, B. B. and Wallis, J. R. (1969a) Computer experiments with fractional gaussian noises. Part 1: Averages and variances. *Wat. Resour. Res.* **5** (1) 228–241.
Mandelbrot, B. B. and Wallis, J. R. (1969b) Computer experiments with fractional gaussian noises. Part II: Rescaled ranges and spectra. *Wat. Resour. Res.* **5** (1) 242–259.
Mandelbrot, B. B. and Wallis, J. R. (1969c) Some long run properties of geophysical records. *Wat. Resour. Res.* **5** (2) 321–340.
Mandelbrot, B. B. and Wallis, J. R. (1969d) The robustness of the rescaled range R/S in the measurement of noncyclic long run statistical dependence. *Wat. Resour. Res.* **5** (5) 967–988.
Mejia, J. M., Rodríguez-Iturbe, I. and Dawdy, D. R. (1971) A model for preserving the covariance function, crossing properties and the Hurst effect in long range hydrologic simulation. Part II: The broken line process as a potential model for hydrologic simulation. *Wat. Resour. Res.*
Rodríguez-Iturbe, I. (1967) The application of cross-spectral analysis to hydrologic time series. *Colorado State University Hydrology Paper, No.24:* Fort Collins, Colorado.
Rodríguez-Iturbe, I. (1969) Estimation of statistical parameters for annual flows. *Wat. Resour. Res.* **5** (6) 1418–1421.
Rodríguez-Iturbe, I. Dawdy, D. P. and Garcia, L. E. (1971a) Adequacy of Markovian models with cyclic components for stochastic streamflow simulation. *Wat. Resour. Res.*
Rodríguez-Iturbe, I., Mejia, J. M. and Dawdy, D. R. (1971b). On the estimation of floods with high return periods. *J. Hydraul. Div., Proc. Amer. Soc. civ. Engrs.*
Rodríguez-Iturbe, I., Mejia, J. M. and Dawdy, D. R. (1971c) A model for preserving the covariance, crossing properties and the Hurst effect in long range hydrologic simulation. Part I: A new look at Markovian models, fractional gaussian noise and crossing theory. *Wat. Resour. Res.*
Scheidegger, A. E. (1970) Stochastic models in hydrology. *Wat. Resour. Res.* **6** (3) 750–755.
Wallis, J. R. and Matalas, N. C. (1970) Small sample properties of H and K estimators of the Hurst coefficient h. *Wat. Resour. Res.* **6** (6) 1583–1594.
Yevjevich, V. M. (1963) Fluctuation of wet and dry years. Part I: Research data assembly and mathematical models. *Colorado State University Hydrology Paper 1:* Fort Collins, Colorado.

DISCUSSION

R. L. Kashyap and A. R. Rao:

A narrow view of the role of stochastic processes in hydrology is apparent from the comments in general and on our paper in particular, by the general reporter. It appears more propitious to comment on some of these general comments before answering the specific questions which have been raised on our paper.

In the initial remarks in the general report, the hydrological problems are divided into two categories, as the short range and the long range problems followed by the statement, 'short range analysis arises mostly in the input—output studies, where the deterministic approach is most useful'. However, the justification for the claim that the deterministic approach is most useful has not been presented at all. Deterministic methods, such as those based on unit hydrographs or the Fourier methods, yield models whose value in prediction of runoff is small. In view of such poor performance of deterministic models, the apparent reluctance to even consider the stochastic difference or differential equation is surprising.

Our surprise is heightened by the fact that is is over forty years since stochastic differential equation models (the Ohrnstein—Uhlenbeck equation) were introduced in physics to analyse problems which could not be properly explained by purely deterministic models. One of the purposes of our paper is to show the utility of stochastic models for short-term forecasting. We do not know of any deterministic models which yield forecasts whose average prediction error is of the same order of magnitude as the prediction error of the stochastic prediction schemes of our paper. We welcome the general reporter to furnish references in which the deterministic models are shown to yield better forecasts than the stochastic models.

The comments of the reporter on the long range analysis are equally vague. The phrase 'equally-likely' is well defined in probability and statistical literature and means equally probable (Gnedenko, 1900). A perusal of any standard textbook on probability theory or mathematical statistics reveals that it is not appropriate to speak of equiprobable sequences when dealing with sequences of continuous-valued random variables. An appeal to the spectrum of these sequences does not reduce the vagueness of the comments.

We will now answer the specific questions raised by the reporter after clarifying certain aspects of our prediction scheme. Our scheme consists of two phases. First of all, we postulate a stochastic difference equation model for the variate to be predicted, in which certain parameters are undetermined. We develop a method for estimating these parameters and update the estimates as new observations about the variate arrive. For prediction, we will use the time varying difference equation in which the parameters are replaced by their latest estimates. In classical methods of prediction, the undetermined parameters are replaced by their estimates computed on the basis of a fixed amount of data. The resulting time invariant difference equation is then used for prediction. We may add that our method usually leads to a smaller average prediction error than the classical method since the estimate of the parameters at each instant is based on all the available information.

Next, we consider the question of the number of parameters in the model. We have clearly mentioned in the paper that the optimal model L3 yields a mean square prediction error which is *less* than or equal to those obtained with other models having a larger number of parameters than L3. We believe this reason to be sufficient for preferring the model L3 for prediction since prediction is the principal purpose of our paper. Contrary to the reporter's general assertion, we can achieve estimates (measured by the variance of the estimate) with any desired accuracy by using a sufficiently large number of samples.

The confusion which the reporter has experienced in our paper between simulation and prediction can be attributed to his semantic misunderstanding.

It is undoubtedly correct to state that models like equation (2) are mostly used for simulation. It is, however, wrong to state that these models are not used for prediction. Further, to claim that 'if this were the case then there would not exist a stochastic philosophy behind them', is to betray an ignorance about the development of prediction theory itself. The development of prediction methods, as we have indicated in the paper, is indeed necessary for the operation of systems, just as the simulation of synthetic traces is useful in the optimal design of water resource systems. In view of this, to claim that 'simulation and not prediction is the main purpose of these models and we do not worry about updating their parameters at every time step', amounts not only to take a limited view of the uses of these models, but also to ignore economical computational procedures in parameter estimation of stochastic models. However, we agree with the general reporter in his claim that if these models are to be used for simulation, then they must preserve the correlations and other characteristics such as the average of flows in logarithmic models. We point out that logarithmic models are used only with data which do not have zero values in them, and that this is standard practice in engineering whenever using logarithmic transformations.

REFERENCE

Gnedenko, B. V. (1900) *The Theory of Probability*, 2nd edition, p.26: Chelsea Publishing Co., New York.

P. E. O'Connell:

Some comments and clarifications are necessary on the general discussion by the reporter on the above paper. For the purpose of synthetic hydrology, the hydrologist is mainly interested in generating synthetic sequences which preserve historic record estimates of ρ_1, the first autocorrelation coefficient, and of the Hurst coefficient. Low-order Markov models fail to do just this and I dispute the reporter's classification of the ARIMA (1, 0, 1) model with such models.

Wallis and Matalas (1971) have found that to model an estimated H value of 0.82 on the Colorado River at Lees Ferry, they have to use a first-order Markov model with $\rho_1 = 0.80$ whereas the observed value was $\hat{\rho}_1 = 0.21$. Hence, $\hat{\rho}_1$ was not preserved, resulting in a gross misrepresentation of the high frequency nature of the historic series. The ARIMA (1, 0, 1) model on the other hand, allows the preservation of $\hat{\rho}_1$ and $\hat{h}$ together by suitable choice of ϕ and θ. A number of combinations of ϕ and θ exist which allow preservation of $\hat{\rho}_1$ and $\hat{h}$ but allow different lengths of 'effective memory'.

A few comments about the nature of the autocorrelation structure of the ARIMA (1, 0, 1) model for lags greater than 1 seem pertinent in the light of the reporter's remarks. This is given by

$$\rho_k = \rho_1(\phi)^k \qquad k \geqslant 1$$

so that the decay of the autocorrelation function is determined by ϕ.

High values of ϕ were chosen for the simulations to illustrate that large effective memories could be achieved if required, but smaller values of ϕ could be chosen as already mentioned. The observed agreement with Hurst's law suggests that the autocorrelation function used, is, in fact, of the right form. The reporter's claim that the autocorrelation function should be one 'which, although indistinguishable from zero by statistical tests, does not become effectively zero until a large amount of time lags' does not have any supporting evidence on a yearly time scale for which Hurst's results were observed. Observed correlograms of the usual limited samples available tend to suffer from large sampling fluctuations for large lags as observed by the reporter himself. This variability is accentuated in the presence of long-term persistence where observed correlograms often bear no resemblance to the underlying autocorrelation structure. On the basis of such supporting evidence, it would appear impossible to distinguish between alternative underlying autocorrelation structures on the basis of statistical tests based on the sample correlogram. In fact, it would appear that the only real proof of a correct autocorrelation structure is agreement with Hurst's law together with preservation of $\hat{\rho}_1$, and the ARIMA (1, 0, 1) achieves this.

REFERENCE

Wallis, J. R. and Matalas, N. C. (1971). Correlogram analysis revisited. *Wat. Resour. Res.*

James R. Wallis:

As a result of the pioneering work of Box and Jenkins, the large family of stochastic processes has been an available tool for practising hydrologists to both use and misuse, and, presumably, both have been done within the last few years. The present paper by O'Connell, represents the first and only serious attempt to convert the 'ARIMA's elegant mathematic simplicity' to the level of a practical engineering tool. It is an heroic undertaking and his results are most encouraging. We hope that he continues with his Boswellian role, and that he extends the work so as to include the multivariate situation, small sample bias corrections for ρ_1, v and H, and a mundane documentation of the sampling properties of the estimators of ϕ and θ. When finished, we may well have a quick, usable way of generating synthetic hydrological sequences that preserve all desirable statistical properties.

·Minor caveats which we suggest here in the hope of preventing a premature rush to an ARIMA bandwagon are of three kinds. First, problems with computer random number generation are more widespread than are generally recognized. These can be serious and should be solved before embarking upon expositions work of this nature. For instance, in other work, we found that most existing Markov generating computer programmes do not preserve the variance expected for given N and ρ. The most common cause of this v truncation is in the choice of the gauss transform algorithm, occasionally aggravated by the use of single precision arithmetic. Our solution was to take a well-tested uniform random number generator (Lewis *et al.*, 1969), and write our own Fortran IV double precision Box–Muller transform.

The result was a programme that gives a very good representation of a normal independent process. The average of 1000 H estimates by least squares for sample functions of 10,000 using this programme is 0.532 if the fit is from time 10 onwards, and

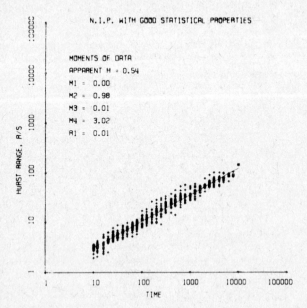

Figure 1. *R/S* box diagram of sample of 10,000 points from a normal independent process. Fifteen uniformly spaced starting points were used and estimates of *R/S* made at times 10, 12, 16, 20, 25, 31, 40, 490, 630, 780, 1000, 1250, 1600, 2000, 2500, 3100, 4000, 4900, 6300, 7800, 10, 000. Least squares fit for mean *R/S* versus time was used as the estimator of *H*.

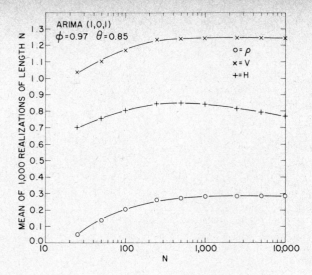

Figure 2. Mean of 1000 realizations of lag one correlation, variance, and H for an ARIMA $(1,0,1)$ process with $\varphi = 0.97$ and $\theta = 0.85$.

0.519 if the fit is from time 49 onwards. An example is shown in Fig.1. The average value of H is given by O'Connell (Table 2, vol.1, p.181) for his N.I.P. process was 0.576 which is surprisingly high in terms of our results.

Secondly, the elegant mathematic simplicity of ARIMA models may have as a corollary some practical difficulties in application. From Fig.2, it can be seen that ARIMA models have small sample biases in ρ and v. In addition, for the single ARIMA model which we investigated, it is evident that estimates of H are an unusual function of N, it will be an unpleasant complication if such relationships are found to be general throughout the ARIMA domain.

A third concern of ours is the blanket statement that fractional noises are 'not commercially viable'. Surely, the decision as to whether or not a given mechanism is commercially viable depends upon the size of the projected hydrological investment, as well as the computer hardware and software configuration that is available to the hydrologist. We agree that filtered fractional noises demand much number crunching, but we feel that they may still be useful. For instance, using an IBM 360-91 running under OS version 18, using a Fortran IV programme compiled under Fortran H opt. 2, we found that to generate 1000 points of a filtered fractional noise with $H = 0.8$, $M = 10,000$ and $p = 10$, took 7.75 s. Core storage required during execution was 170 K characters. The same programme running under CP-CMS on an IBM 360-65 took 128 s of execution time. We would assume that a carefully reprogrammed assembler language routine could result in a considerable reduction in these C.P.U. times.

Current rentals for an IBM 360-65 vary from about \$700 h^{-1} during prime time to a low of \$150 h^{-1} or less for the early morning hours and the slack weekend periods. We believe that computer costs of this magnitude may deter graduate students from selecting filtered fractional noises as dissertation topics (Mandelbrot's fast fractional noise generation procedure provides a possible alternative for those with access to only small and slow computers), but they are not so excessive as to completely exclude their use by a state agency concerned with a major investment optimization for a large river system

REFERENCE

Lewis, P. A. W., Goodman, A. S. and Miller, J. M. (1969) A pseudorandom number generator for the System/360. *IBM Syst. J.* **8**, 136–146.

N. C. Matalas:

I would like to make two comments. Number one refers to the paper by Wallis and myself concerning bias. I cannot offer you a simple explanation why we obtained the results that we did. The particular estimators of the correlations yielded biased estimates, and these biases tend to increase as persistence increases. The bias is downward; that is, the estimators tend to yield estimates whose values are less than the underlying correlations. Thus, observed sequences are seemingly more independent than is actually the case.

My other comment refers to the matter of persistence that has been discussed here today. Fractional noise models have been offered as alternatives to Markov models. In doing so, what we are discussing is not whether the world is Markovian or non Markovian, but whether the world is Brownian or non Brownian. In a fractional noise world, which does not belong to the Brownian domain of attraction, $H \neq \frac{1}{2}$.

The integral of the correlation function does not converge. The fractional noise models, introduced to hydrology by Mandelbrot and Wallis, are not 'mere moving average processes'. To think of them as such is to fail to understand them.

I. P. Druzhinin:

On behalf of all the authors of our paper, I would like to express our gratitude to Dr. Rodríguez-Iturbe for a thorough analysis of our work and for his highly interesting questions.

As I have already mentioned, about 10 years ago we initiated in Irkutsk some studies on the models of the multi-annual oscillations of the river flow. Very soon we came to the conclusion that our knowledge of the basic laws and properties of this process was inadequate. We have no more than 25–30 per cent of all the necessary information.

Consequently, we had another problem concerned with the type, properties and the amount of information which can be obtained from these sources. And, finally, we had the problem of obtaining proper tools enabling retrieval, processing and the use of various information originating from numerous sources.

The paper by Grinevich indicated that complementary information can be obtained from historic records. This source is far from being completely exhausted. Various information can also be obtained from the relationships with the geophysical, cosmic and other processes. Omission of these sources would be, at least, unreasonable.

For instance, we found the cause of the increase or decrease of the annual river runoff for about 50 per cent of the analysed series. These oscillations were caused by the rapid change in solar activity.

Oscillations of the annual runoff can be considered, therefore, as a simple change in the series of years with the increased or decreased water volume. The change of these series is unusually interesting. At present, we can assume that we know the cause of this phenomenon for about half of the analysed cases. There are also some other causes for these oscillations. For quite a long time many authors have discussed the periodic components. We decided to check this hypothesis, and we found that it has some justification. Recently, we have made one additional and precise test by the so called WMO method. Some of the results are presented in our paper. The next test concerns the coherence. Here we also found some serious justifications. We encountered some interesting properties which are related to the atmospheric circulation periods lasting for 30–40 years. When a given type of circulation period begins, it controls some part of the oscillations of the river runoff. Next, there is a period of another circulation and we have a different pattern of oscillations and a different type of transformation of the external factors – among them also the solar activity. The multi-annual observation series are genetically nonhomogenous from this point of view.

This fact should be taken into consideration in the choice of the length of the controlling series.

The above information provides an answer to the remark made by the general reporter about the evaluation of the harmonics with periods of 10–11 years for a comparatively short sample. There is no possibility of extending the sample. To achieve greater certainty, one can apply group analysis. We have made this analysis and it confirmed our conclusion very well. Other remarks made by the general reporter are also very interesting, and, I think, that here also we have an answer. In particular, relationships between the coherence index and the squares of the correlation coefficients for the definite frequencies are just certain approximations. None of the authors consider them completely accurate. For that reason some differences are quite possible. The opinion expressed by the general reporter about the linear and nonlinear filters is also very interesting. It can be said that just one of the comparative series was filtered.

I. V. Khomeriki:
First of all, I would like to express my thanks to the general reporter for the work accomplished by him.

Now let me present some explanations concerning my paper. Although spectral analysis found broad application in the structural analysis of the hydrological series, I think that some of the best forms of the mathematical description of spectral analysis have not been used so far in hydrology. I would like to mention here the Grenlander–Rosenblatt scheme presented 10 years ago. We decided to try this scheme. Moreover, we wanted to prove that the Monte-Carlo method, beside its potential use in modelling, can also be used for some other purposes. It can be used, for instance, in structural analysis for the description of the runoff cycles or for the evaluation of the influence of the periodic components in the hydrological series on the results of the water management computations. I would also like to comment on the remark made by the general reporter about the additive models which are, in his opinion, abstract, i.e. random functions plus deterministic functions when the periodic component has periods of 10, 15 or 20 years. We also consider these models abstract – we did not claim that they have a real value. We have proposed these models for the well defined purpose of analysing the dislocation of the parameters in the hydrological series as well as for proving that the evaluation of the storage reservoir parameters on the basis of some general nomograms is ill-founded.

Closing remarks by the general reporter
I wish to thank the authors very sincerely for their interesting comments and I want to add some closing remarks at this point. I have found O'Connell's presentation especially attractive. He mentions at one point in his discussion that he does not see any evidence for my claim of a stronger decrease of runoff correlograms than the one that he has in his model. I believe that this evidence exists when we look especially at the stochastic part of monthly and daily flows after the annual cycle has been removed. I also think that there are great differences in the crossing properties of processes which obey correlograms like the ones given by O'Connell and those processes with correlograms which last the same as the previous ones but that decay much more sharply. I also want to thank the other authors for their discussions. With respect to the points raised by Rao, I still believe that models like the ones used in his paper have *a major role* in hydrological simulation and not in prediction as such. For simulation, the updating of the coefficients at every time step is not very sensible. Also for hydrological purposes, the evaluation of the models should be made with criteria somehow different to those used in this paper.

Session V   27 July 1971

General Reporter   Dr. S. Dyck, Professor of Hydrology, Technical University of Dresden, Dresden, GDR

Chairman   Dr. A. Ramachandra Rao, Professor, School of Civil Engineering, Purdue University, Lafayette, Indiana, USA

Structural analysis of hydrological sequences — V
Statistical techniques

S. Dyck

The papers discussed in this general report are the following. All are published in volume 1 of the Symposium Proceedings.

1. *U. Maniak*   Flood frequency studies of rivers in the Federal Republic of Germany
2. *J. Winter*   Stochastic analysis of short-time rainfalls
3. *I. Zsuffa*   Loi de probabilité à deux variables des basses eaux
4. *A. B. Kasansky*   A statistical filtration model for the inseepage of melt water into frozen ground
5. *I. Bogárdi*   Determination of the degree of protection offered by flood control systems on the basis of distribution functions
6. *V. Stanescu and C. Catana*   Statistical analysis applied to the study of the components of a flood wave
7. *P. Guillot*   L'utilisation des modèles mathématiques pour le prévision en hydrologie et météorologie

This group of papers deals with various aspects of applying frequency analysis to hydrological events. In a rigid sense, none of the papers is in full conformity with the subject of this session, as none of them deals with sequences. The papers reflect the broad field of the application of frequency analysis in hydrology.

The paper by Winter applies frequency analysis to rainfall events. Though the term 'stochastic analysis' is used in the title of the paper, it is not a random process of rainfall that is considered, but a statistical independence of the events. For 15 rainfall stations, a depth—duration—frequency analysis is performed. The durations are from 10 min to one day. The well-known fact is stated that a good fit to empirical short-time rainfall distributions is obtained by the lognormal distribution function.

The paper by Maniak is concerned with the development of a method to determine the magnitude and frequency of momentary peak discharges for catchments with an area less than 300 km^2 in the Federal Republic of Germany. In his paper, Maniak presents results obtained with unadjusted areal peak runoffs of 121 gauging stations for the logarithmic Pearson type-III distribution.

Seventy-three per cent of the used observation series are shorter than 30 years, only three series are longer than 50 years. At first, therefore, the influence of the size of the data set on the reliability of the parameter estimation is investigated. Naturally, the skewness shows the greatest variation, both in the comparison of different stations

(Fig.1, vol.1, p.222), and in partial-duration series (Fig.2, vol.1, p.223). Figure 3 (vol.1, p.223) shows that the error is unacceptably large for short series. With that, the uncertainty of the parameter estimation is obvious. Maniak then tries to find statistical relations between the skewness and the characteristics of the basin (area, average slope of the watercourse, and shape of the basin). The results show that no statistical relation can be observed (Fig.4, vol.1, p.224). This result is not surprising, as the parameters and the characteristics of the catchments can be estimated only with limited accuracy. Moreover, the factors used for the basins may not be sufficient to describe the very different behaviour of small basins. Better results may be obtained if the territory is divided into flood-hydrologically homogeneous regions and if statistical characteristics of storm-rainfall are introduced in a multi-regression equation.

For the purpose of separating the factors affecting flood peaks, a common period of time for all records is desirable. This eliminates the variability with time, so that the effect of other factors on flood peaks may be analysed more easily. Flood peak data are often faulty, therefore, a very careful adjustment is necessary in many cases before any frequency analysis is performed.

The object of the interesting paper by Bogárdi is to develop a method for estimating the actual state of the degree of protection offered by flood levee systems along streams having a wide flood plain. It is divided into sections according to the non-uniformity of the flood protection within a system. A method is presented for estimating the degree of protection of flood control sections and a procedure is suggested for the determination of the degree of protection for a system consisting of several sections.

Bogárdi shows that the degree of protection is a random variable (Fig.1, vol.1, p.247). It is defined as the flood exposure, x_i, which the levee at the point i is capable of resisting without special reinforcement. x_i depends on the stage and on the duration of the flood. As shown in Fig.2, (vol.1, p.248), x_i results mainly from one of four possible causes for the rupturing of the levee. For a particular section along the full length, the degree of protection is determined at cross sections selected by random point spacing or by a grid method. The characteristic x_i value is obtained at the lowest of the stages leading to a rupture of the levee by one of the different causes. The empirical distribution of the degree of protection can be approximated by an extreme value distribution with an upper limit or by an exponential distribution with lower and upper limits. The probability of an original degree of protection lower than the considered peak flood is defined by the distribution function along the levee section. If the flood fighting activity at the weak points of the levee is taken into account, the non-uniform degree of protection is made more uniform. To determine the limiting value of the load bearing capacity of the particular flood control section, the maximum flood fighting activity must be estimated in advance.

For describing the combined degree of protection offered by several levee sections, the probabilities pertaining to these stages are introduced. The resulting function indicates, for floods of different probability, the probable length of the levee over which the degree of protection is smaller than the flood under consideration. It seems to me that the careful practical application of the method to larger systems is very laborious. The paper by Bogárdi is a valuable example for the possible opening of new fields of application of the probability theory.

In the following two papers, multi-dimensional distribution functions are applied to flood waves and to low flow events. Stanescu and Catana investigate attributes of flood waves originated by rainfalls in the basin of the Jiu River (Roumania), especially the relations between peak discharge and volume at a station in the outlet of the basin and the peak discharges and volumes at upstream stations. Because of the difficulties of a genetical interpretation (owing to the variability of rainfall in space and time, vegetation, soil moisture, etc.), a statistical analysis is given employing multivariate and conditional distribution functions.

At first, a test is made on the hypothesis of a two-dimensional normal distribution of peak discharge and volume at one station, and of discharges and volumes at different stations, using logarithmically transformed values of the variates. As the hypothesis is not rejected the conditioned distribution is a one-dimensional normal distribution with specific parameters, and the two-dimensional distribution can be written as the product of a conditional and a one-dimensional marginal distribution.

The three-dimensional probabilities of the volumes at three stations at the main river and one of its tributaries can also be split into marginal and conditional distributions. By repeatedly applying this method, the volume at the outlet station is partitioned into the most frequent contributions of 11 upstream stations and of the areas of the basin between two neighbouring stations.

Seeking analogous assessments of cross sections without direct measurements, interconnections between the correlation coefficients of peak discharges and volumes at different stations and also of morphological characteristics of corresponding basins are established.

In the paper the symbol of probability, p, is not used in a unique manner (being greater, less than, or equal to x). Therefore, it is not fully clear that if, in the conditional distributions, the condition is the exceedance of a given value or the equality to it. The used simplification of the conditional distributions is possible only in the last case. The authors are asked to explain this, and also to illustrate applications of the regional considerations, especially in the graphs given, without discussion, at the end of their paper. The idea of the authors is a valuable contribution to the analysis of the formation of flood waves in a basin.

The paper by Zsuffa outlines a possibility for constructing the two-dimensional distribution function $F(x, t)$, describing periods of low flows by the probability that x will be equalled or not exceeded by the discharge Q during a time interval $T \leqslant t$.

The time within which direct runoff occurs is not considered, but it is assumed to be less, by an order of magnitude, than the following depletion period. Therefore, the real hydrograph can be substituted by a model consisting of subsequent recession curves, with the following properties:

(1) The duration θ of a depletion period, i.e. the time between two runoff events, is distributed exponentially: $p(\theta \leqslant x) = 1 - e^{-\lambda x}$

(2) The depletion curves are represented by $Q(t) = Q_0 e^{-at}$

(3) The starting value Q_0 is constant or fixed, or it is also distributed exponentially: $p(Q_0 \leqslant x) = 1 - e^{-\lambda x}$. This decision has to be taken by observation.

The first step is to determine the one-dimensional distribution function of the lowest discharges Q_{min} of the depletion periods, given by $p(Q_{min} \leqslant x) = F(x, 0)$. This distribution follows exactly from those of θ and Q_0; in the case of Q_0 = const., it is a power function of x, in the case of Q_0 being a random variable it is a combination of power and gamma functions of x.

The generalization to the two-dimensional distribution function $F(x, t)$ is accomplished in the case of Q_0 = const. by substituting x by xe^{-at} in $F(x, 0)$, and in the other case by shifting the curves representing $F(x, 0)$, by e^{-at}.

Because the surface and subsurface runoff is not included in the model, it is taken to be valid for $F(x, t) \leqslant 0.5$.

Proceeding from physically motivated assumptions, and employing comparatively easy mathematical tools, Zsuffa has derived results which are much more precise than those obtained from conventional methods in low flow statistics. Only the transition from the one-dimensional to the two-dimensional distribution should be elucidated a little more, for the case when Q_0 is a random variable.

Guillot reports on linear regression models used in France for predicting low flows, flood waves and precipitation, and reviews the application of mathematical models in hydrology and meteorology.

(1) Low flows: the discharge volume to be predicted for the months April to August is written as $E = I + P$; I represents the initial conditions linearly composed of precipitation and discharge values of preceding months, and P is a quantity linearly composed of precipitation and temperature values of the months within the prediction period. Using the probability of P, one can determine the probability of E for each known value of I.

The author maintains that no deterministic 'principle of depletion' can yield results applicable for the prediction of low flows, since there are a great number of recession curves, additionally disturbed by intermittent rainfalls. Purely statistical statements, e.g. based on the distribution of extreme type III's, have failed too.

(2) Flood peaks and discharge volumes: the regression of these quantities is made on variates such as depth, duration and intensity of precipitation, discharge before the flood wave, season, soil moisture, etc.

The shape of the flood wave is represented by a regression on the preceding discharge values at the station under consideration as well as at upstream stations, and on various precipitation values. Guillot points out that deterministic methods such as the IUH concept apply to extreme events only, and are very incorrect due to baseflow separation, determination of direct runoff, and the nonlinearity of the system. For very extreme events far beyond the observed reach, the regression as well as the extrapolation of distribution functions is too inaccurate; here the Gradex method is recommended.

(3) Precipitation: Using a sample of weather situations analogous to the present one, which has been selected from 2000 stored situations, the precipitation is computed by regression on atmospheric pressure values at various heights.

Guillot emphasizes the futility of physically motivated deterministic models for prediction, and the expediency of regression models. He expects substantial progress in hydrology by fitting such regression models to many basins and comparing the results. I believe it will be as difficult to draw conclusions from the comparison of the anonymous regression coefficients as of the parameters in physical laws. It must be borne in mind that, to obtain linear correlation and homoscedasticity among the variables, the regression has frequently to be made on transformed values (by square roots or logarithms). In this way the relations between the original variables are again of the nonlinear type equivalent to the physical laws underlying the process.

The paper by Kasansky deals with the infiltration of meltwater into frozen ground. Starting from the micro-picture of the process, the essential macroscopic characteristics may be found by averaging over a porous medium. A formula for the averaged velocity of filtration is used as the statistical model (equation (2): vol.1, p.242). Kasansky has used a model proposed in previous publications to derive equations of the transfer of heat and impurities through porous media. After that, he cites an equation of infiltration into frozen ground (equation (5), vol.1, p.243).

It is possible to regard the process of the passage of a fluid particle through a porous medium as a stochastic process. However, this needs some limitation. In one pore channel, the path of an individual particle is entirely determined by the prevailing boundary conditions and the equations of motion. In the statistics of velocity vectors a multivariate probability distribution should be used.

Only in the case of a homogenous and isotropic porous medium and a homogenous field of forces is the matter made simpler. Without detailed assumptions about the probability distribution and the field of forces, it is not possible to derive additional relations. Thus, it is very difficult to recognize a physical foundation to the subsequent equations. The resulting infiltration model (equation (13), vol.1, p.245) shows the time dependence of the rate of infiltration. It includes the parameters $\aleph$ and $\aleph_1$, whose experimental estimation should be very difficult. Moreover, equation (13) includes the structural coefficient F, depending on tortuosity. Without precise assumptions about the form of pore channels, this coefficient cannot be determined. Last of all the initial porosity of the porous medium has to be estimated.

From this point of view, the practical applicability of this infiltration model seems to be difficult. I must emphasize that my remarks are based on my own interpretation of the paper and I am not certain that I have understood the paper correctly.

Three questions are posed for discussion:

(1) The true frequency distribution of hydrological events (e.g. floods) is not known, several theoretical distributions are fitted to the data. What experience has been gained on the construction of frequency distributions starting from the physical background of the hydrological processes?

(2) The few known multivariate distributions have rather complicated and unexplored features with the exception of the multivariate normal distribution. Is there any experience with non-normal multivariate distributions in hydrology? And what transformations to normality have proved to be most suitable for hydrological variables?

(3) The third question, connected with the second, is, strictly speaking a question to the mathematicians. It is well known that, in the case of a joint normal distribution of variables, the linear combination of them (e.g. the sum) is also normally distributed, with parameters easy to compute. But, if normality has been achieved by any transformation, is it possible to get analogously simple statements for the original variables?

DISCUSSION

I. Bogárdi:

In my paper a mathematical statistical method is applied for flood control developed along streams having a wide flood plain.

The model involves the hydrological, meteorological, geological, soil-mechanical features of the levee system, since all these factors influence the degree of protection. We have to combine the stochastic characters of:

(1) the flood protection along the levee line, that is, as a function of space, and
(2) the flood occurrences as a function of time.

For item (1), the distribution function of protection was introduced; for item (2), the distribution function of yearly maximum stages of flood exposures (Bogárdi, 1968) was used. While I agree with the general reporter that it is hard work to carry out the investigation in the case of large streams, I have to underline its necessity, because:

(a) flood levels are generally increasing due to human interferences (deforestation, leveeing, etc.),
(b) the levee earth material is ageing, and
(c) the value of the protected area is increasing.

As a first step to form a sound flood control development programme, the present degree of protection of the system must be determined.

In Hungary, the 4200 km-long levee system protects 25 per cent of the country's area. The method proposed was used to determine the present degree of protection of this large system. A great part of the work was effected by computer, similar to the dynamic programme which gives the optimal development and sequence in the case of a limited budget.

Programme results are not however, the last step in the decision making process, since we are often bound to deviate from this economic optimum. Anyhow, in such cases we can also compute the 'costs' of the deviation and, by examining the pros and cons, we can produce an optimal decision.

REFERENCE

Bogárdi, I. (1968) Flood exposure recommended as a parameter for describing the fatigue loading on flood control structures. *Bull. int. Ass. scient. Hydrol.* **XIII** (3).

Wolfgang H. O. Trau:
I should like to give a contribution to the paper presented by Maniak. The planning engineer in our country (Federal Republic of Germany) is often faced with the problem of determining the design flood for hydraulic structures, such as weirs and bridges in small basins, by applying simple methods of computation. The question had to be examined whether simple statistical methods provide a useful tool. Evidently this is true only for long-term homogenous records.

The influence of man, however, on runoff conditions in small basins situated in fairly highly industrialized regions often has to be taken into account. For instance, in parts of the Ruhr district, the runoff conditions are changing every year. There is often not sufficient time for runoff and rainfall measurement. The question arises — which statistical methods can be applied and whether, in these cases, deterministic methods can be used. The change of peak runoff with time can be tested by means of regression analysis. Often, however only small values of the coefficient of determination are computed. Therefore, other criteria, such as the shape of the floodwave, might be taken into account. An additional adjustment of the flood records may be achieved by methods which admit to estimate the relation between baseflow and peak flow, and to eliminate the baseflow. In small basins, the baseflow often only amounts to a very small percentage of the total runoff. Comparing several flood events, the individual baseflows only show slight differences, as it was checked for several rivers in the Federal Republic of Germany. Therefore the error is relatively small.

Finally, I would like to discuss a question which is of some interest to the designing engineer. Hydraulic structures have a limited lifetime. Experience shows that the lifetime, even of larger plants, is steadily decreasing.

The chance of the recurrence period of a design flood during the lifetime of a structure can be calculated. This gives a prospect of calculating the hydrological risk for a certain type of structure, thus limiting the error caused by differences in the lifetime of the individual structures of flood protection measures.

I. Zsuffa:
M. le prof. Dyck dans sons rapport a touché le problème essentiel de notre méthode: c'est à dire l'élargissement du loi de type II (d'un loi gamma special) des valeurs minima au loi *à deux variables*.

Pour répondre je rapelle au modèle mathématique (Fig.1, tome 1, p.234). Il se compose d'une série des courbes de tarissement avec les périodes et les débits initiaux qui sont des variables aléatoires; par les hypothèses adoptées nous avons dérivé les deux types de loi pour la répartition des valeurs minima. Il faut mentionner qu'on peut utiliser les hypothèses plus générales mais la résolution d'un tel modèle général ne peut se faire que numériquement calculer par les matrices respectives.

Quant au problème posé par M. le professeur Dyck il faut remarquer qu'on caractérise par la même probabilité la valeur $Q = xe^{a\tau}$ que $Q_{min} = x$, au moins dans le cas du loi du type 1. C'est a dire quand Q_0 est fixée. Quand Q_0 est une variable aléatoire, p.e. avec une répartition exponentielle, il faut considérer que la relation

$$F(x, t = 0) = \psi(x) = F(xe^{a\tau}, \tau)$$

n'est valable que si

$$xe^{a\tau} < Q_0$$

Etant que, dans le cas du loi de type II, la répartition de Q_0 est exponentielle

$$G(y) = p(Q_0 < y) = 1 - e^{-\lambda y}$$

la fonction a deux variables cherchée dans le cas de type II

$$F(Q < x, t < \tau) = \psi(xe^{-at})(1 - e^{\lambda x}) =$$

$$(1 - e^{-\lambda_2 x})^2 + (\lambda x)^{\lambda/a} \Gamma\left(1 - \frac{\lambda_1}{a}\right)[1 - \Gamma_{(1-\lambda_1/a)}(\lambda_2 a)] * [1 - e^{-\lambda x}]$$

P. Guillot:

Un des points signalé par M. le rapporteur pour la discussion, est celui de la distribution adoptée pour les valeurs extrêmes des débits; Monsieur le rapporteur signale à juste titre qu'il existe un très grand nombre de formes analytiques, théoriques, qui sont proposées pour la distribution des débits extrêmes.

Je vais faire un dessin au tableau: Sur le graphique classiquement attribué à Gumbel, on porte en ordonnée le débit de crue et, en abscisse, le double logarithme de la fréquence. La distribution empirique des débits est connue jusqu'à une durée de retour de dix ans lorsqu'on a quelques dizaines d'années d'observations, au delà nous sommes dans un domaine d'extrapolation hypothétique, jusqu'à cent ans et mille ans de durée de retour, la loi de Gumbel s'extrapole par une droite, la loi de Gumbel des logarithmes par une exponentielle, la loi de Galton par une parabole concave, la loi de Gauss par une parabole convexe ce qui fait que nous avons un très vaste éventail d'hypothèses qui ne sont pas plus défendables les unes que les autres si on se place sur le plan purement statistique. Il est donc nécessaire et inévitable de proposer une hypothèse physique qui s'appuie sur des idées le plus simples possible. Et notre méthode, que je voudrais vous proposer consiste à extrapoler au delà de la crue décennale par une droite parallèle à droite de Gumbel des précipitations, en effet, nous supposons premièrement que la loi de distribution de la précipitation sur le bassin versant est raisonablement extrapolable par la loi de Gumbel (ce qui comporte encore une part l'hypothèse, malgré le grand nombre d'observations) et indépendants recueillis dans le monde entier.

Deuxième hypothèse – qui parfaît en tout cas prudente et physiquement raisonable, c'est que, au delà de la crue décennale, quand il pleut 10 mm de plus, il coule dix millimètres de plus. Cette méthode est recommandable, à notre avis, pour les bassins versants imperméables, en tout cas à faible retention et ayant des temps d'écoulement et des temps de concentration qui ne dépassent pas trois à quatre jours.

Je pense que pour un grand nombre de petits bassins versants de l'ordre de mille kilomètres carrés ou moins, cette méthode est assez simple et elle évite cette énorme incertitude des formes analytiques.

Un deuxième point, sur lequel le rapporteur désire qu'une discussion s'instaure est celui de la forme des distributions multivariables. Je pense qu'il n'est pas nécessaire que la distribution des variables aléatoires soit gaussienne dans toute leur étendue pour pouvoir appliquer les méthodes de calcul de la corrélation multiple entre les variables aléatoires. En vérité, la forme normale, multivariable, n'est que la partie principale de la distribution à plusieurs dimensions, elle est utilisable dans la zone centrale des distributions et il n'est évidemment pas question de l'extrapoler aux valeurs extrêmes. Il est seulement question de l'utiliser dans le domaine des valeurs connues ou à proximité du domaine des valeurs connues. Dans ces conditions, tous les calcule et toutes les théories qui ont été faites par l'école anglaise de Cambridge sur la distribution multivariable Gaussienne, sont utilisables au moins comme règle pratique comme critère de liaison qui suffit pour la plupart des problèmes de prévision.

M. S. Gruszewskij:

I would like to comment on the papers presented during the former session, since time was then too short for my intervention.

It is worth mentioning that scientists draw different conclusions from the analysis of such an important problem as the presence of a periodic component in the annual river runoff series. Kartvelishvili and his collaborator have applied a very interesting method, and have analysed runoff series from rivers flowing in different parts of the world. Their conclusion was that the existence of a significant periodic (deterministic) component has a very low probability. This conclusion is also shared by our general reporter, Dr. Rodríguez-Iturbe. However, Druzhinin and his associates arrived at a different conclusion, proving that the variability of solar activity exerts an influence upon the volume of the annual river runoff. This idea was also reflected in the presentation made by Roche. In my opinion, further research is needed to eliminate this contradiction and to explain both the limitations and scope of applications of the methods used for the analysis of this phenomenon. Consequently, the existence or the lack of a significant periodic component would be clarified.

At this point, I would like to make one more general remark. We are dealing with one nature which obeys both the stochastic and deterministic laws. For that reason, there is a great future in the investigations which apply simultaneously the methods already discussed as well as some other methods. Besides, there are some other investigations of this type. For instance, those carried out by Vinogradov from Alma-Ata who is present at this Symposium. The results of these investigations have not been discussed yet.

I. Zsuffa:

Je suis très content de voir que les méthodes françaises sont en conformité avec l'évolution récente de nos prévisions hydrologiques, en Hongrie. En constatant ce caractère paralèlle il me semble qu'il y s'agit des idées plus ou moins générales.

Voici les analogies: en 1955 nous sommes parti aussi de livre Linsley–Kohler–Paulhus et nous avons commencé les travaux par la méthode des relations coaxiales, mais il faut avouer, avec très peu de succès: Après les premiers essais infructueux M. Szesztay introduisait une méthode simple: au lieu d'une relation a n variables il utilisait $m = n/2$ relation à trois variables:

$$Y = F(x_1, x_2, \ldots, x_n) = p_1 f_1(x_1, x_2) + p_2 f_2(x_3, x_4) \ldots p_m f_m(x_{n-1}, x_n)$$

où $f_i(x_{2i-1}, x_{2i})$ sont les fonction à trois variables construites par simple voie graphique, et p_i sont les convenable poids choisis subjectivement. Pour améliorer la pondération j'ai élaboré alors une méthode basée sur les moindres carrées. Mais, c'était aussi un biais: elle exige aussi l'ordinateur, donc la corrélation multiple doit la déplacer.

Quant aux problèmes des prévisions des étiages nous avons aussi constaté que s'appuyer sur l'équation des courbes de décrue, considérés comme une loi extrapolable, n'ont jamais abouti à aucun résultat pratique pour les *prévisions* des étiages. (Quant aux problèmes de *l'évaluation de la probabilité* des étiages nous utilisons de courbes de tarissement!.) Étant, qu'en effet, que la vitesse de décrue dépend de l'état initial d'humidité du sol et qu'il existe une population de courbes de tarissement possibles: nous avons développé en série Fourier des courbes de tarissement observées et nous avons cherché les relations entre les données hydrométéorologiques de l'hiver et les coefficients Fourier.

De l'année 1965 les ordinateurs sont introduit aux calculs hydrologiques. Après l'élaboration d'une méthode (presque identique à l'example de Drac au Sautet de Guillot) pour la simulation des séries longues des débits des petits cours d'eaux partant des données météorologiques et se basant sur la corrélation multiple, nous avons commencé le remaniement des relations des prévisions des crues de Danube. Pour sélecter les variables nous avons élaboré une programme de l'ordinateur pour la corrélation multiple qui répète automatiquement le calcul si l'un des coefficient de corrélation partielle est négative en omettant la variable en question (étant donné que par raison physique il y faut avoir des relations positives).

Maintenant pour les prévisions continues nous avons élaboré une méthode plus complexe:

(1) On a constaté que les prévisions faites sur la base des graphiques construites des données de 60 années ont servi parfois les résultats pires en comparaison avec les 'prévisions' des techniciens basées sur leurs expériances et sur les données *récentes*. Ainsi, pour opprimer cette contradiction c.à.d. pour éliminer l'influence des variations du lit et de la rugosité nous élaboré le programme suivant.

Nous n'utilisons que les données des deux dernières années, l'index i désigne le numéro d'ordre en sens invers (c.a.d. $i = 1$ désigne les données d'aujourd'hui $i = 2$ les données de l'hier), les index n désignent des stations (p.e.: $n = 1$ Krems, $n = 2$ Vienne, $n = 3$ Pressburg, . . . , $n = 8$ Baja, $n = 9$ Mehács). Quant à variable à prévoir nous nous bornerons aux sept stations hongroises $n = 3$ Pressburg, . . . , $n = 9$ Mohács.

Nous prévoyons les différences journalières. Ainsi l'ordinateur calcule ces differences

$$X_{2n}^i = X_{2n-1}^i - X_{2n-1}^{i+1}$$

où on désignes par X_{2n-1}^i le niveau observé à la station n au jour i, et par X_{2n}^i la différence des niveaux respectifs.

(2) Sélection parmis des k laps de temps possibles entre les observarions resp. des différents couple des stations. Pour ce fin on utilise de fonction de corrélation croisée. L'ordinateur calculs les coefficients de correlation

$$r_{2n,\,2m}(k)$$

oú n, m désignent les stations, $(3 \leqslant n \leqslant 9$ stations a prévoir; $1 \leqslant m \leqslant n + 1$ stations servant des données des variables indépendants) et k le laps de temps. Pour chaque prévision à l jours d'échéances et chaque n, m couple des stations l'ordinateur cherche

$$\max_{k \leqslant l}\{r_{2n,\,2m}(k)\}$$

(3) Sélection parmis les données de deux années dernières pour construire l'échantillon statistique. Puisque pour la prévision des données d'une étiage les niveaux des crues ne servent guère information, il faut faire une sélection convenable. A ce fin l'ordinateur calcule pour chaque index les differences suivantes

$$S_n = \sum_{1 \leqslant m \leqslant n+1} \left(X_{2m-1}^1 - X_{2m-1}^y \right)^2 + \sum_{1 \leqslant m \leqslant n+1} [X_{2m-1}^{1-k(m)} - X_{2m-1}^{y-k(m)}]^2$$

où $k(m)$ désigne que le laps de temps optimal se varie selon n, m. Ensuite l'ordinateur sélecte les $N = 100$ plus petits valeurs S et le programme se continue avec les données de ces $N = 100$ jours ainsi choisis.

(4) Sélection parmis les variables indépendants. L'ordinateur calcule pour chaque n station et pour chaque l jour d'échéance les correlations multiples et il donne les coefficients de corrélation partielle aussi. Le programme se répete si quelques uns des coefficients de corrélation partielle sont négatifs

$$\rho^l_{2n,\,2m} \leqslant 0 \qquad \text{(les index pairs indiquent qu'il s'agit des relations entre les différences des niveaux)}$$

Étant donné que c'est absurde de point de vue physique, l'ordinateur élimine les données X^i_{2m} et il calcule une corrélation multiple nouvelle.

Il continue ce processus jusque chaque $\rho^l_{2n,\,2m}$ est positif. Si dans la derniere étape ils restent plus que 10 variables indépendant, le programme élimine les variants

$$|\rho^l_{2n,\,2m-1}| < 0.1$$

les variants des niveaux aussi.

(5) L'ordinateur calcule la relation à deuxième et troisième ordre.

$$X^{i-e}_n = A_1(X^{i-l+k_1}_{m_1}) + B_1(X^{i-l+k_1}_{m_1}) + C_1(X^{i-l+k_1}_{m_1})^3 +$$

$$+ A_2(X^{i-l+k_2}_{m_2}) + B_2(X^{i-l+k_2}_{m_2})^2 + C_2(X^{i-l+k}_{m_2})^3 +$$

$$+ A_3(X^{i-l+k_3}_{m_3}) + B_3(X^{i-l+k}_{m_3})^2 + C_3(X^{i-l+k}_{m_3})^3 + \ldots$$

L'ordinateur donne les résultats après deux schémas. Selon le schéma large nous recevons sur l'affiche de l'ordinateur:

(a) les coefficient des corrélations croisées,

(b) les numéros d'ordre choisi par les sommes des carrés des différences,

(c) les résultats des calculs des corrélations multiples et les coefficients des corrélations partielles,

(d) les équations linéaires finales,

(e) les équations polynomiques,

(f) l'ordinateur donne les prévisions pour chaque n station et pour l journées prochaines,

(g) le 'plotter' de l'ordinateur déssine les hydrogrammes à prévoire pour chaque, $3 \leqslant n \leqslant 9$ station et les coups longitudinaux à prévoir pour chaque $1 \leqslant l \leqslant 8$ jour.

Après ça l'ordinateur *emmagasine les données actuelles* et *élimine les données plus anciennes* et change les numéros d'ordre i par un pas.

Dans une situation critique c.à.d. pendant une grande crue ou pendant une étiage basse l'ordinateur accomplit le programme entier et si la situation critique dure longtemps l'ordinateur ne donne que les résultats finals [voir et (f) point ci-dessus]. Dans une situation neutre on prépare des prévisions par calculs manuels selon les polynoms les plus frais mais l'ordinateur emmagasine les données nouvelles (arrivant en outre par telex ainsi aptes pour l'ordinateur). Mais dans une telles situation même au moins par 6 semaines l'ordinateur calcule les polynomes actuelle selon le schéma décrit.

La première forme de notre programme est près et elle est en service. La plus nouvelle, décrit ci-dessus et au cours de l'épreuve (la différence entre eux est la sélection des dates des données servant l'échantillon statistique, dans le programme en usage, l'échantillon se compose des données de dernières 100 journées).

P. Guillot:

Je suis extrêmement intéressé d'apprendre que nos collège hongrois ont reconcé aux méthodes de corrélation graphique pour établir les formules de prévision, parce que je pense qu'effectivement les méthodes de corrélation calculée sont sonsidérablement supérieures aux méthodes de corrélation graphique. A mon avis, la méthode des relations coaxiales qui a été publiée largement pendant les années cinquante aux Etats-Unis, et très largement utilisée dans tous les pays de l'Est européen depuis également cette époque, est une aberration regrettable. L'expérience montre, en effet, que cette méthode ne permet pas de décider si une variable explicative est ou non utile à conserver pour la prévision. La seule réponse rationnelle à cette question c'est le calcul du coéfficient de corrélation partielle, même si – et ceci est très important – les relations ne sont pas linéaires. Je crois que cette remarque est en relation avec les soucis de M. Dyck. Beaucoup d'hydrologues considèrent que les calcule de corrélations multiples, classiques ne sont pas utilisables à partir du moment où le modèle naturel n'est pas linéaire. Eh bien! Même si le modèle est très loin d'êtres linéaire, la méthode de corrélation linéaire est un outil extrêmement puissant et toujours valable de sélection des variables explicatives les plus utiles. Ensuite, naturellement, une fois cette sélection opérée il faut examiner les graphiques de corrélation partielle, et les graphiques qui montrent la correspondance entre valeur calculée et la valeur observée, et faire des transformations de variables, si nécessaire, pour se ramener à des liaisons à peu près linéaires et homoscédastiques dans la zone d'utilisation, mais cela n'est pas du tout une faiblesse des calculs classiques de corrélations multiples. J'insiste beaucoup sur cette idée, parce que je crois que cette crainte de la non-linéarité est peut être la raison majeure du faible emploi ou du mauvais emploi par les hydrologues des méthodes de corrélation calculée.

G. S. Cavadias:

When we discuss multiple regression as a forecasting tool in hydrology, we must make a distinction between two types of models, In the first type of model, the independent variables are not random variables but fixed numbers and the only source of randomness for the dependent variable is the error term. In this case we do not have to assume linearity with respect to the dependent variables but only with respect to the coefficients.

In the second type of model, the dependent variables may also be stochastic. I agree fully with Guillot's remarks in the first case where it is not only desirable but also possible to consider non-normal disturbances. In the case of stochastic dependent variables, however, the necessary sampling theory is not available for making statistical inferences in the non-normal case. The situation is even more complicated when the dependent and the independent variables are measured with error.

Consequently, the distinction between the two types of models is necessary when we consider extensions to non-normal regressions between hydrological variables.

M. Roche:

Puisqu'on a évoqué le problème de la linéarité des régressions, entre autres en a parlé, et également au sujet de l'homoscédasticité, je voudrais signaler que, dans l'esprit de beaucoup, les propriétés de linéarité et d'homoscédasticité sont liées au fait que la variable dépendante et la variable indépendante, dans le cas par exemple d'une distribution à deux variables, sont toutes deux des variables normales. S'il est exact que dans la distribution normale à deux variables on a effectivement régression linéaire et homoscédastique, l'inverse n'est pas vrai. Rappelons qu'il y a homoscédasticité lorsque la distribution de probabilité conditionnelle d'une variable, sachant que l'autre a une valeur connue, est indépendante de cette valeur. En particulier, ce n'est pas parce qu'on a constaté que les distributions marginales de deux variables sont normales que la régression qui les lie est obligatoirement homoscédastique, ni même obligatoirement linéaire; par contre, on peut très bien envisager des régressions linéaires sans qu'aucune des répartitions marginales soit normale. On peut s'amuser ainsi à créer des distributions qui jouissent tout à fait indépendemment de l'une ou l'autre de ces propriétés. Il n'en reste pas moins vrai que, dans la nature, on observe assez souvent, lorsque les distributions marginales sont normales, ou que par transformation des variables ou par anamorphose on peut les rendre normales, qu'on se rapproche d'une régression linéaire et homoscédastique. Mais ce n'est pas du tout obligatoire et c'est une chose qui demande à être vérifiée.

La deuxième chose que je voudrais dire, c'est que, depuis le début de ce colloque, on entend beaucoup parler de modèles déterministes, stochastiques, et je me demande parfois si on ne mélange pas un petit peu tout.

Finalement, qu'est-ce qu'on entend actuellement par modèle stochastique? Pour les anglo-saxons, c'est une opération qui consiste à faire l'analyse d'une chronique, ce qui revient à examiner comment les valeurs successives de cette chronique sont liées les unes aux autres et quelles sont les lois des résidus probabilistes non expliqués par ces liaisons. C'est un point de vue, une règle du jeu, qui a été adopté en particulier par le groupe d'hydrologie stochastique de Yevjevich; je ne sais pas si on l'a assez dit au reste des spécialistes. On a l'impression que l'on considère parfois comme modèles stochastiques, à bon droit du reste, des choses tout à fait différentes comme par exemple certains modèles utilisées par transformer en débits des données climatologiques. C'est un autre point de vue, car si, dans un modèle stochastique au sens anglo-saxon, on peut avoir des passages de chroniques climatiques à des chroniques de débits, en passant notamment par l'intermédiaire de processus, il n'en reste pas moins qu'on s'en tient à l'analyse structurale de ces chroniques.

Il semble enfin que ceux qui essayent de rechercher des relations entre les éléments climatologiques et les éléments hydrologiques par l'intermédiaire des régressions, procèdent d'un autre point de vue. Il conviendrait peut-être de séparer nettement ces relations de régression des modèles stochastiques à proprement parler.

Viennent enfin les modèles dits conceptuels ou déterministes. Là encore, il s'est crée un langage bien particulier qui est né, je pense, du groupe d'hydrologie paramétrique, en marge de l'AISH, et qui semble être resté une langue entre gens d'une même société, discutant entre eux de choses qu'ils connaissent parfaitement pour avoir mis un certain nombre d'années à les élaborer. Je ne suis pas certain que ce langage soit compris par tous les hydrologues. Il conviendrait que les spécialistes de ce groupe se donnent la peine de bien expliquer ce qu'ils entendent par leurs différentes définitions. On retrouve en particulier une certaine confusion lorsque à l'intérieur des modèles dits déterministes, il s'agit d'ajuster les paramètres en faisant appel à des méthodes qui effectivement relèvent de la statistique. Ces méthodes ont été évoquées hier après-midi; je pense notamment au problème de l'optimisation de ces modèles par l'intermediaire de critères dits objectifs. De tels critères ne sont en fait objectifs que dans l'esprit des gens qui les créent; en effet, si une fois créés ils peuvent effectivement servir à un jugement mathématique objectif, au moment où on les créé, ils dépendent et doivent du reste dépendre du résultat qu'on cherche à atteindre, et par conséquent restent éminemment subjectifs par rapport à celui qui utilise le modèle. Je vous remercie.

Closing remarks by the general reporter
This session has shown that statistical techniques are applied under various aspects of

hydrology. Examples are given for the possible opening of new fields of application of the probability theory. Professor Cavadias already stated in the previous session, it is to be noted that in most areas of the application of statistics, special methods have been developed and adapted to the nature of the variables encountered. It is to be hoped that hydrologists will not only take over useful elements from existing mathematical theory, but will continue developing specific methods adapted to the nature of hydrological variables. When applying statistical methods to hydrological events, it must be borne in mind that the methods of statistics were developed to process large data samples. In hydrology, we often have to deal with very small sample sizes, with very short observation series. We often have to deal with faulty data. Therefore, we always have to take the greatest care with the selection of suitable methods. I think, as has already been said in the previous session, the shorter the observation series, the greater has to be the expense in statistical evaluation. We have to use all the available additional information to obtain reliable results to build up the appropriate models for frequency analyses, for forecasting, or other problems.

General Reporter   Professor Dr. G. G. Svanidze, Head of the Chair of Hydrology, State University of Tbsilisi, USSR

Chairman   Dr. A. Svoboda, Institute of Hydrology and Hydraulics, Slovak Academy of Sciences, Bratislava, Czechoslovakia

Stochastic analysis of water storage problems

G. G. Svanidze

The papers reviewed and discussed in this general report are listed below. The papers are all published in volume 1 of the Symposium Proceedings.

1. *I. Kontur*   Application of stochastic hydrological models in the design of lake level control
2. *I. Zsuffa*   Quelques relations arithmétiques de matrice dans le calcul des réservoirs
3. *S. K. Srinivasan*   Emptiness of a finite dam—time dependent theory
4. *K. Nacházel*   Multiple Markov processes in hydrology and their importance for stochastic analysis of water accumulation problems
5. *G. G. Svanidze*   Mathematical models of runoff for the computation of storage reservoirs
6. *W. G. Strupczewski*   Reservoir operating policy in relation to inflow prediction
7. *Erich Weber*   Model of a lake reservoir
8. *I. Tzvetanov*   A probabilistic model for an annual optimum control of a complex reservoir
9. *Zdzisław Kaczmarek*   Some problems of stochastic storage with correlated inflow.

The contemporary methods of the streamflow storage theory, as well as various methods of water management computations, *can be divided into two basic groups.* The first group embraces the composition methods (including also the analytical methods), the second includes all the applications of the Monte Carlo method. Accordingly, we can divide the papers presented at this Symposium and discussed here into two groups.

The first one includes the papers by Srinivasan, Zsuffa, Tzvetanov, Kaczmarek, and Strupczewski, and the second group includes papers by Nacházel, Kontur, Weber, and Svanidze. The method of chronological series which is still used in the design practice, can be considered a primitive variant of the Monte Carlo method. In this case all computations are based on a single and usually short realization of the streamflow process recorded in the past. For this reason the results of such computations cannot be considered conclusive.

Moran (1959), one of the creators of the analytical computation methods, has made a comparison between two groups of methods with the conclusion that, in the case of more than two reservoirs, analytical methods are not sufficiently powerful, and the application of the Monte Carlo method is recommended (the independent inflows are taken into consideration, i.e. the model proposed by C. Sadler). This method allows solution of the problem with the avoidance of many analytical difficulties. Nine years later, Kritsky and Menkel (1968) confirmed the validity of this thesis writing that '. . . in the case of the probabilistic analysis of such systems (i.e. complex water resources systems), the Monte Carlo method is the only one which can help us in our search for the solution'. Although there are many interesting works aimed at the perfection of the composition methods, including the analytical methods of computations, these methods, in fact, still do not allow for the solution of such problems as the operational control of a system of storage reservoirs which includes more than two reservoirs. On the contrary, the Monte Carlo method has been substiantially developed and now it is an important tool enabling the solution of the complex problems both for the single multi-purpose reservoir and systems of mutually-dependent reservoirs (Svanidze, 1964; Reznikovszky 1969).

We should not exclude the possibility, however, that the analytical methods will dominate in the future, when the number of simultaneously analysed storage reservoirs will grow and, consequently, the size of the problem will expand. For this reason, the improvement of the analytical methods shall be also considered as one of the actual scientific problems.

We shall begin with the first group of papers. Srinivasan has proposed in his paper a new analytical method for the computation of the storage reservoir with finite capacity and stochastic inflow. The problem is solved by a Laplace transform technique and the results are in the form of the probability distribution function of filling the reservoir during the flood season. It is also possible to obtain the explicit form of the probability distribution function governing the reservoir water levels at any arbitrary time.

The paper is interesting as one more successful trial aimed at the formulation of the analytical method for the computation of the storage reservoir. It should be noticed, however, that the paper takes into consideration a very simple case (an isolated reservoir with steady fixed outflow), which is extremely rare in engineering practice.

The paper by Zsuffa gives equations which allow preparation of algorithms and matrices for the computation of one or two reservoirs. In the first case, the work is based on the well-known model of Moran (1959) with independent inflows in the neighbouring time periods. If there is a cascade of two reservoirs, additional lateral inflow to the lower reservoir is assumed to be proportional to the inflow to the upper reservoir. This assumption is valid, of course, only for very small rivers with small catchments. Equally doubtful is another assumption that there are no stochastic dependences between streamflows in the neighbouring time periods.

The problems presented in both of the above-mentioned papers have already been solved by analytical methods, for instance by Kartvelishvili (1967). However, these problems can also be solved by the Monte Carlo method, in such a case the solution is more complete and simpler. Regardless of this fact, it must be admitted that investigations along this line have good prospects. The only requirement to be observed is that the models should be more adequate to the real conditions.

The paper by Tzvetanov is concerned with the probabilistic model for the determination of the optimal operation policy for the control of a multi-purpose storage reservoir (within-the-year storage). As usual, the three following components have been taken into consideration: (a) input process — river inflow, (b) control operator — regime of the oscillations of the actual storage volume in the reservoir, and (c) output process — total target outflow with due consideration given to evaporation, seepage and free discharge losses.

The outflow consists of the controlled component — water supply to all consumptive and non-consumptive water users including the free discharges, and the uncontrolled component – evaporation and seepage losses. The second component is defined as the elementary stochastic process of type $j(t) = h\phi(t)$, where $\phi(t)$ is a known function and h is a random variable. From the economic viewpoint, the aim is to analyse the optimal operation of the reservoir as a system which is serving all elements of the common goal. A penalty function is introduced.

The distribution of water resources is accomplished according to a hierarchic structure with three control levels. The recurrent equation of dynamic programming is defined for the first control level.

In the future, the solution of this problem should be completed giving, at the same time, some practical examples illustrating the advantages of the method proposed.

Kaczmarek proposes in this paper the probabilistic method for the solution of some river flow control problems, taking into consideration stochastic relationships between the mean inflows (Q_i) in the separate time intervals. He proposes that the outflow from the reservoir should be determined as being dependent on the inflow and the rate of storage at the beginning (V_{i-1}) and the end (V_i), of a given time period.

In the general solution of the problem, Kazmarek makes use of the principles formulated by W. I. Romanovskij in his well-known work about the Markov chains (1948). According to these principles, it is possible to construct a vector on the basis of its former value and the transition probabilities matrix. Two possible schemes of such constructions are given. However, he then proposes the original approximate method which provides a solution of the problem with sufficient accuracy for practical purposes.

Kaczmarek writes that, if the discrete inflow process forms a lag-m Markov chain, then the process of oscillations of the rates of storage (V_i) can be described as a lag-$(m + 1)$ Markov chain. This thesis should be somewhat clarified. Some special investigations concerned with this subject have proved that this thesis is actually valid for the two-dimensional process (Q_i, V_i).

The determination of the optimal control policy for the reservoir can prove to be a difficult problem without the determination of the operational plan or without the application of the optimization methods. In a case of similar computations for a system of reservoirs, Kaczmarek indicates the limited applicability of such generalizations, mainly due to large computational difficulties.

In the paper by Strupczewski, consideration is given to the operational problem concerning optimal control of the river flow with the aid of storage reservoirs, assuming that a probabilistic prediction of the inflow is available. Although there are a number of works (Kartvelishvili, 1967; Svanidze, 1967) where the authors have tried to describe the river flow as a continuous stochastic process, solutions of the respective problems have not yet been brought to such a stage as to yield concrete results. On the contrary, if we substitute continuous models by discrete ones, many engineering problems of river flow regulation can be solved with good results.

At present, the stochastic dynamic programming model, with inflow and time described in discrete form, shall be considered the best tool for the operational optimization of the operation of the storage reservoir. In the work presented by Strupczewski, the problem is analysed exactly in such a form. Moreover, this work shall be considered to be a successful attempt at such a formulation of the presented model as it takes care of the inflow prediction, which is one of the elements of available information. Such an approach is a logical continuation of the traditional Polish investigations concerned with the probabilistic forecasting of hydrometerological processes, among which should be mentioned the works of Kaczmarek (1961, 1969).

As underlined by Strupczewski himself, his complete model is very complex and two proposals are made for its simplification. The first one is based on the substitution of a sequence of independent variables (predictors) by a single aggregated variable. The second proposal assumes that, in the case taken the reservoir volume often takes its extreme values, the introduction of one extended prediction after each maximum or minimum filling of the reservoir can be justified without expanding the scale of the problem. It is obvious that both proposals lead to some loss of available information, but, temporarily, this defect cannot be avoided.

The construction of the statistical conditional distribution functions $F(Q_j|Q_{j-1})$ (for instance, for the mean monthly flows) with the aid of the fragments method (Svanidze, 1964) could be recommended.

Let us now consider the second group of papers, in which the Monte Carlo method is used for the computation of river flow regulation.

In the paper by Nacházel, the author deals with the problem pertaining to the determination of the multi-annual component of the usable capacity of the regulation storage reservoir, in the case when synthetic hydrological series are applied. Stochastic relationships between annual runoffs in different years have been analysed with the aid of a higher-order Markov chain.

The initial statistical characteristics for the River Elbe are evaluated on the basis of historical series, and a simple analytical relationship has been applied for the correlation function. By the application of well-known methodology for the generation of synthetic hydrological series (Svanidze, 1964; Reznikovszky, 1969, Maass et al., 1962), it is possible to determine the relative capacities of the reservoir (β_v) at various release levels (a) and various guarantee levels (p).

The author concludes that the application of a higher-order Markov chain leads to a more economical solution, i.e. at the same guarantee level, the reservoir of smaller capacity is required. This is true for the River Elbe since curves $\beta_v = f(a, t)$, at different numbers of years, t, are located in between curves $r_1 = 0$ (no correlation between the series elements) and $r_1 \neq 0$ (correlation computed according to the lag-one Markov chain). In the case of some other rivers, for instance the River Ob (city Kamien profile), River Naryn (Uch-Kurgan), River Selety (village Iljinskoe profile), the situation is quite different; the higher the order of the Markov chain (for $t = 2, 3, \ldots$) the larger the necessary capacity of the reservoir.

There are, however, some cases which confirm the conclusion of the author. For instance, the Ural (Kushum) (Reznikovszky, 1969) and other rivers. We can also encounter such rivers for which the capacity of the reservoir is irrelevant or just slightly dependent on the lag number (Irkutsk storage reservoir). The solution of the problem also depends on the absolute values of a and β_v, as well as on their mutual relationships; it depends also on the choice of the date of the beginning of a year (since in many cases the correlation function itself depends on this assumption), and on many other factors.

Some preliminary estimates can be made depending on the relative location of the correlation function for the recorded historical series and the correlation function for the lag-one Markov chain ($r_t = r_1^t$). If the second curve is located below the first one, the corresponding capacity of the reservoir should be smaller, and vice versa.

The paper by Kontur (Hungary) deals with the regime of the Lake Balaton water levels. This regime is determined on the basis of water balance computations with the application of the Monte Carlo method for the statistical modelling of the lake inflow and some other elements of the water balance. Mean data describing individual elements of the water balance are taken from the work by Szesztay (1966). It was found that observations of the variability of the lake volume recorded during the period 1963–1966 are not from a single general population, therefore, the most recent observations have been taken as a basis for the determination of the statistical characteristics. The analysis has been made by comparison of the spectral functions.

The correlation relationships between rainfall and the increase of the lake volume, as well as similar relationships among other hydrometeorological parameters have been investigated. For individual months, the conditional functions of distribution of probabilities associated with filling the lake bowl have been constructed. This is a scientific problem of great theoretical and practical significance, therefore, it would be advisable to apply various computational methods for its thorough analysis. Especially recommendable would be the construction of a set of conditional realizations of the water balance elements described by the higher-order Markov chain, by extrapolating the historical data recorded in the last 10–15 years for the future 10–20–50 years (see the paper by Weber). This approach is justified by the assumption concerning the change of the stationarity conditions (water transfer from neighbouring basins, intensification of the water use, etc.).

Conditional distribution functions describing the process of the lake level oscillations should be constructed on the basis of a statistical analysis of available data, made for the time periods specified for the future (e.g. one, two, three years, etc.).

The paper by Weber describes a mathematical model for a balance-type computation of the regulation of the 'Muritzseen', which is the largest lake system in the GDR.

Particular elements of the water balance have been modelled by the Monte Carlo method according to the scheme of the lag-one Markov chain. The statistical characteristics have been determined from the 46-year period of record; the last have been defined by an analysis of available information concerning lakes outflow, rainfall and evaporation from the lakes' surface with due consideration given to water withdrawals and discharges. The modelling of the synthetic hydrological series is effected by a simple regression for the monthly time periods.

The Pearson type III distribution has been used as the inflow distribution function allowing, however, for some flexibility in the choice of the probability distribution functions for rainfall and evaporation. The curves have been graphically built on probability paper according to 46 available points, without using any other information about these processes and without introducing any hypothesis as it is usually done in such cases. Following the evaluation of the results of the analysis, the recommendations are to adopt a computational guarantee level for the outflow of 92.5 per cent.

It has been underlined in the paper by Svanidze that two principal cases may occur in calculating river flow regulation:

(1) stationary regime – inflow and water use conditions do not change in the calculation period (regime is static), and

(2) nonstationary regime – inflow or water use conditions (or both of them) do change (regime is dynamic). The conditions of modelling the hydrological series (input processes) by the Monte Carlo method should reflect these aspects.

In case (1), a hydrological series of considerable length (n = 1000 years and more) is modelled, taking into account within-the-year flow distribution, and the unconditional probability distribution functions of the target release, reservoir volume and the free release, are structured by the balance technique. In case (2) the inflow is assumed to be a conditional random process and an aggregate of conditional realizations of this process is modelled for the forthcoming limited calculation period of the operation of a water economy installation (n = 20–50 years). The processing of the statistical material according to the output functions is carried out not along the realizations [as in case (1)], but for the cross sections of all the realizations at the same moment of time (e.g. one, two years, etc.). The whole concept of this method, together with its theoretical base and practical assumptions, is presented in a more detailed way elsewhere (Piranashvili, 1969; Svanidze, 1969; Svanidze and Khomeriki, 1970; Svanidze and Piranashvili, 1970). The author describes a number of steps which should be taken for the choice of the most suitable one-dimensional probability distribution function, with the aid of the Pearsonian graph. Taking as an example about 200 rivers from all over the world, the conclusion was reached that, in the majority of cases, the most suitable is the beta-distribution, i.e. the Pearson distribution, type I.

On the basis of data given by Kalinin (1968), the author presents empirical formulae for the determination of extreme values of the modular coefficient of the mean annual discharge.

For the minimum modulus

$$K_{min} = e_\nu^{-4C}$$

For the maximum modulus

$$K_{max} = \begin{cases} 1 + 4.45C_\nu, \text{ at } C_\nu \leqslant 0.9 \\ 5 + 7.5(C_\nu - 0.9), \text{ at } C_\nu > 0.9 \end{cases}$$

The application of the Johnson distribution is recommended generally.

Passing to the question of synthetic hydrological series modelling by the Monte Carlo method, the author proposes the application of a unified methodology, which is valid for any one-dimensional distribution. For this purpose, the so called N-class processes (Piranashvili, 1970) are isolated by means of the N-hypothesis introduced by Kartvelishvili (1967) for a random vector.

This hypothesis reduces to the assumption that any multidimensional distribution of a finite number of correlated random values, whose one-dimensional distributions are normal and the correlation matrix is non-negatively defined, is also normal. The introduction of the N-hypothesis means that the modelling of any process may be reduced to modelling a normal process with the parameters $(0, 1)$.

Next, the author discusses recurrent dependences for the modelling of the lag-one Markov process, the higher-order Markov process and the multi-dimensional lag-one Markov process.

After this brief analysis of the papers presented, the following questions have been formulated:

(1) What are the further prospects concerning the application of analytical methods for the computation of the streamflow regulation in complex systems of mutually dependent storage reservoirs?

(2) What is the most rational form of the probabilistic forecast of the river flow for its application in the stochastic dynamic programming model?

(3) Can the Markov chains be used for the mathematical modelling of river flow in flow regulation problems? If not, what would be the justified substitute for the Markov chains?

(4) In the cases when Markov chains are used, the required capacity of the storage reservoir changes, depending on the lag-number taken into consideration. What criteria should be applied for the resolution of such problems?

(5) There are opinions, that the probability distribution functions limited at their ends (distribution beta or Pearson type I, Johnson distribution S_B, and others), are, in most cases, more suitable for the description of the annual oscillations of the river flow than other distributions which are widely used at the present time (e.g. Pearson type III or its special case gamma distribution, lognormal distribution, and others). To what extent are these opinions justified?

REFERENCES

Kaczmarek, Z. (1961) Statisticheskie metody v gidrologicheskikh progiozakh. *Doklady konferentsii po gidrologicheskim prognozam*: Budapest.
Kaczmarek, Z. (1969) *On the Methods of Selection of the Forecasting Variates:* Wrocław.
Kalinin, G. P. (1968) *Problemy Globalnoi Gidrologii*: Gidrometeoizdat, Leningrad.

Kartvelishvili, N. A. (1967) *Teoriya Veroyatnostnykh Protsessov v Gidrologii i regulirovanii rechnogo stoka*: Gidrometeoizdat, Leningrad.

Kritsky, S. N. and Menkel, M. F. (1968) Ob osnovakh teorii regulirovaniya rechnogo stoka. *Trudy GGI*, Vyp. 160.

Maass, A., Hufschmidt, M. M., Dorfman, R., Thomas, H. A. and Marglin, S. A. (1962) *Design of Water Resource Systems*: Harvard University Press, Cambridge, Massachusetts.

Moran, P. A. P. (1959) *The Theory of Storage*: Methuen, London.

Piranashvili, Z. A. (1969) O predstavlenii uslovnykh sluchainykh protsessov. *Trudy Instituta prikladnoi matematiki TGU*, t.I.

Piranashvili, Z. A. (1970) Nekotorye voprosy statistiko-veroyatnostnogo modelirovaniya nepreryv-nykh sluchainykh protsessov. *Sb. Voprosy issledovaniya operatsii*. Izd. AN GSSR, Tbilisi.

Reznikovszky, A. Sh. (editor) (1969) *Vednoenergeticheskie raschety metodom Monte Karlo*: Izd 'Energiya' Moscow.

Svanidze, G. G. (1964) *Osnovy rascheta regulirovaniya rechnogo stoka metodom Monte-Karlo*. Izd. AN GSSR 'Metsniereba', Tbilisi.

Svanidze, G. G. (1967) River runoff as a stochastic process and its mathematical modelling. *International Hydrology Symposium, Fort Collins, USA*.

Svanidze, G. G. (1969) Nekotorye voprosy modelirovaniya gigrologicheskikh ryadov. *Doklady mezhdunarodnogo simpoziuma po regulirovaniya rechnogo stoka (Sofiya-1967)*. Izd. AN Bolgarii.

Svanidze, G. G. and Khomeriki, I. B. (1970) General scheme for calculating the water balance of closed inland seas and lakes by the method of statistical modelling. *Symposium on World Water Balance*, pp.289–294: IAHS Publ. No.93.

Svanidze, G. G. and Piranashvili, Z. A. (1970) O matematicheskikh modelyakh rechnogo stoka i ikh ispolrzovanii dlya gidrologicheskikh raschetov i prognozov stoka gornykh rek. Tashkent.

Szesztay, K. (1966) Some problems of lake hydrology. *Publ. Research Institute for Water Resources Development in Foreign Languages*, No.4, Budapest.

DISCUSSION

Z. Kaczmarek:

In the first place, I would like to express my appreciation of the paper presented by the general reporter in which the author puts to the fore a number of problems that in my opinion, are most relevant to the stochastic theory of storage. I would like to make a few remarks with reference to the paper, and to what has been presented in the individual works. First of all, I share Professor Svanidze's view that the methods which he called analytical, as well as the Monte Carlo methods, should be developed and applied according to the tasks they are expected to serve. I think personally that, at present, it would be difficult to decide unequivocally in favour of any method of that group. Another problem which I would like to raise concerns my paper presented at this Symposium. This work, as Professor Svandize said, takes into account the inter-dependence between the inflows to the reservoirs in successive time periods. However, I would like to point out that the approach set out in my paper makes it also possible to take into account the nonstationarity of the inflow. Therefore, the method can also be applied to problems connected with flow regulation for periods shorter than a year. The paper by Professor Svandize contains a remark which concerns the charac-ters of the interdependence of the rate of storage, marked by the symbol V. My paper presents a thesis which is substantiated by some reasoning to the effect that, if the inflow to the reservoir forms a lag-one Markov chain, or simple Markov process, then V_i, V_{i+1}, etc., constitute a lag-two Markov process. In other words, if the inflows bear the character of the lag-N random process, then the rate of storage forms a lag-$(N + 1)$ random process. I think that, under specific assumptions, this thesis is cor-rect. One final remark which is in close connection with the general report, and, strictly speaking, with the questions posed by Professor Svandize. You may be interested to know that a work concerning stochastic dynamic programming, in my opinion a very interesting work, has been published in Poland. This is a paper by Dr. Kornatowski which those colleagues of ours who are interested in these problems may find useful. It is available as a publication in one of our scientific periodicals.

I. P. Tzvetanov:
After the thorough analysis of my paper made by the general reporter, I shall not dwell again on the content of the paper, not even on the summary. However, I shall take the opportunity of expressing my opinion on two basic problems associated with the control of complex reservoirs.

First problem. The complex reservoir as a complex control plant with three hierarchic levels
When analysing the generalized block diagram of the complex reservoir given by Fig.1 (vol.1, p.423), it is seen that the complex reservoirs are complex systems with three hierarchic control levels:

level μ_1 — the level on which the reservoir itself is controlled,
level μ_2 — the level on which the water is distributed between the basic types of water users (hydroelectric power stations, fish industry, sports bases, etc.) and consumers (agriculture, industry, etc.),
level μ_3 — level on which the water is distributed between consumers and water users of a given type.

In the paper, I give the solution of the problem only for the level μ_1 without considering the complexity of the entire problem. This is because I obtained the complete solution of the problem later.

Since it is not possible to formulate and solve this complex problem within five minutes I am pleased to say that its complete solution may be found in a paper which I shall present at the Vth IFAC Congress in Paris, June 1972.

Second Problem. Probabilistic description of the flow (the input) of a given reservoir
When forming the flow (the input) of the complex reservoir, a number of random factors such as rainfall, air temperature, moisture absorption of the soil, etc. have their effect. This fact permits us to consider the flow as a random process. The investigations of authors like Kartvelishvili, Svanidze, and others allow us to assume the hypothesis that, for practical purposes, the flow is described accurately enough by the Markov process. This hypothesis was also confirmed by our investigations carried out with my colleagues. Hence, the flow may be predicted accurately enough by a Markov process with 'incomes' (in this case the 'incomes' are the discharges ξ_{ij} flowing into the reservoir when the Markov process passes from the state i into the state j). I have no more explanations since this problem is developed in detail in my paper.

S. K. Srinivasan:
I am in full agreement with the report presented by Professor Svandize, particularly with regard to the analytical solution of finite dams. Our attempt at the solution of finite dams is part of a general programme, involving a larger class of inventory problems in which the inputs are very much correlated. After the contribution was submitted to the Conference, we have taken up and completed investigations relating to other types of dam as well. Other papers soon due for publication will meet some of the points raised by Professor Svanidze (Srinivasan, 1972; Srinivasan and Ramani, 1971). Besides these, there are other attempts, particularly by Cohen (1969) and his collaborators from the Netherlands and Kalashnikov (1967) from Russia on the general solution of inventory problems. I fully share Professor Svanidze's views regarding dams in series. Perhaps in this case, we may have to use both simulation and analytical methods together to throw light on some of the important hydrological problems.

REFERENCES

Cohen, J. W. (1969) *Single Server Queues*: North-Holland.
Kalashnikov, V. V. (1967) Some results in inventory theory. *Kybernetika* **3** (1), 58–62.
Srinivasan, S. K. (1972) Analytic solutions of a finite dam governed by a general input. *J. appl. Probab.*
Srinivasan, S. K. and Ramani, S. (1971) A continuous storage model with alternative random input and output. *J. Hydrol.*

I. Zsuffa:

Permettez moi de répéter le titre de mon rapport: 'Quelques relations arithmétiques de matrice dans le calcul des réservoirs'. Ainsi je n'ai voulu pas ici donner rien de nouveau à la base théorique. Je ne veux que démontrer les avantages techniques de l'utilisation des mêmes différences ΔQ pour chaque valeur. Le modèle est le modèle de Moran, la méthode, avec quelques modifications et quelques complément, est issu de la littérature française (Bernier, Roux). Les relations algorithmiques des matrices transistoires permettent telles simplifications, avec lesquelles on peut élaboré une programme de l'ordinateur très vite servant les répartitions des volumes emmagasinés dans le réservoir et des volumes superflus passés par l'évacuateur pour chaque K capacité et chaque M volume d'eau exigé.

Quant au remarque sur les hypothèses je note qu'il n'y a pas modèle sans hypothèse. Naturellement il faut examiner et justifier ces hypothèses. Selon le rapporteur même, la relation des débits des bassins versants voisins est constant pour les petits bassins versants. Il me faut remarquer que la Hongrie entière est relativement petite en comparaison avec les bassins versants de l'URSS, et nous n'avons travaillé que sur les bassins versants à surface 100–600 km^2. Quant à l'indépendance des débits des périodes voisines je pense que ce n'est pas une question hydrologique: ils existent des tests statistiques exactes convenables pour donner la réponse: tests Wald–Wolfowitz, même le test χ^2. Il faut avouer que justement c'est le manque de l'indépendance pourquoi nous n'utilisons pas cette méthode dans le cas des petits réservoirs dits réservoirs, saisonniers. Pour ces réservoirs nous utilisons les méthodes plus primitives, basées sur la répartition des débits mensuels, bimensuels, trimensuels etc.

Quant à la répartition des volumes d'eau passer par l'évacuateur, la dimensionnement de deux réservoirs n'est qu'une simple exemple. On peut aussi montrer un autre exemple. Notre plus grand lac c'est le Balaton. En l'envisageant comme un réservoir nous avons utilisé cette programme. Nous avons reçu pour chaque K capacité du lac et chaque M eau exigé par l'irrigation, les répartitions des niveaux de l'eau et des volumes d'eau évacué.

Naturellement les graphiques de ces répartitions ne sont qu'illustration, les répartitions sont données par les vecteurs de rang des matrices des résultats de l'ordinateur. Ces vecteurs servant les données de base aux calculs de l'usine optimal du lac au point de vue de récréation, de navigation, de défense contre inondation, de l'irrigation et de la production d'energie.

Quant à la question philosophique de M. le rapporteur il me semble aussi que c'est le domaine de calcul des réservoirs où on peut utiliser en premier lieu des méthodes de simulation, parmis eux la methode 'Monte Carlo'.

Mais c'est fait que par simulation on ne peut pas acquérir plus d'information: ce n'est que les informations des données pures qui sont à notre disposition. Ainsi, de point de vue, de précaution, il me semble que les modèles directes, p.e. le modèle de M. Moran, malgré leur domain d'application plus restreint, sont plus convenables.

W. G. Strupczewski:

In a stochastic reservoir operation model we want to base the decision reached at the beginning of the ith period on the reservoir volume V_i and some hydrometeorological variables $\{X_i\}$ which are known at that time.

These variables serve as predictors for the future inflows:

$$Z_i = Z_i(V_i, \{X_i\})$$

while

$$\{X_i\} = \{Q_i^-\}, \{Y_i\}$$

where Q_i^- stands for past inflows according to the number of conditions which appear in the Markov process, and $\{Y_i\}$ denotes other variables. It is difficult to give an *a priori* estimation of the degree to which a particular predictor or a collection of predictors bears on the decision Z_i. Therefore we do not know of which components $\{X_i\}$ should consist. Indeed, we know the case where the standard policy determined by the queues theory is the best and the knowledge of future inflows $Q_{i+1}, \ldots$ is immaterial for making the optimal decision.

In general, it is purposeful to make the decision dependent upon $\{X_i\}$. This is a *sine qua non* condition for a reservoir model of dynamic programming to be of practical applicability. On the other hand the numerical side of the problem requires us to minimize the dimension of the X_i vector.

Our experience indicates that the predictor vectors $\{X_i^k\}$ which, in various time periods $i + k, k = 1, 2, \ldots$, are optimal for inflow forecasts, are all different, and this would mean a high dimension of the $\{X_i\}$ vector. Therefore, instead of operating with predictors $\{X_i\}$, it seems reasonable to introduce some multidimensional functions of $\{X_i\}$ which I would like to call the aggregated predictors. Let us express the aggregated predictors by the equations for average conditional inflows in future periods

$$\{\tilde{Q}_i\} = \{\tilde{Q}_i^i, Q_i^{i+1}, \ldots\}$$

Then $\{X_i'\} = \{\tilde{Q}_i\}, \{Q_i^-\}$

With an assumed horizon of a forecast one can reduce the dimension by applying the forecast only to selected time periods e.g. those with large instability of inflows. One can also consider the possibility of omitting $\{Q_i^-\}$.

The application of aggregated predictors allows for an immediate use of inflow forecast schemes which were developed for our reservoirs, in order to determine the reservoir operating policy. To a certain extent, the above approach was indicated in my paper.

The usefulness of Markov processes as inflow models in the problem of water system control

In problems of this type, some apparently simple cases may well extend beyond the possibilities of computers. To depart from Markov processes would even more complicate the solution.

The estimation of computational possibilities will clearly affect the number of relations assumed in the inflow process and the range of the forecast. Unfortunately, the usefulness of an assumed inflow model can be estimated after the relevant problem of water system control is solved. In view of the fact that the model is realistically inadequate one should not identify an average theoretical effect, determined by solving the control model, with an average effect resulting from the operation based on relevant decision functions. The usefulness of an inflow and forecast model is estimated by comparing the effects resulting from decision functions corresponding to various models during an observational series of inflows.

Such an approach also allows an estimation of the usefulness of theoretically inconsistent deterministic models which admit average values as a future inflow. The study of an inflow model should not be separated from the context of a problem which is to be solved.

K. Krajewski:

I would like to answer briefly a few of the questions posed by the general reporter. One question concerned the likely future of deterministic methods and methods based on modelling by the Monte Carlo method. This question has already been answered in part by Kaczmarek. I would like to add that it is possible to formulate the answer in a different way, namely, that such methods which are simply based on the solution of the actual operation problems are likely to become increasingly useful. Another question posed by the general reporter concerned the forms of forecasts which could be utilized to the best advantage of the reservoir operation.

For many years now, forecasts, both in this country and elsewhere, have been so formulated that they indicate a conditional distribution of an anticipated quantity. This conditional distribution is, in fact, the sole objective we should aim at. Having at our disposal a conditional mean value and a conditional error of forecast, we can use these quantities for all water management calculations.

There is one more thing that I would like to mention. The previous discusser said that, in his own opinion, the best reservoir policy is such a policy which is only connected with an appropriate forecast. I think that my learned colleague simply forgot to add that this is not only his opinion, but one which is shared by many of his colleagues.

Closing remarks by the general reporter

I have no more comments since the authors of the reviewed papers have not protested against my remarks and wishes concerning their work. Only Kaczmarek stands by his thesis concerning change of the Markov chain lags. I take Kaczmarek's word for it; however, I brought with me a very basic research paper written by Abramishwili and Chitashvili, two young mathematicians from Tbilisi. I have given them exactly the same problem to clarify — how the Markov clain lags change in river flow regulation. The authors arrived at some well specified conclusions and I have given this paper to Kaczmarek. I think there were no other reservations. Once again I would like to express my gratitude to all discussers for their patience.

Rapporteur général Professeur Jacques Bernier, Laboratoire National d'Hydraulique, Electricité de France, Chatou, France
Président Dr. I. Zsuffa, Direction des Eaux, Baja, Hongrie

Processus stochastiques multidimensionnels en hydrologie — I

J. Bernier

Les communications analysées dans ce rapport général sont les suivantes. Celles-ci sont publiées en tome 1 du colloque.

1. *R. Gras et J. P. Henry* Un modèle stochastique de simulation de la température d'un cours d'eau
2. *J. Déri* Analyse de la concomitance stochastique de certains processus hydrologiques
3. *Timothy Doak Steele et Nicolas C. Matalas* Principal-component analysis of streamflow chemical quality data
4. *N. C. Matalas et J. R. Wallis* Correlation constraints for generating processes
5. *E. P. Chemerenko* Objective analysis of snow cover fields
6. *Ignacio Rodríguez Iturbe et Carl F. Nordin* Frequency domain analysis of hydrological systems with monthly data

Introduction

L'analyse statistique multidimensionnelle traite de données collectées en plusieurs dimensions sur le même individu. En hydrologie on considère généralement des unités de temps tels que l'année, le mois, le jour etc. pendant lesquelles sont mésurées des grandeurs multidimensionnelles comme la précipitation ou le débit relatifs à un ensemble de stations dans une région ou un bassin donnés. Dans un domaine connexe comme celui de la qualité des eaux, les multiples paramètres de pollution mesurés dans un site donné peuvent constituer les éléments d'un vecteur à multiples dimensions.

La nécessaire prise en compte des modèles multidimensionnels tient à plusieurs facteurs. En premier lieu les problèmes posés par l'exploitation des ressources en eau doivent être résolus à l'échelle régionale dans la mesure où les localisations des ressources et des besoins diffèrent et où l'importance croissante des demandes nécessite la diversification de ces ressources. Eu égard à l'aspect qualité des eaux, les conséquences étendues à l'aval d'un réseau hydrographique, d'un rejet polluant doivent être examinées à l'échelle régionale. Dans ce cas, les aspects divers que peut prendre la pollution imposent des critères de qualité non pas basés sur un seul paramètre mais sur un ensemble de paramètres biochimiques, salins, minéraux thermiques, radio-actifs etc. qui possèdent de multiples interactions entre eux et avec l'hydrologie du milieu.

En second lieu l'information hydrométéorologique disponible est limitée en quantité et en qualité, les durées d'observations peuvent différer grandement d'une station à l'autre. L'utilisation de modèles probabilistes multidimensionnels permet une amélioration importante de la précision des estimations de ressources par rapport à la méthode unidimensionnelle d'étude de chaque station séparément. En effet les processus hydrologiques spatiaux sont très fortement structurés et les dépendances régionales sont généralement étroites. La prise en compte de ces structures est d'ailleurs nécessaire si on veut représenter avec réalisme les lois de probabilité de grandeurs hydrométérologiques pour les utiliser dans des modèles de simulation notamment.

Les phénomènes hydrométéorologiques présentent des variabilités temporelles et spatiales importantes. Ces deux types de variabilité ne sont pas de même nature. Ce fait n'a pas été reconnu par certains hydrologues qui (notamment par la méthode dite des stations années) ont essayé de compléter un échantillonnage temporel insuffisant par un échantillonnage spatial en mélangeant les observations de stations différentes avec ou sans correction. Ces méthodes ne peuvent permettre une estimation convergente des lois de probabilités des variables étudiées. En effet la prise en compte de l'échantillonnage temporel est fondée sur la loi des grands nombres qui assure la convergence des paramètres statistiques calculés sur les échantillons temporels vers des quantités certaines. Cette loi n'est généralement pas valable pour les échantillonnages spatiaux Kalinin et Nikolskaya ont précisé l'hypothèse d'ergodicité conditionnelle qui fonde l'estimation de probabilité en échantillonnage spatial. Cette hypothèse n'est généralement pas valable pour des bassins d'étendue moyenne où se posent les problèmes d'exploitation des ressources. A ce niveau les structures de dépendance régionales sont trop fortes et il importe de définir un modèle de processus multidimensionnel:

Etant donné un vecteur X dont les constituants $X_1, \ldots, X_n$ sont des variables aléatoires dont les réalisations sont les observations à n stations d'un bassin, ce vecteur X_t peut dépendre du temps et constituer alors un processus stochastique. Pour une époque t fixée, la loi de probabilité:

$$F_t(x_1, x_2, \ldots, x_n, \theta) = \mathrm{Prob}(x_{it} \leqslant x_1, \ldots, x_{nt} \leqslant x_n/\theta)$$

définit la structure spatiale.

Les paramètres non aléatoires θ caractérisent les différences spatiales. En d'autres termes la variabilité temporelle aléatoire est prise en compte par la loi de probabilité; la variabilité spatiale est prise en compte par les variations déterministes des paramètres de ces lois pour chaque station et par la structure de dépendance.

Les communications analysées dans ce rapport ont en commun l'utilisation à des degrés divers, d'un point de vue multidimensionnel. En dehors de ce point commun, leur diversité quant aux problèmes traités et aux méthodes d'approche illustrent l'étendue et la fécondité de l'approche multidimensionnelle en hydrologie. Nous étudierons dans l'ordre: la communication de Chemerenko qui traite de mise en forme préliminaire de l'information spatiale en vue des prévisions, les communications de Déri, Rodríguez-Iturbe et Nordin qui traitent des méthodes de recherche des modèles plus spécialement bidimensionnels, la communication de Matalas et Wallis qui traite d'un problème technique sur les contraintes imposées par la structure de corrélation et leurs conséquences sur l'estimation des paramètres dans le cas d'information incomplète. Nous discuterons enfin des deux communications de Steele et Matalas, et Gras et Henry dont le domaine d'application concerne la qualité des eaux.

Objective analysis of snow cover fields: par Chemerenko

En vue de préparer l'information hydrométéorologique pour des prévisions numériques, on doit disposer des valeurs aux noeuds d'une grille régulière disposée sur une zone geographique donnée. Le but de l'analyse objective est de reconstituer par interpolation les valeurs aux noeuds à partir des des données des stations voisines irrégulièrement disposées, cette procédure automatisée permet le contrôle de l'homogénéité des observations spatiales.

Les méthodes d'interpolation exposées, reposent sur les relations lineaires:

$$f_0 = \sum_{i=1}^{n} p_i f_i$$

exprimant la valeur interpolée f_0 en fonction des valeurs aux stations i. La recherche des pondérations p_i est basée sur la minimisation de la moyenne quadratique de l'erreur d'interpolation et donne le résultat:

$$\sum_{j=1}^{n} p_j \mu_{ij} = \mu_{i0}$$

où les μ_{ij} sont des fonctions de corrélation entre les stations i et j, μ_{i0} la function de corrélation entre les stations i et le noeud interpolé.

Passant en revue diverses méthodes (méthode des moyennes, méthode des moyennes pondérées, interpolation polynomiale, interpolation optimale) l'auteur donne diverses formules permettant le calcul des μ_{ij} et donc des pondérations p_i.

Les méthodes sont illustrées et comparées à propos du traitement des données sur la couverture neigeuse (hauteur et équivalent en eau) qui présentent une variabilité spatiale considérable sur les 700 stations situées dans la partie européenne de l'URSS.

Chemerenko décrit la méthode de relation automatique des stations pour l'interpolation et donne quelques résultats numériques sur la comparaison des méthodes à partir du critère de l'écart quadratique moyen entre valeurs calculées et observées:

$$\hat{\sigma} = \sqrt{(f_0 - f_c)^2}$$

Exprimé en pourcentage de l'écart quadratique moyen de la hauteur de neige. $\hat{\sigma}$ est égal à 0.20–0.25; il monte à 0.30–0.35 pour l'équivalent en eau. La méthode d'interpolation optimale donne des résultats légèrement meilleurs.

La précision augmente avec la densité du réseau mais atteint la saturation pour un station par 4000 km^2.

Cette interessante méthode d'interpolation numérique n'est pas de nature probabiliste car elle ne prend pas en compte les observations comme réalisation de variables aléatoires. Le paramètre $\hat{\sigma}$ fait intervenir à la fois des fluctuations aléatoires et des variations spatiales systèmatiques.

La méthode objective de sélection et d'interpolation est certes validée dans des régions géographiquement homogènes comme les grandes plaines d'URSS. On peut discuter son application à des zones de relief plus tourmenté et à des grandeurs comme les précipitations soumises à ce relief.

Enfin l'auteur pourrait peut être fournir des indications complémentaires sur l'interpolation optimale dont il donne les résultats sans préciser les principes de base.

Analyse de la concomitance stochastique de certains processus hydrologiques: par Déri

Déri passe en revue diverses méthodes permettant d'apprécier la probabilité de l'occurrence simultanée des états critiques (des points du vue, crue, étiage ou richesse) de deux processus hydrologiques.

Dans le cas de processus stationnaires les propriétés des coefficients de corrélation simple, sériels (autocorrélation) et fonctions de corrélation croisée (fonction d'intercorrélation) sont rappelées ainsi que leur utilité pour résoudre le problème de l'extrapolation stochastique. Ceci est fort classique.

Une idée intéressante est l'étude de la concomitance au moyen de fonctions indicatrices attachées aux processus, soit pour les cours d'eau 1 et 2:

$$I(t)_1 = \begin{cases} 1 \text{ si } Q(t)_1 < KQ(t)_1 \\ 0 \text{ si } Q(t)_1 \geqslant KQ(t)_1 \end{cases}$$

$$I(t)_2 = \begin{cases} 1 \text{ si } Q(t)_2 < KQ(t)_2 \\ 0 \text{ si } Q(t)_2 \geqslant KQ(t)_2 \end{cases}$$

Ces fonctions indicatrices, d'étiages dans le cas présent, dépendent du choix des fonctions sérielles KQ, constantes ou dépendant du temps, il y aura concomitance pour une période t_i entre les deux processus si

$$I(t_i)_1 = I(t_i)_2 = 1$$

$$I(t_i)_1 = I(t_i)_2 = 0$$

Déri donne la formule suivante pour la probabilité conditionnelle:

$$\text{Prob}[I(t_i)_i = 1 | (t_i)_2 = 1] = p + r(1 - p)$$

où

$$p = \text{Prob}[Q(t_i)_{1,2} \leqslant KQ(t_i)_{1,2}]$$

et r un paramètre dont la signification n'est pas précisée mais qui semble relié au coefficient de corrélation des valeurs des fonctions indicatrices entre les stations.

Dans la deuxième partie de sa communication, Déri expose les prémisses d'un modèle bidimensionnel stochastique markovien pour l'étude de la concomitance de la sècheresse définie à partir d'un critère sur l'humidité du sol

$$S(t_i) \leqslant S_0 \text{ (seuil)}$$

et de l'étiage dèfini par un critère sur le débit

$$Q(t) \leqslant KQ(t)$$

Ces critères permettent la définition de nouvelles variables comme les durées des sécheresses ou des étiages, le nombre d'épisodes secs, les volumes déficitaires et les besoins en eau maximaux dont les distributions statistiques ont été ajustées par des lois exponentielles. de Poisson, et de Pearson.

Pour l'étude de la concomitance on peut définir un système à quatre états:

E_1: sécheresse et étiage (définis par les critères précédents)
E_2: sécheresse et abondance en eau (définie comme le contraire de l'étiage)
E_3: état humide du sol et étiage
E_4: abondance en eau dans le sol et dans le cours d'eau

Les transitions d'un état E_i à un autre entre deux époques successives sont définies par une matrice de probabilité de transition caractérisant un processus markovien d'ordre 1 pour l'ensemble des états E_i.

Cette communication donne un état préliminaire de méthodes intéressantes. Il restera notamment à utiliser la théorie des processus markoviens pour le calcul des probabilités de concomitance utiles pour la prévision.

Frequency domain analysis of hydrological systems with monthly data: par Rodríguez-Iturbe et Nordin

Voici une présentation heuristique et des illustrations de techniques très en faveur actuellement auprès des hydrologues à savoir: les analyses spectrales undimensionnelles et croisées. Les données utilisées sont les séquences d'anomalies mensuelles (écarts de débits mensuels à leur moyenne) des débits liquides et solides de deux stations situées sur le Rio Grande dans le Nouveau Mexique. Les séquences d'anomalies sont supposées stationaires.

Les auteurs donnent les expressions approchées des représentations spatiales des deux processus $x_1(n)$ et $x_2(n)$ des débits solides et liquides pour le mois n, on a

$$x_1(n) = \sum_{j=0}^{\infty} a_j \cos (w_j n \Delta t) + b_j \sin (w_j n \Delta t)$$

$$x_2(n) = \sum_{j=0}^{\infty} c_j \cos (w_j n \Delta t) + d_j \sin (w_j n \Delta t)$$

Les fréquences w_j sont mesurées en radians par mois. Les auteurs considèrent les a, b, c, d comme des réalisations de variables aléatoires multinormales. Ceci est une hypothèse qui demanderait vérification et qui ne peut être valide en toutes circonstances pour les débits mensuels. Cette hypothèse n'est d'ailleurs pas néssaire pour la validité de la représentation spectrale pour laquelle on suppose généralement que ces coefficients sont des variables aléatoires indépendantes possédant des variances fonction des fréquences w_j.

Etudiant les séries dans le domaine du temps, les auteurs donnent des estimations des fonctions d'autocorrélation:

$$a_{x_1 x_1}(\tau) = E[x_1(n)x_1(n-\tau)]$$

et fonctions de corrélation croisée

$$a_{x_2 x_1}(\tau) = E[x_2(n).x_1(n-\tau)]$$

Il apparaît que le débit liquide peut être mieux prédit à partir de son propre passé que son homologue solide (les autocorrélations du premier ordre étant respectivement de 0.43 et 0.26).

Les auteurs mettent cependant l'accent sur la représentation spectrale des corrélations simples et croisées dans le domaine des fréquences en utilisant la notion de spectre de puissance:

$$\Gamma(f) = 2 \int_0^{\infty} a(\tau) \cos 2\pi f \tau \, d\tau$$

Dans le cas du spectre croisé ou a est la covariance croisée $a_{x_2 x_1}(\tau)$, on sait que $\Gamma(f)$ est une fonction complexe:

$$\Gamma x_1 x_2(f) = c_{x_1 x_2}(f) + i Q_{x_1 x_2}(f)$$

dont les parties réelles et imaginaires permettent la définition du spectre d'amplitude croisée

$$C_T(f) = \sqrt{C^2 + Q^2}$$

et du spectre de phase:

$$\theta(f) = \text{arc tan} \frac{Q}{C}$$

Les figures de spectres unidimensionnels montrent dans chaque cas une contribution très importante (70 à 60 pour cent) des basses fréquences à l'explication de la variance totale du phénomène. Cette constatation peut être faite également sur le spectre d'amplitude croisée. Le spectre de phase indique les retards entre les oscillations corellées des deux processus.

Rodrigues Iturbe et Nordin utilisent ces spectres pour ajuster un modèle au système des deux processus $x_1(t)$ et $x_2(t)$:

$$x_1(t) = dx_1(t - 1) + \eta_1(t)$$

$$x_2(t) = x_t + \eta_2(t)$$

où x_t est un processus mixte explicitant la relation entre les débits liquides et solides:

$$x(t) = ax(t - 1) + bx(t - 2) + cx_1(t)$$

où les $\eta_1(t)$ et $\eta_2(t)$ sont des bruits blancs (variables aléatoires indépendantes). Ce modèle bidimensionnel à l'advantage de permettre une simulation conjointe des deux processus et de tenir compte des écarts de phases entre les séries.

Cette illustration des méthodes spatiales est particulièrement intéressante par la discussion très complète de l'estimation des modèles. On notera que l'analyse spectrale est utilisée ici en supposant les processus stationnaires, il reste à prouver que la seule transformation additive opérée en retranchant le débit moyen permet d'éliminer la nonstationnarité éventuelle notamment pour les structures de dépendance des processus. Des expériences différentes tendraient à montrer une évolution saisonnière des corrélations entre les débits mensuels.

Correlation constraints for generating processes: par Matalas et Wallis

Les auteurs montrent les contraintes imposées par les structures algébriques des matrices de covariance des modèles probabilistes aux méthodes d'estimation adaptées à l'information disponible.

Pour générer par modèle probabiliste les débits à n stations ou saisons, la matrice de covariance C entre les n composantes du vecteur de variables doit être définie positive ce qui entraîne un certain nombre de conséquences notamment le déterminant de C doit être positif.

La nature des contraintes imposées est illustrée sur deux cas: Le premier cas concerne les modèles markoviens exprimés sous forme matricielle:

$$X(t) = AX(t - 1) + B\epsilon(t)$$

où les éléments de ϵ sont indépendants de ceux de X et de ϵ des époques antérieures. Dans ce cas où $n = 2$ la contrainte imposée aboutit à la condition

$$R^2 \leqslant \frac{(1 - \rho^2)(1 - r^2)}{(1 - r\rho)^2}$$

où R, ρ et r désignent respectivement les coefficients de corrélation entre $X_1(t)$ et $X_2(t)$, $X_1(t)$ et $X_2(t - 1)$, $X_2(t)$ et $X_2(t - 1)$ (X_1 et X_2 sont les deux composantes du vecteur).

Lorsque les paramètres R, ρ et r sont estimés sur des séries observées de longueurs différentes, cette condition peut n'être pas remplie. On doit alors contrôler si cette situation est due aux erreurs d'arrondissement dans le calcul ou à la nature des corrélations croisées, dans ce dernier cas le modèle markovien ne peut être utilisé sans ajustement des paramètres du modèle.

Un autre cas étudié est celui des modèles de bruit fractionnaire initialement proposés par Mandelbrot et Wallis sous la forme undimensionnelle:

$$X(t) = \left(h - \frac{1}{2}\right) \sum_{u=t-M}^{t-1} (t - u)^{h-3/2} \epsilon(u)$$

Matalas et Wallis proposent une généralisation bidimensionnelle de ce modèle.

Dans ce cas le coefficient de corrélation R doit obéir à une contrainte du type $R^2 \leqslant K$ où K a une forme assez complexe.

D'après les auteurs il semblerait que la contrainte imposée à R soit plus large dans le cas des modèles à bruits fractionnaires que dans le cas markovien.

Dans tous les cas on constate que la limite supérieure de R est égale à 1 si les processus marginaux sont décrits par des modèles homogénes avec paramètres identiques.

.Cette communication illustre quelques unes parmi les précautions à prendre pour ajuster des modèles multidimensionnels. On pourrait demander aux auteurs la raison pour laquelle ils utilisent un modèle markovien particulier où les écarts ϵ sont tous mutuellement indépendants.

Un modèle markovien où les ϵ sont corrélés serait peut être plus réaliste dans certains contextes. Dans ce cas il serait utile d'adapter les contraintes imposées aux paramètres du modèle.

Principal-component analysis of streamflow chemical quality data: par Steele et Matalas
L'analyse en composantes principales est, avec l'analyse spectrale multidimensionnelle, une des techniques d'analyse statistique multivariable les plus en faveur parmi les hydrologues. Elle réalise la décomposition d'un vecteur X de n variables aléatoires sous la forme.

$$X = AW$$

où W est un vecteur de m variables non corrélées (covariances nulles) m étant le rang de la matrice des covariances $C = E[XX^T]$ des éléments de X.

Les variances des composantes principales que sont les coordonnées de W sont les valeurs propres de la matrice C, soit $\lambda_{ii}(i = 1, m)$.

Ces λ_{ii} constituent les éléments d'une matrice diagonale λ vérifiant:

$$\lambda = A^T A$$

$$\sum_{i=1}^{m} \lambda_{ii} = \sum_{j=1}^{m} C_{jj}$$

où les C_{jj} sont des éléments de la diagonale principale de C. La recherche des composantes principales est appliquée par les auteurs à un problème de qualité chimique des eaux des rivières où les n coordonnées de X sont les concentrations de n constituants chimiques.

Chimiquement ces n concentrations sont liées par une relation:

$$\text{meq}/1.(\text{cations}) = \text{meq}/1.(\text{anions})$$

Cette équation exprime l'égalité de la concentration totale des cations (ou ions chargés positivement) et de la concentration totale des anions (ou ions chargés négativement). Les concentrations de chaque constituant doivent être exprimées en poids équivalent:

$$\text{meq}/1. = \frac{\text{mg}/1}{G}$$

où G est une constante de proportionnalité spécifique à chaque constituant. Par ailleurs les concentrations peuvent être exprimées par des relations linéaires stochastiques en fonction de la conductivité spécifique Z

$$C_i = a_i + b_i Z + \epsilon$$

Ainsi la conductivité spécifique Z peut être un indicateur de la quantité d'ions inorganiques présents dans un cours d'eau.

Le cas présenté concerne l'étude des concentrations moyennes annuelles de 6 composants: Ca^{+2}, Mg^{+2}, Na^{+1}, HCO_3^{-1}, SO_4^{-2}, Cl^{-1} dont les concentrations sont significatives à 8 stations situées en Arkansas, Texas et Nouveau Mexique.

Les variables X_i utilisées sont les concentrations normalisées

$$X_i = \frac{C_i - \bar{C}_i}{S_i} \quad (\bar{C}_i: \text{moyenne}, S_i: \text{écart-type})$$

pour éliminer l'effet d'échelle.

Selon les cas la première composante principale exprime entre 56 et 84 pour cent de la variance totale exprimée par

$$\sum_{j=1}^{n} C_{jj}$$

égale à 6 dans le cas présent.

La dernière composante a une variance pratiquement nulle ce qui exprime la condition chimique d'égalité des quantités d'anions et de cations. Par ailleurs le coefficient de corrélation entre la première composante et la conductivité spécifique varie entre 0.90 et 0.99.

Les auteurs essaient d'expliquer les premieres composantes en fonction des constituants chimiques les plus fortement corrélés avec ces composantes. Il faut remarquer qu'une présentation géométrique non utilisée par les auteurs pourrait faciliter cette recherche. Dans un espace à 6 dimensions ou chaque vecteur X est repéré par ses 6 constituants chimiques les 6 composantes constituent un repère orthogonal. On peut projeter les points observations sur le plan principal engendré par les deux premières composantes. Le fait que ces deux composantes expliquent entre 81 et 94 pour cent de la variabilité totale, nous montre que le nuage de points dans R^6 (espace à 6 dimensions) est très peu dispersé autour du plan principal qui permet ainsi de visualiser le phénomène. Par ailleurs le cosinus de l'angle de la projection des axes originaux sur le plan principal avec chacun des axes principaux est égal aux corrélations entre les constituants qui représentent l'axe original et la composante principale.

Steele et Matalas terminent leur communication par une analyse de sensibilité en éliminant les constituants dont les corrélations avec les autres concentrations sont les plus faibles.

Un modèle stochastique de simulation de la température d'un cours d'eau: par Gras et Henry

Pour étudier le réchauffement des rivières lié à l'utilisation de l'eau comme fluide réfrigérant par l'industrie, les auteurs proposent un modèle de simulation basé sur l'établissement du bilan énergétique par tronçon défini par l'équation:

$$\rho ch \frac{dT}{dt} = F(T, t)$$

h est la profondeur, ρ et c la masse et la chaleur spécifique de l'eau.

La fonction d'échange d'énergie $F(t, t)$ où intervient la température de l'eau contient notamment des termes de flux de rayonnements atmosphériques et solaires, la chaleur latente de vaporisation de l'eau qui avec les paramètres d'humidité absolue de l'air permet d'exprimer les échanges par évaporation.

Cette méthode du bilan énergétique pour le calcul de la température naturelle d'un cours d'eau tombe en défaut là où on ne dispose pas de séries de températures naturelles nécessaires not amment pour les rivières à forte pente dans les régions montagneuses.

Les auteurs proposent un modèle stochastique de simulation dont l'estimation peut être effectuée sur des données météorologiques disponibles.

Ce modèle repose sur les principes suivants: La fonction d'échange peut être exprimée en fonction de la température naturelle T_n:

$$F(T_n, t) = a(t) - \beta(t)(Lq_e(T_n) - \rho'C_p T_n) - \sigma'T_n^4$$

où L est la chaleur latente de vaporisation de l'eau. q_e l humidité absolue de l'air, C_p la chaleur spécifique de l'air à pression constante.

Les processus $\beta(t)$ et $a(t)$ représentent respectivement les conditions d'échanges particulières au site et les conditions climatologiques régionales.

Les trois processus a. β et T_n sont liés entre eux de telle sorte que seuls deux d'entre eux sont indépendants.

Gras et Henry donnent un modèle de simulation des deux processus $T_n(t)$ et $\beta(t)$ sous la forme suivante où t est exprimé en jours:

$$T_n(t) = T_0 + T_1 \cos\left[\frac{2\pi}{365}t + \varphi\right] + T_2(t)$$

$$\beta(t) = \beta_0 + \beta_1 \cos\left[\frac{2\pi}{365}t + \varphi\right] + \beta_2(t)$$

Le résidu aléatoire $T_2(t)$ est représenté par un processus markovien stationnaire du premier ordre.

Dans l'exemple exposé le coefficient d'autocorrélation du résidu est très élevé de l'ordre de 0.989.

La terme aléatoire β_2 semble avoir une structure relativement complexe; l'hypothèse de bruit blanc adopté n'est qu'approximative. La fonction de répartition de β_2 a été ajustée sur l'exemple traité par une loi de Pearson type III à trois paramètres.

Gras et Henry illustrent leur modèle en comparant les fonctions de répartition naturelle et perturbée par une centrale thermique calculées sur 10 ans suivant les deux modèles déterministe et stochastique dans un cas où la disponibilité des températures naturelles permettait la mise en oeuvre du premier.

Le faible écart entre les deux courbes montre la validité de l'approche stochastique dont la mise en oeuvre ne demande qu'un nombre limité d'années d'observations météorologiques.

Cette méthode présente un très grand intérêt mais elle se situe encore à un stade exploratoire. Pour la rendre opérationnelle dans le cas d'information métérologique limitée, il faudrait s'assurer de la validité des hypothèses de structure du modèle: processus de Markov, bruit blanc, loi de Pearson en traitant divers cas de rivières possédent des régimes thermiques différentes.

Conclusions

La statistique mathématique met à la disposition des hydrologues un arsenal de techniques d'analyse multidimensionnelle très étendu. Outre l'analyse en composantes principales dont les applications commencent à être développées dans le domaine de l'hydrologie, il existe des méthodes comme l'analyse discriminante, l'analyse des correspondances, les corrélations canoniques, l'analyse factorielle dont un inventaire peut être trouvé dans Morrison (1967) et dont les applications pourraient être fructueuses. Matalas et Reiher (1967) ont avancé un certain nombre de critiques à l'égard des techniques d'analyse factorielle appliquées notamment par Wallis (1965). Ces critiques concernent surtout l'interprétation phénoménologique trop stricte des facteurs obtenus. Cependant une approche récente de ces méthodes conçues comme des outils de description des données indépendamment de tout modèle probabiliste, description basée sur des interprétations géométriques du style de celle présentée dans ce rapport et concernant les composantes principales, permet de répondre à certaines des critiques avancées. Il n'est pas douteux que ces techniques pourraient être utiles en hydrologie. Malgré leur diversité les communications présentées dans ce rapport ne présentent qu'un échantillonnage assez limité des méthodes statistiques et probabilistes multidimensionnelles en hydrologie.

BIBLIOGRAPHIE

Kalinin et Nikolokaya (1967) Spatial and temporal analysis of maximum runoff. *International Symposium on Floods and their Computation,* Leningrad: AISH Publ. No.84.
Matalas, N. C. and Reiher, B. J. (1967) Some comments in the use of factor analysis. *Wat. Resour. Res.* 3 (1).
Morrison, D. F. (1967) *Multivariate Statistical Methods*: MacGraw-Hill
Wallis, J. R. (1965) Multivariate statistical methods in hydrology. *Wat. Resour. Res.* 1 (4)

DISCUSSION

J. Déri:

Tout d'abord je voudrais remercier notre rapporteur général pour son grand travail, valable. Je suis particulièrement reconnaissant pour les remarques concernant les détails de mon rapport numéro deux. Personnellement je voudrais apporter quelques considérations techniques concernant le calcul des paramètres des fonctions hydrologiques. Dans le cadre de notre symposium il s'agit de l'application des modèles mathématiques en hydrologie. Mais alors il faut tout de suite ajouter qu'il s'agit d'une application de ces modèles dans le domaine de l'hydrologie dont les processus sont de plus influencés et modifiée dans les diverses activités humaines. En vue de l'intensification de ces activités, de ces interventions humaines il faut toujours tenir compte que les séries historiques en hydrologie, ne sont pas valables à la description des phénomènes hydrologiques, naturels. Cela veut dire que les paramètres hydrologiques, surtout les paramètres de l'écoulement, sont modifiés sous l'effet des activités techniques et économiques. Parlant très sommairement on peut supposer que de plus en plus grandes régions hydrologiques ont des séries d'écoulement où l'hypothèse de la stationarité et de l'ergodicité ne sont plus valables. Pour illustrer les modifications et les changements structuraux des séries chronologiques on peut trouver quelques résultats méthodologiques brillants. A ce titre je pourrais citer les analyses de Monsieur le Professeur Yevjevich qui a démontré dans son étude la réduction des fonctions d'autocorrélation sous l'effet des prélèvements pour l'irrigation et des dérivations de l'eau. Moi aussi, j'ai analysé les séries d'écoulement modifiées par l'homme dans les régions agricoles et on a justifié de ces résultats mentionnés précédemment. En somme je voudrais souligner l'importance de l'analyse des effets humains sur les processus naturels et je voudrais également souligner que les méthodes présentées dans mon rapport numéro deux ne sont valables que dans les conditions non influencées, c'est-à-dire les analyses de simultanéité des séries chronologiques exigent beaucoup de précautions et des analyses spéciales. En ce qui concerne la question posée par M. le rapporteur général je peux dire que le calcul de probabilité conditionnelle utilise ou bien emploi un coefficient de corrélation simple, reliant les deux fonctions indicatrices pour un intervalle de temps choisi. En ce qui concerne la probabilité P figurant dans mon rapport c'est une probabilité de l'occurence d'un événement aléatoire concernant l'un des deux processus analysés, en remarquant que dans le cas des deux processus simultanés les probabilités d'occurence d'un événement donné ont la même valeur de probabilité. C'est une hypothèse utilisée dans ce rapport, dans le calcul. Merci pour votre attention.

R. Gras:
Je dois d'abord remercier M. Bernier de la manière fidèle avec laquelle il a traduit ma pensée. Je ne suis ni hydrologue, ni mathématicien. Je suis en plus un fervent partisan des modèles dits déterministes. J'ai été amené à cause de l'incapacité de ces modèles déterministes à attaquer un modèle stochastique pour étudier les questions de qualités des eaux, et ce modèle est encore à l'état — disons — exploratoire, je le reconnais, et c'est plus en tant que méthodologie que j'ai présenté ce papier ici. Je voudrais attaquer le point sur lequel Bernier a demandé des explications, qui est celui du coefficient béta. Le coefficient béta fait partie, est une partie disons interne de la fonction d'echange d'énergie dans l'atmosphère et cette fonction d'énergie. . .

Le coefficient béta est, sans entrer dans les détails physiques de la chose, est une fonction plus ou moins linéaire d'un terme qui est la vitesse du vent et sur lequel les physiciens discutent encore beaucoup qui est le coefficient d'échange des coefficient d'évaporation ou le coefficient de convection, mais le problème n'est pas lá aujourd'hui, le problème est de savoir si la représentation en tant que bruits blancs et la loi de Pearson adoptée, est valable dans le modèle choisi. A priori l'étude analytique a montré que le vent ne suivait pas un bruit blanc, il avait une certaine structure, et on s'est heurté à la difficulté de trouver cette structure. Je reconnais que je ne suis pas arrivé à une solution convenable et j'ai adopté cette hypothèse, qui est très simplificatrice et très commode dans le calcul de bruits blancs.

Mais, physiquement il y a un phénomène qui sauve cette hypothèse, qui est l'inertie thermique d'une masse d'eau. Du fait de la linéarité du coefficient béta en fonction du vent, d'une part, et de la linéarité des phénomènes d'échange quand la température de l'eau varie peu, on peut admettre que même si l'introduction de quantités de vent se fait selon une structure qui n'est pas la structure réelle, si sur une sommation de termes les quantités introduites ont une somme égale, il se trouve que l'effet d'échanges d'énergie sur la température est strictement le même. Ce qui veut dire que pratiquement, le modèle ne pourra être utiliser et donnera des températures vraisemblables qui si les distances de parcours des particules d'eau, c'est-à-dire les tronçons, sont de dimensions telles que les quantités de vent introduites en sommation, sont peu dépendantes finalement de la structure du vent, c'est-à-dire sur des périodes très longues on a les mêmes quantités introduites. Alors je ne pense pas discuter davantage sur des questions plus physiques, mais qui ne font pas l'objet de ce symposium, aussi, je vous remercie, je passe la parole au suivant. (*Note*: The intervention of M. R. Gras has not been authorized.)

E. J. Plate:

It is customary to smooth data for coherence estimates with suitable filters, and I assume that Nordin and Rodríguez also used a smoothing filter. But coherence estimates might be influenced by the filters that were used. I would like to suggest that Nordin and Rodríguez append the details of their data preparation to their paper for future reference.

G. Cavadias:

Je voudrais addresser une bréve question à Déri au sujet de sa communication.

La quantité donnée par l'équation (6) est un probabilité. Or, si le coefficient de corrélation a une valeur négative et la probabilité p est très petite, la quantité donnée par l'équation est négative. Déri pourrait peut-être donner quelques explications supplémentaires au sujet de cette équation.

L. S. Kuchment:

I would like to draw your attention to the paper by Chemerenko – the paper concerned with investigations on the statistical properties of hydrometeorological fields and their objective analysis. Our general reporter presented an excellent and very detailed summary of the basic theses of this paper. I would like, however, to underline once again some of the results obtained by Chemerenko and the methodology adopted for the investigations. Although three general reports concerned with distributed parameter models are presented at this Symposium, in practice none of them is concerned with these models in the strict sense.

If we seriously begin investigations on the application of distributed parameter models, no doubt we will be faced with the problem of an objective analysis of the hydrometeorological fields. It has been known for a long time that the numerical methods of weather forecasting were unsatisfactory, mainly because of the lack of an infallible method allowing for an objective analysis of various hydrometeorological elements.

Next, if we consider a catchment as a certain system showing significant variability of its parameters which should be statistically described, the statistical properties of various hydrometeorological elements must be investigated. We can seriously approach the task of structuring a sound lumped parameter model only in the case when we know the distribution of such catchment characteristics as various storage capacities, roughness, infiltration coefficients, etc.

In the study I accomplished with Koren, we have tried to obtain the lumped parameter models making use of the statistical distributions of various characteristics. In this way, we came to the conclusion that, in practice, we know very little about the statistical distributions of the hydrometeorological characteristics. The methodology applied by Chemerenko is based on very extensive field experiments and objective meteorological analysis. In his case, this methodology led to very promising results. When we meet again at one of the next conferences on mathematical models, I do hope we shall have a general report on the objective analysis of various hydrometeorological fields. It may even happen that all sections will discuss this problem.

G. de Marsily:

To answer a question raised by the general reporter about other applications of multi-variable analysis in hydrology, I would like to mention the work completed by Roux (1900) about 'Correspondance factor analysis' (referred to as AFC) on piezometric data. This method can be used, among other things, to optimize existing observation networks. This method was originated by Benzecri (1900) for other applications.

The aim here was to group, in a few families, a large number of piezometers observing the level of water in an aquifer. These piezometers had been observed weekly for several years, and a decision on the reduction of the number of observations had to be made. Having classified the piezometers in a few families having the same behaviour throughout the year, one or two peizometers of each family could be observed only, and, if necessary, the measurements of other unobserved piezometers could be generated by correlation.

This method can also be used to choose significant sets of observations on which to build a correlation analysis to reconstruct missing data on any observation network.

AFC is in some features, similar to principal component analysis (referred to as PCA), but, in others, quite different. We shall briefly compare them.

In both methods, the data collected are disposed in an array of p lines (for instance, the p stations of observation) and q columns (the q observations made in each station, for instance, at q different dates). However, with AFC, the original data k (i, j) are normalized in order that the sum of all elements of the array is 1:

$$f(i, j) = \frac{k(j, i)}{\sum\limits_{1}^{p} \sum\limits_{1}^{q} k(i, j)}$$

In both methods, two modes of analysis are successively used, by considering the array as either p vectors, each of them having q components (line analysis), or q components (column analysis). Let us only consider the former, for instance.

The goal is to visualize more easily those p vectors, which belong to a space of q dimensions; this will be done by projecting those vectors on a sub-space of lower dimension, the sub-space will be determined as the one loosing the least possible information during the projection. Generally speaking, this sub-space is of dimension 1, 2 or 3, and its graphical representation can be easily given, as Dr. Bernier has pointed out.

With PCA, this will be achieved by alloting a uniform weight (generally $1/p$) to each point representing each of the p vectors in the q dimensional space, by using Euclidian distance, and by determining the first axis of inertia of this set of p points. Thus, this single axis, for which the moment of inertia of the p points is minimum, is the one best representing those p points in a one-dimensional space: it will be oriented in the direction where the points are the most scattered.

If a second dimension is needed, the projections of the p vectors on the sub-space of dimension $(q\text{-}i)$ orthogonal to this first axis are computed, and the second axis of inertia is determined in the same way and so on.

With AFC, before any analysis, the 'profiles' of each vector (or each station) are considered, that is to say, the set of the q measures made on each station. In order that two stations having identical profile except for a coefficient of similarity, should have the same representation, each profile will be normalized; the sum of the measures of each profile will be 1. The array will become:

$$g(i, j) = \frac{f(i, j)}{\sum\limits_{j=1}^{q} f(i, j)}$$

However, each of the p points representing the p stations in the space of q dimensions will be alloted a weight equal to the sum of a profile before normalization:

$$\omega(i) = \sum_{j=1}^{q} f(i, j)$$

This gives relatively more importance to the points having major fluctuations than the ones of a lower variation. Then, the distance between two points i and i' among the p stations will be defined as

$$d^2(i, i') = \sum_{j=1}^{q} \frac{[g(i, j) - g(i', j)]^2}{\sum_{k=1}^{p} f(k, j)}$$

that is to say the χ^2 distance between the two laws $g(i, j)$ and $g(i', j)$, centred on the law

$$\sum_{k} f(k, j)$$

In other words, this distance is null for stations having similar profile, or increases as the differences, for the two stations, between the measurements of each datum, balanced by the average value of this measure for each data, and for all the stations. Then, inertia analysis is carried out as for PCA.

In both methods, the two types of analysis (line and column) are made successively, with simultaneous graphical representation: a first set of points represents the stations, and a second represents the dates.

I would like to raise another question concerning Chemerenko's paper. His aim is to determine the value of a regional variable on a regular grid, using irregularly spaced data. I would like to know if, in his excellent review of interpolation methods, he has considered 'universal kriging', a method developed by Matheron (1969) which can fall in his category of optimum interpolation, but has actually been demonstrated as being 'optimum'. This theory considers the effective experimental data as a realization of a random function $Z(x)$ defined in three or two-dimensional space. If the number and the localization of the data allow the calculation of the moment of order 2 (covariance or variogram) of the spatial law, the theory of 'universal kriging' gives the optimal estimate (of minimal variance) of:

(1) the so-called 'trend' or 'derive' which is the first moment of the spatial law: expectation $E\{Z(x)\} = m(x)$

(2) the value of the variable at any point included in the area of definition of the spatial law:

$S(x)$ being an estimator of $Z(x)$, the theory guarantees that

$E\{Z(x)\} = Z(x)$ for any x: unbiased estimation

$E\{Z|(x) - Z(x)|^2\}$ minimum: optimal estimation

The variance of the estimation can also be determined, which is most important.

REFERENCES

Benzecri, J. P. (1900) Analyse des données et autres publications due *Laboratoire de Statistiques Mathématiques, Faculté des Sciences de Paris,* Quai St-Bernard, Tour 45, Paris 5.

Matheron, G. (1965) *Les Variables Regionalisées et leur Estimation:* Masson et Cie, Paris.

Matheron, G. (1969) Le krigeage universal. *Cahiers due Centre de Morphologie Mathématique* (Fascicule 1), Ecole des Mines de Paris, Fontainebleau.

Roux, P. (1900) Apport de l'analyse factorielle des correspondances a l'optimisation des réseaux de piézomètres. *Revue de l'Institut Francais du Pétrole,* Rueil-Malmaison, France.

Roux, P. (1971) Note summaire sur l'analyse factorielle des correspondances. *Rapport, Ecole des Mines, Laboratoire d'Hydroléologie Mathématique.*

I. Rodríguez-Iturbe:

Plate has a question about the calculation of the coherence in our paper and I want to say that we used a Tukey window to obtain the individual spectra and no other filter or similar thing was involved in obtaining the coherence spectrum.

Ch. Kluge:

As to the last question of the general reporter about other multivariate methods applied in hydrology, I should like to mention the paper of Dyck and Kluge (vol.1, pp.39–46) presented in session II. In this paper the well-known formulae for the distribution of simple functions of stochastic variates were applied, e.g. if we know the joint distribution function of basin precipitation x_1, and basin retention x_2 for storm events, we can compute the distribution of the difference $Y + X_1 - X_2$ according to the integration formula given in our paper, and so we get the distribution of the resulting runoff volumes of the storm events. Because the mathematical form of the joint distribution of D_1, X_2 is often unknown, we have given also an approximation formula using a sample of storm events. If X_1, X_2 has a bivariate normal distribution, then Y has a univariate normal distribution with mean value and variance very easily computed. However, in this case, a problem remains. If not X_1, X_2 but a function of them, e.g. $\log X_1$, $\log X_2$ are normally distributed, we get the parameters only of the difference $\log X_1 - \log X_2$ in this easy manner, but not those of the flow volume $Y = D_1 - X_2$. I think this is the very problem raised by Dyck as the last question in his general report (session V).

Finally, I should like to mention that we also tackled the problem of finding the distribution of a variable A represented by a linear regression equation $A = b_1X_1 + b_2X_2 + b_kX_k + \ldots + b_pX_p$. Because of the dependence of the X_k, their joint density function will be rather difficult to handle mathematically, and therefore we recommend the use of the principal components of the X_k, both for the regression and for computing the distribution of A. Due to the independence of the principal components, the density function of A will be found by successive convolution of their one-dimensional density functions.

J. Déri:

J'ai reçu une question de Cavadias concernant l'équation (6). Je pourrais répondre en prenant l'hypothèse de travail à l'égalité suivante: $p_1 = p_2 = p$ [(voir l'expression (5)] . Dans ce cas il est possible justifier que les valeurs de $[p + r(1 - p)]$ sont toujours positives pour un intervalle de temps t choisi.

N. T. Kottegoda:

I would like to ask a question of the authors of the papers on multi-variate analysis. My problem arose from the use of multi-variate analysis as opposed to a simple two station regressional analysis for generating some missing hydrological data at stations with very short records.

Consider the basic equation for a first-order Markov process:

$$Z^p_{t+1} = AZ^p_t + B\eta_{t+1}$$

where Z^p_{t+1} is the standardized flow at station p at time $t + 1$; Z^p_t, at the same time t; A and B are constants and η_{t+1} is a vector of identically distributed standardized random variates. It has been shown by Matalas that $A = M_1 M_0^{-1}$ where M_0 is the cross-correlation matrix and M_1 is the lag-1 covariance matrix. I have found that very low values of serial correlation such as 0.1 and 0.2 which are denoted by the diagonals of M_1 give a downward bias to the cross-correlation. In other words, the M_0 matrix of the synthesized data tends to be lower than that of the historical data. This may be due to the very low values in A arising from the low values in M_1 though B is also reduced simultaneously. This apparent shortcoming made me resort to simple two station regression analysis when the serial correlation was low.

P. Guillot:

Je voudrais mentionner au moins deux problèmes pratiques de prévision et d'estimation de valeurs extrêmes en hydrologie où l'analyse multivariable peut rendre des services. Le premier problème est l'estimation des étiages extrêmes c'est à dire des plus faibles débits à craindre. L'étiage dans un bassin versant est provoqué par la concomittence de conditions initiales sèches, et par de faibles pluies pendant la période d'étiage. Etant donné qu'on peut admettre qu'il y a independance entre les évenèments météorologiques antérieurs et les évènements météorologiques régnant pendant la periode de sécheresse on a par là un moyen de mieux evaluer la probabilité des étiages en combinant deux lois de distribution qui ne sont pas très bien connues, je m'explique par un schème sur le tableau. Ceci est le graphique représentant la relation entre l'écoulement (l'écoulement par example d'un mois) les conditions initiales au début du mois et la pluie du mois. On connaît la fonction de répartition des réserves initiales du bassin versant jusqu'à une probabilité de 1/10, la queue de la distribution est mal connue, comme toutes les distributions empiriques. De même pour la pluie du mois, on connaît la distribution dans la zone centrale, la queue de la distribution est mal connue. Néanmoins il suffit de connaître chacune des 2 grandeurs 'réserves initiales' et 'pluie du mois' jusqu'à une probabilité de 1 chance sur 10, pour en déduire la grandeur 'écoulement du mois' jusqu'à une probabilité voisine de 1 chance sur 100.

L'autre problème est la prévision calculée de la pluie journalière sur un bassin versant: cette pluie dépend du type de temps, c'est-à-dire de la forme du champ de pression en altitude et au sol. La méthode la plus puissante pour constituer avec un ordinateur un échantillon de journées météorologiquement analogues est la 'reconnaissance des formes' par les composantes principales des champs de pression. Nous estimons qu'il y a là une voie très prometteuse pour faire progresser les prévisions météorologiques graduées en probabilité, qui sont indispensables pour la prévision hydrologique.

G. Cavadias:

I would like to point out that discriminant analysis could be applied to forecasting when the variable can be classified in different categories. It may turn out that it is advantageous to forecast the category or the class in which a future value of a hydrological variable will fall. The criterion in this case is the probability of misclassification rather than the minimization of the sum of the squared forecasting errors.

This type of forecasting has already been applied in meteorology and I think that there may be some advantage in using this approach in hydrological forecasting.

Multi-dimensional stochastic processes in hydrology — II

N. C. Matalas

The papers discussed in this general report are given below. The papers are all published in volume 1 of the Symposium Proceedings.

1. *Subin Pinkayan and Raziuddin Khaleel*   A multivariate mathematical model of monthly river flow
2. *S. N. Kritsky and M. F. Menkel*   Finding regularities in long-term streamflow fluctuations by analysis of hydrological ensembles
3. *G. A. Alexeev*   Optimal spatial interpolation and its errors in the case of various distances between observation points or watershed centres
4. *M. G. Elgizawa*   Use of cross-correlation techniques for determining the relationship between groundwater levels and sources in the Nile delta area
5. *J. Bernier*   Modèles probabilistes à variables hydrologiques multiples et hydrologie synthétique
6. *Chester C. Kisiel and Lucien Duckstein*   Time and frequency-domain identification of a causal bivariate stochastic process

Introduction
Of fundamental importance is the description of the multivariate structure of hydrological processes. A mathematical formulation of this structure is needed for the solution of such problems as predicting hydrological events in time and space, generating multisite, multi-season synthetic streamflow sequences, transferring information from longterm to short-term sites, and designing data collection systems. Correlation analysis has been used extensively to assess and model the spatial—temporal multivariate structure of hydrological processes, particularly streamflow. To a lesser extent, spectral analysis and various techniques of multivariate analysis have been used.

For all practical purposes, we are concerned with the space—time domain of hydrological processes, and because correlation analysis operates in this domain, it is, in a sense, a natural form of analysis. Moreover, other analyses are essentially dependent upon a transformation of the correlation structure among and within hydrological processes. Spectral analysis operates within the frequency domain, entailing a Fourier transform of a correlation structure. There are various techniques of multivariate analysis, but of these, the two which have received the most attention in hydrological studies are principal-component analysis and factor analysis. Each of these techniques is based on an orthogonal set of variables derived from a transformation of an intra-correlated set of variables.

While the various analyses are interdependent, via transformations of correlation structures, each seeks to provide additional insight into the multivariate structure of hydrological processes. Ultimately, all insights are transformed into the space–time domain of hydrological variables. It is in this domain that the mathematical models of the multivariate structure of hydrological processes must be operated.

In the following paragraphs, a brief review is given of six papers dealing with the multivariate structure of hydrological processes. Of the six papers, two deal with the construction of mathematical models for generating streamflow sequences, three with prediction and estimation problems, and one with the design of data collection systems.

Streamflow simulation
In the design of water resources systems, the historical sequences of flows pertaining to the various sites in a basin provide but a single response to the system. This response is of limited value because the historical sequences are unlikely to be repeated for the period corresponding to the economic life of the system. Based on the multivariate structure of the historical flow sequences, models can be constructed for generating synthetic flow sequences whose multivariate structure is, to some extent, identical, in a statistical sense, to that of the historical sequences. The synthetic flows are not predictions of future flows, rather the synthetic flow sequences of a given length represent realizations from a stochastic process having the modelled multivariate structure. Sets of synthetic flow sequences may be used to determine a set of system responses from which a basis may be formed for evaluating alternative system designs.

Under the assumption of a multivariate Markovian structure for monthly flows, various investigators have developed models for generating synthetic monthly flows for one or more sites. The various models are dependent upon there being sufficient data at each site to allow reliable estimates to be made of the model parameters. In many cases, a sufficient data base is not available and in more extreme situations, non-existent. To generate synthetic flows at data deficient sites, it is necessary to transfer information to the sites from others having more extensive data sets.

The problem of information transfer to allow the generation of synthetic flows at data deficient sites is considered by Pinkayan and Khaleel in their paper. The authors begin with a general formulation of a time series where the stochastic variable of interest, discharge, is assumed to be composed of the sum of a trend, a periodic, and a random component. The trend is attributed to irrigation diversions, and is eliminated simply by the addition of the diversions to the measured monthly flows.

With the de-trended series, the authors seek to 'remove' the periodic component. Several steps are involved in the removal procedure. First, the data are transformed in log-space. The transformed data are expressed as standardized deviates where the means and standard deviations relate to the annual sequences for each of the months and are expressed as a Fourier series. The mean and standard deviation of the standardized deviates, taken over all monthly annual sequences, are used to form another sequence of standardized deviates which are taken to be the variate values of the random component.

Given the random component for two sites, a bivariate first-order autoregressive Markov model is used to relate the variate values for the two sites. Here, one site is assumed to have a longer time series than the other. Via matrix algebra, the random component for the short-term site is expressed in terms of the random component for the long-term site. The inverse of the transforms for the removal of the period component used to obtain a mathematical model for generating synthetic flows at the short-term site is expressed in terms of the random component for the long-term site. The information transferred from the long-term to the short-term site is reflected in the parameters of the generating model which are functions of the moments describing the bivariate structure of the flow sequences for the two sites.

In general, we are concerned with generating synthetic flow sequences at several sites in a manner that preserves the correlations exhibited between the historical sequences and the serial correlations for each of the sequences.

Bernier's paper describes the analytical procedures involved in using a multivariate first-order autoregressive process for generating synthetic flows at several sites. This problem was first approached by Fiering (1964) and later by Matalas (1967).

Fiering's approach was based on transforming a set of interdependent variables into an orthogonal set by principal-component analysis. Pertaining to each transformed variable, synthetic sequences were generated independently by a first-order autoregressive model. Via the inverse transform, the synthetic flow sequences were derived. While this procedure maintained the historical cross-correlations in the synthetic sequences, it failed to do so for the serial correlations. Matalas considered a multivariate first-order autoregressive model and showed that the model generates synthetic flow sequences which are identical to the historical sequences in terms of cross-correlations and serial correlations.

Bernier presents a thorough review of the mathematical operations involved in using the multivariate first-order autoregressive model for generating multi-site synthetic flow sequences. The differences between this model and that proposed by Fiering are pointed out. The multivariate first-order autoregressive model involves two matrices whose elements are functions of the cross-correlations between and serial correlations within the historical flow sequences. Matrix algebra leads to a direct determination of the elements for one of the matrices. The other matrix times its transpose is shown to be a function of the space—time (lag-zero and lag-one) correlations for the historical sequences. The use of principal component analysis for determining the elements of this matrix is illustrated.

Prediction and estimation

Estimates of the statistical parameters characterizing streamflow sequences are subject to sampling errors whose magnitudes vary inversely with the length of the sequences. To improve reliability, the sequence length must be extended either directly through the passage of time or indirectly via the transfer of information from other sequences. In their paper; Kritsky and Menkel address themselves to the problem of indirectly increasing the reliability of statistical estimates.

The authors point out that one of the main difficulties faced in information transfer is the transformation of an ensemble of sequences into an ergodic set. With such a set, time and space are interchangeable. That is, averages taken across the ergodic ensemble at any point in time equal the averages over time for any member of the ensemble. The transformation process, generally referred to as regionalization, involves the 'removal' of meteorological and physiographical controls on the statistical estimates.

Suppose a particular statistic is obtained for each of several streams in a given basin. The average value of the statistic is subject to two error sources — time and space (referred to as random and geographic, respectively, by the authors). The geographical component error is attributed to the non-uniformity of the statistic. While the random error may be reduced by increasing the number of streams, the geographic error tends to increase as the increased number of streams involves more diverse coverage of the basin.

How wide a coverage of streams is needed to provide a good estimate of the average (regional) value of a statistic may be determined by the relative magnitudes of the random and geographical errors. The authors take as their criterion of goodness that the geographical error is less than the random error. Another criterion is proposed to measure the goodness of the regional estimate relative to the estimate of the statistic for a given stream. They note that the random error associated with the regional estimate is equal to the time error of the estimate for a given stream. Thus. it is possible to form a weighted average of the regional and at-site estimates and, determine its variance.

In certain cases, it is possible to determine the errors theoretically under simple assumptions about the probability distribution of the statistic and the correlation structure for the set of streamflow sequence. However, the general case requires some procedure for estimating the variances of the errors. It is well to note that the correlation structure among the estimates is a function of that among the sequence. In the case where the estimate is of the mean, the two structures are the same.

The association between two inter-dependent sequences may not be noted by the lag-zero cross-correlation coefficient. This is the case if one sequence is, in effect, the response of the other and there is a response time lag. Thus, to assess the degree of association between the two sequences, it is necessary to consider the cross-correlation coefficients for lag-times which bound the response time. Elgizawi in his paper seeks to determine the predictive capability of groundwater levels from levels for the Nile River.

Time series consisting of observations separated by an interval of five days were considered. The cross-correlations between the series were calculated for various lags. Because groundwater levels were assumed to be responsive to previous river levels, the lags were considered for one direction only. From this analysis, the author reports that the maximum cross-correlation is equal to 0.87 with a time lag of 39, which is approximately 6.5 months. Three conclusions are drawn: (1) the river is the main contributor to the aquifer, (2) river levels provide a good basis for predicting groundwater levels, and (3) a basis is provided for designing a system of observation wells.

In their paper, Kisiel and Duckstein have carried out an exploratory bivariate time series analysis of flows at two stations 12 miles apart on the same stream. Very little inflow occurs between the stations, and there are no losses through the stream bed. The data used are not time averaged events, but rather actual events as recorded by either a digital or analog recorder.

The sample time series are assumed to be realizations of a causal bivariate stochastic sequence with upstream disturbances producing downstream disturbances. The time series analysis carried out consisted of determining autocovariance functions, cross-covariance functions, phase spectra, coherence spectra, and gain functions. The effect of using either 15-min or 2-h sampling intervals was investigated. Only flows in the late spring and late summer were used.

The maximum lag for the 15-min data was 480 min, while for the 2-h data it was 48 h. The maximum 8-h lag used to obtain the cross-covariance function for the 15-min data leads to a much smaller cross-covariance function than does the 48-h maximum lag used with the 2-h data. The reason given is that long sequences of summer storm runoff lead to 'noisy' cross-correlograms because of the averaging effect induced by cross-correlating diverse inputs with subsequent inputs.

The various analyses indicate the plausibility of a combined nonlinear delay and linear attenuation model to describe the transformation of flows in the reach between the two stations. Regression models for predicting downstream flows for different segments of the record are given having associated squared correlation coefficients ranging from 0.73 to 0.99. One of the regression models is an autoregressive one with squared correlation equal to 0.88.

Network design

The configuration of a network may be characterized by one or more parameters such as type of stations, number of stations per unit area, frequency of observation, and spacing between stations. The spacing between stations is considered in the paper by Alexeev.

The author considers a network of observation points at each of which some hydrometeorological parameter ϕ has been observed for a number of time points, say N_j for the jth observational point. Based on these observations, the statistical characteristics of ϕ are estimated. The sample correlation coefficients r_{jk} are plotted against a_{jk}, the distance between the jth and kth observational points. A function $\mathcal{T}(a)$ is obtained which gives the variation in r_{jk} as a function of distance.

The variances of the observations are assumed to be equal for all sites, and the spatial homogeneity of correlations is tested by means of Fisher's logarithmic transform. For any points in the area, the correlation between pairs of points is estimated by $\tilde{r}(a)$. The parameter ϕ is estimated at a non-observation point by means of multiple regression with the observed values at the observational points. The variance of the regression estimate is a function of only distances between points.

Let $\dot{S}_0$ and $\bar{S}_0$ denote the maximum and average relative mean square error for non-observation points between the observation points, and let δ denote an estimate of the common variance. The criteria given by the author for determining optimal distances between observation points are

$$\delta/\sqrt{l} \leqslant \dot{s}_0 \leqslant 0.5$$

$$\delta/\sqrt{l} \leqslant \bar{s}_0 \leqslant 0.4$$

An iterative procedure is used to determine the optimal distances. The author recommends using regressions on three observational points, since additional points only marginally reduce $\dot{s}_0$.

REFERENCES

Fiering, M. B. (1964) Multivariate technique for synthetic hydrology. *Proc. Amer. Soc. civ. Engrs* **H45** (September).
Matalas, N. C. (1967) Mathematical assessment of synthetic hydrology. *Water Resour. Res.* **3**(4), 937–945.

DISCUSSION

S. Pinkayan and R. Khaleel:

First of all, we would like to thank the general reporter for summarizing our paper and presenting all the points so well.

From our experiences in working in developing countries, we would like to share with you several of our problems which may lead to some discussions.

Problem of inadequate data for the estimation of the parameters of the models
In developing countries, many water resources projects have been built and must be developed on the basis of the analysis of a short period of hydrological record. One solution to this problem which we tried in our paper was to generate hydrological sequences from data available at other sites.

Problem of data subject to errors
The errors may be due to measurement and deduction of data. Measurement errors may be introduced by equipment, methods, and personnel, while the error from deduction of data is caused randomly by personnel.

We do not know how to solve this problem yet. However, we have started some studies concerning the effects of data subject to errors on some estimated statistics such as mean, variance, and the first serial correlation coefficient.

General comment
This comment is concerned with a previous discussion. Some of the delegates divided the river flow data into several periods of observation and kept some of the data for checking against the generated flow sequences. We who work with short periods of record say that with 10 to 15 years one cannot afford to do that.

To answer the question posed by the general reporter: how did we evaluate the model?
Since we were using a short period of record, say 15 years, only the first moment such as the mean was tested. In other words, the generated means were checked by the observed means as to whether they came from the same population. The higher moments were not tested because of small sample sizes.

A. Reznikovszky
Kritsky and Menkel were unable to attend our Symposium, therefore, they have asked me to present some explanations about their paper. First of all I would like to express my gratitude to our general reporter for a detailed presentation of the basic theses of this paper, which of course makes my task a much easier one.

I would like to concentrate on one issue related to this paper. In the design or operation of water resources systems, we are modelling random hydrological and meteorological processes — I would like to stress that these comments as well as my comments presented yesterday are concerned with such processes only. In such a case, regardless of the type of the model applied, we have to make a quantitative evaluation of some parameters to be used in the computations. As we know, evaluations made on the basis of short samples are inaccurate. To increase the accuracy, Kritsky and Menkel propose that the evaluation of individual parameters based on short samples should be combined with the group evaluations which are being called by the authors a 'geographic component' of the evaluation. The group evaluations are carried out on the basis of observation data concerning streamflows in rivers which are similar from the physical and geographical viewpoints.

A considerable part of the paper is concerned with the fact that streamflows in rivers situated near each other can be substantially synchronized since conditions on which they depend are analogous. In such cases, there may be a distinct correlation between the series used in the computations. One of the paragraphs in the paper by Kritsky and Menkel is solely concerned with the problem of how to consider this correlation in the group analysis.

To conclude I would like to mention that the method presented in the paper by Kritsky and Menkel has been recently applied by the authors and their associates for the practical evaluation of the distribution parameters of hydrological series. They carried out, among others, such combined evaluations of the variability and asymmetry coefficients for maximum flows in many regions of the Soviet Union.

E. Plate:
I would like to point to an important shortcoming of the model employed by Kisiel and Duckstein. The fluid mechanics of the flood flow problem is not treated correctly by using a model consisting of a pure lag model plus some linear or nonlinear system. Not only does the flood wave change with time, but also with distance, as represented for example by a kinematic wave, which obeys the diffusion equation. A suitable input—output relation must take cognicance of this behaviour.

C. C. Kisiel and L. Duckstein:
We would like to make the following points in response to questions raised by the general reporter and discussers:
(1) As indicated by the general reporter, the maximum lag should be 480 h and

not 240 h as noted properly in Fig.3(b) (vol.1, p.367).

(2) The question is raised as to how far apart might the regression relations hold. It is plausible that the strength of association between upstream and downstream flows may decrease linearly or exponentially as first approximations; but we have not evaluated this. The answer would depend on changing the topography of the river, in particular, the curvature of the channel.

(3) Tukey–Hanning weights were used to obtain spectral estimates.

(4) Concerning the question on confidence intervals for the phase, coherence and gain spectra given in Fig.4 (vol.1, p.369) we have computed a few confidence intervals based on the theory developed by Jenkins and Watts (1968). Thus for $u = 61, N = 334$ data points (sample size), 2-h data points, a Tukey–Hanning window, and the degrees of freedom $v = 29$, we find the 95 per cent confidence limits, given in Table 1.

TABLE 1

Frequency (cycles/h)	Phase (fraction of a circle)	Coherence Confidence limits	Coherency	Gain
0.01	–	–	–	1.55 ± 0.35 or (1.20, 1.90)
0.04	-0.26 ± 0.03 or (−0.23, −0.29)	(0.49, 0.85)	0.7	0.90 ± 0.30 or (0.6, 1.2)
0.20	-0.98 ± 0.08 or (−0.90, −1.06)	(.002, 0.49)	0.22	0.40 ± 0.38 or (0.02, 0.78)

From the brief analysis, we feel that the general form of our conclusions holds. Confidence limits are not an absolute substitute for knowledge of the theoretical forms of the coherence, phase and gain functions. Our analysis is exploratory and serves as a basis for hypotheses to be corroborated by further empirical and theoretical study.

Concerning Fig.4(b) (vol.1, p.369) the ordinate is the coherence function and *not* that of the phase.

We must correct ourselves by stating that the results given in Fig.4 are not based on alignment. Subsequent to submission of the paper, alignment produced a much smoother coherence function with the same general trend.

(5) Jenkins and Watts (1968) answer the question about the effect of filtering on the estimates of the phase coherence and gain functions. Linear filtering does not alter the coherence and phase but the cross amplitude spectrum that enters into computation of the gain is altered by $G^2(f)$, the square of the gain of the filter. The latter result is based on the filtering operation being applied to both input and output processes. These theoretical results are based on the assumption of a truly linear system.

(6) As suggested by Plate, aligning the data to obtain good frequency domain results becomes necessary only when filtering is used; furthermore, the method used here to obtain the lag time between the two time series, namely, measuring the maximum of the cross-covariance function, may not be the best. Use of the phase plot may be preferable.

(7) Finally, as a result of this discussion, it was decided to pursue the analysis of the San Francisco River data in a model building framework as a joint effort between Erich Plate in Karlsruhe, Germany and the authors in Tucson.

REFERENCE

Jenkins, G. and Watts, D. G. (1968) *Spectral Analysis and its Applications:* Holden-Day, San Francisco.

J. Bernier:

Je voudrais faire quelques commentaires sur mon papier, quoique parfaitement bien résumé par M. Matalas. Ceci pour parler d'un paradoxe intéressant qui apparaît dans la modélisation à plusieurs dimensions. Si on considère cette équation comme un modèle à une dimension, ceci est un exemple d'un modèle markovien, si le résidu épsilone est indépendant de xt. Une généralisation toute naturelle à plusieurs dimensions consiste à prendre x comme un vecteur à p dimension. Mais à ce moment là il faut bien voir le fait que si le vecteur est considéré comme markovien selon ce modèle, les marges, c'est-à-dire chaque composante du vecteur, n'est plus markovienne, c'est-à-dire que le processus marginal n'obéit pas à la propriété de Markov. Il y a un trouble, parce que la structure du modèle marginal dépend à ce moment de la dimension du vecteur, ce qui peut être gênant en hydrologie, puisque en fait on aurait un processus marginal dont la structure dépendrait du nombre de stations considérées dans une région. Mais en fait difficulté n'est qu'apparente parce que en hydrologie, à l'échelle régionale, les débits spatiaux sont très structurés et précisément cette structure est très bien représentée lorsqu'on fait une analyse des débits à l'échelle régionale par les composantes principales, et dans le papier je montre comment: en analysant chaque vecteur pour chaque époque t par les composantes principales, on peut éviter la difficulté que je soulevais en adaptant un modèle de cette sorte, non plus sur les débits directement, mais sur les composantes principales.

G. de Marsily:

I would like to make a comment on the very interesting paper by Elgizawi who has used a stochastic process to link groundwater levels and Nile levels. As shown by Pinder *et al.* (1969) a deterministic approach can also be used.

Under certain assumptions concerning the aquifer, they show that a linear analytical function exists, which gives the response of a piezometer to a step increment in the level of the river. Using the convolution relation, they can determine the movement of the piezometer for any movement of the river.

We have used the deconvolution procedure presented in a later paper by Emsellen *et al.* (deconvolution et identification automatique de paramètres en hydrologie (vol.2, pp.709–735) to determine experimentally this function, assuming only its existence and linearity, without any *a priori* on its shape or analytical expression. This was tested on the River Adour, with the cooperation of the Compagnie d'Amenagement des Coteaux de Gascogne (Mm. Bourdelle, Demay & Mutin). The result was the actual existence of such a relation, but essentially for the ascending parts of the hydrograms: in this area, flood in the river was usually linked to simultaneous rain on the alluvial plain where the piezometers were observed. The rapid rise in the level of the river could be followed on the piezometers, but the fall in the river was tempered in the piezometers by delayed vertical infiltration to the aquifer through the semi-pervious initial layer of the soil.

In this way, evidence of both types of recharge to the aquifer (by river and by infiltration) could be brought.

REFERENCE

Pinder, G. F., Bredehoeft, J. D. and Cooper, H. H., Jr. (1969). Determination of aquifer diffusivity from aquifer response to fluctuations in river stage. *Wat. Resour. Res.* **5** (4), 850–855.

I. Rodríguez-Iturbe:

I have two points which I want to ask the authors. The first one is a comment more than a question. I agree with Plate that the phase is extremely linear. I agree with him that the only nonlinearities observed are for large frequencies and of course the variance of the estimates is very much a factor because of the low coherences we have. It will be interesting to put some confidence limits to that phase to see if it is possible to put a straight line through the points observed.

I think the authors notice very well that there is a strong lag in the cross-covariance function in both cases analysed in the paper. They also mention in the paper alignment of the series and that translation of the series is necessary when this lag is observed in order to have meaningful coherence and phase spectra.

The times I have encountered such a situation, I have been completely unable to make an interpretation of the coherence because it depends on so many things.

I would like their comments on the interpretation of coherence and phase when alignment is made.

V. Yevjevich:

Three points are raised in the general report and in the papers on which I would like to comment: design of networks, ergodicity of the multivariate point processes in hydrology (several point processes), and generation of multiple series. The general reporter states that the design of networks is a current hydrological problem all over the world. Kisiel has stressed that all depends on what is the objective function. If the objective function is to obtain maximum information out of given funds, then the observational points of a network should be distributed over a river basin or a region according to the theoretical aspect of the problem, by looking for a space—time optimum in information procurement. However, if the objective function is to obtain maximum information at points of interest, and not to diffuse information all over the river basin, the practical design of networks gets a completely different aspect. It is logical to ask the question how can one know the future points of interest? Because one is not interested in information at all points of a river basin, this creates a dilemma. However, one knows a large number of key points in a river basin at which the information will be needed in the future. The points of future storage capacities, of future important withdrawals of water, of returns of water etc., often can be well anticipated, at least in some limits.

The design of networks must find an optimum between two different positions, the design by using the information theory with maximum information to be produced for the available funds, and the approximate design by using the expected future river basin developments, with points of interest thus defined. These two opposite approaches are rarely properly integrated in a network design.

The paper by Kritsky and Menkel raises the question of ergodicity. For fifty points in a river basin, say of observations of precipitation, runoff, evaporation, groundwater levels etc., their series may be conceived as an ensemble of time series. The annual series show the same ensemble averages, so that they can be considered as stationary processes.

However, looking at the individual station series, their means or variances do not converge to ensemble averages. In that case, one is faced with a stationary, at least for the annual series, but non-ergodic process. By removing periodicities in time series, it is feasible to reduce a non-stationary series to a stationary process; there is a need also to make ergodic the series of a set of points. For many hydrological processes, the means, variances and auto-covariances are parameters which make the ensemble of series non-ergodic, because they change from station to station. However, it seems that the higher the moment the less the process is non-ergodic. If the change of the mean over a region is represented by an orthogonal polynomial function, usually affected by the sampling variation, and similarly for the variance, standard deviation, or auto-covariance, the regional deterministic signals may be thus obtained. This approach does not require the concept of geographically similar to climatologically homogenous regions, because these are unprecise concepts. By developing the regional deterministic function for the mean, the variance, the auto-covariances etc., and by removing them, a second-order regional stationary and ergodic process may be approximated, because the higher-order moments usually vary less over a region. For the ergodic process well approximated, the information of many stations (station-year method in a new form) can be pooled together. As long as the mathematics of stochastic processes is very good for stationary and ergodic processes but limited for non-stationary and non-ergodic processes, at the present stage of development, one is forced or induced, or both, to reduce the non-stationary and/or non-ergodic processes to stationary and ergodic components by inferring and removing the non-stationarity (usually periodic) deterministic time components, and the non-ergodicity in the form of regional deterministic trend surfaces for the main parameters.

The generation of new samples for n series simultaneously by preserving the basic properties of historical samples should start with the definition of what is meant under preservation. The generation of n series simultaneously may be the question of optimization. If one uses the principal-components approach in transforming n non-normal and dependent variables into n normal and independent variables of principal components, and by rejecting the principal components with small variances, the principal-components approach may reduce substantially the number of variables to be generated. However, neglecting principal components with low variances represents a loss of information. By reducing the number of variables to generate, it is cheaper to generate the new samples on a computer, but the loss of information incurred requires an optimization approach.

M. Bonnet:

AprèsYevjevich et en tant que réprésentant du Service Géologique National Français (BRGM) je voudrais insister sur l'importance qu'attachent les services nationaux responsables de la gestion des reseaux hydrologiques permanents au problème dit: 'l'optimisation' de ces reseaux (encore appelé dans la terminologie anglo-saxonne: 'network design'). Si dans sa plus grande généralité le probème est extrêmement complexe et devra sans doute, au moins en partie, être posé en termes économiques on peut toutefois rencontrer quelques cas particuliers plus simple qui présentent pourtant une grande importance pratique. C'est notamment le cas du problème de la 'reduction des reseau surabondants'.

Il arrive souvent en effet que l'étude detaillée d'un système hydrologique ait exigé la mise en place et la surveillance pendant une certain période d'un important reseau de mesures. Une fois achevées les études qui ont motivé l'implantation d'un tel reseau il faut décider quel doit être le reseau permanent a conserver pour continuer à obtenir une bonne information sur l'evolution du système.

S'agissant en géneral de la reduction des reseaux enregistrant des données temporelles il faut donc opérer une double optimisation: dans l'espace (reduction du nombre de points) et dans le temps (détermination des pas de mesure). Pour resoudre ce problème nous avons mis au point et experimenté une méthodologie visée sur l'emploi conjoint de deux techniques de l'analyse multidimensionnelle. L'exposé détaillé de cette méthodologie fera l'objet d'un prochain projet mais je lá resumerai très rapidement: — Par une technique *d'analyse spectrale* on detèrmine tout d'abord le pas de mesure assurant un minimum (fixé à priori) de perte d'information. Puis on echantillone reguliérement à ce pas de mesure, l'ensemble des series enregistrées sur les points de mesure du reseau pendant une certaine période (de longueur déterminée d'ailleurs en vu des résultats de l'analyse spectrale) de la phase d'étude detaillée. On obtient aussi un tableau à n colonnes (les n valeurs echantillonnées) et p lignes (les p points de mesure) ce tableau est alors traité par une téchnique d'analyse factorielle des correspondances (technique du Prof. Benzegri).

Les résultats du traitement sont presentés de façon graphique et permettent de déceler les points fortement correlés et d'opérer une reduction dans l'espace.

Ce n'est la qu'un exemple de traitement d'un aspect plus général. Et il est certain que toutes nouvelles suggestion pour la solution de ces problèmes d'optimisation des reseaux permanents sera très favorablement accueilli par les practiciens.

Session IX 28 July 1971

General Reporter Professor G. P. Kalinin, Moscow State University, Moscow V-234, USSR

Chairman Mr. M. Roche, ORSTOM—EDF (Professeur à l'ENGREF), 19 rue Eugène Carrière, Paris 75018, France

The fitting of conceptual catchment models — I

G. P. Kalinin

The following papers are reviewed and discussed in this report. (All are published in volume 2 of the Symposium Proceedings.)

1. *P. Johnson and M. Clements* Determination of instantaneous unit hydrographs by least squares polynomials
2. *Subhash Chander* Representation of a catchment by a network of reservoirs
3. *S. Ramaseshan and Ravi Shanker Anant* A multiple input system model for the hydrological basin
4. *Jean Michel Dujardin et Pierre Jouhet* Les modèles déterministes de transformation des précipitations en débit
5. *Jacques W. Delleur and Ramachandra A. Rao* Characteristics and filtering of noise in linear hydrological systems
6. *Otto Pfafstetter* A swampland model

The papers discussed in this general report are concerned with a very real problem, therefore, they have great importance both from the theoretical and practical viewpoints. There are two basic problems, more or less completely discussed in the reviewed papers. They are:

(1) Construction of a model for the computation of the flow hydrograph.
(2) Determination of the model parameters.

The following is a brief discussion of the individual papers.

The paper by Johnson and Clements proposes another method of kernel function identification by the expansion of the kernel function as well as the input and output functions into the exponential series (MacLaurin series). Such expansion enables the determination of the equation which relates polynomial coefficients of the kernel function to the polynomial coefficients of the input and output functions.

The methodology was tested with data for three flood events. The polynomials of degree 16 were used. Convolved hydrographs are noted to be good approximations to the actual hydrographs, however, the form of the kernel function is slightly different from the real one. In two cases it has non-zero values at the origins and the tails exhibit waviness. There are some doubts about the stability of the proposed method. Expansion into exponential series requires too many terms for good approximation of usually observed hydrographs and histograms.

The paper by Ramaseshan and Shanker Anant considers a multiple input model with lumped parameters. The basin is divided into L parts according to the number of precipitation gauging stations, and of each station the effective precipitation multiplied by the surface coefficient w_i is transformed by its own kernel function. Each kernel function is a gamma distribution with parameters N_j and K_j. This way, the identification of the model reduces to the determination of $2L$ parameters, and an optimization method of steepest descent is recommended for this purpose.

The model was tested on observation data from a basin of about 2000 km^2 area. The station weights were estimated by the Thiessen method (12 precipitation gauging stations were taken into consideration). It was assumed that runoff losses were the same for the entire basin. The parameters N_j and K_j are initially estimated by the method of moments. Optimization of these parameters was carried out separately for each flood. The sum of squared deviations of computed flows from the natural flows was used as an objective function. There are relatively small differences between the optimal values of the parameters and their values computed by the method of moments, however, they vary significantly from flood to flood. The geometric mean of the estimates from several floods was used as the mean value of a given parameter. Hydrographs computed according to such mean parameters are the same as the hydrographs obtained from the model with one single mean input. It has been confirmed that, in some cases, a multiple input model gives better results than the model with a single input.

It should be underlined, however, that the results of tests carried out for those floods which were not used for the evaluation of the parameters, are not presented in the paper. In such cases, with the adopted methodology for the evaluation of parameters, one can expect that the results will be better than from the single input model.

Chander proposes in his paper to represent the basin by a system of non-linear storage reservoirs. It is assumed that all reservoirs have the same capacity and the storage–discharge relation in the closing profile is $\rho = CQ^B$, where C and B are constant parameters.

Similar to the rivers, the reservoirs are classified according to their upstream connections. The reservoir is called a first-order reservoir if no other reservoir discharges into it, a reservoir receiving water from the first-order reservoir is called a second-order reservoir, etc. The reservoirs network of n_w-order, is a network incorporating a reservoir of the n_w-order.

It has been proved by numerical experiments that the second and third-order reservoir networks yield practically the same results.

The proposed model was tested using observation data from a basin of about 20 km^2 area. The basin was represented as a second-order network of reservoirs. The parameters C and B were estimated by the fitting technique. The model generates in the closing profile a closely comparable hydrograph to the naturally observed one.

Pfafstetter considers in his paper a model intended for the computation of the transformation of hydrographs of a large river flowing through a very wide valley, with a pronounced backwater effect and significant water losses by evaporation and infiltration. The model is developed for a basin of about 45,000 km^2 area which belongs to the system of the Paraguay and Cuiabá Rivers.

The transformation computations were carried out for three river reaches each about 60 km long. It is assumed that each reach is characterized by one average stage–discharge rating curve. The parabollic function was used for the approximation of the rating curve, and the relation between storage and water level elevations is represented by the piece-wise linear approximation corresponding to five values of these elevations. This way, these two relationships require 13 parameters. Next, the backwater effect coefficient is introduced to the stage–discharge relationship. This coefficient incorporates the water surface slope and a new parameter. The infiltration, evaporation and inflow due to precipitation are taken into consideration in the next phase of the computations. Separately considered also is the mutual relationship between surface and groundwater resources. All together the model incorporates 32 parameters, from which 23 are determined from a preliminary analysis of the observation data and some *a priori* assumptions. The remaining parameters are defined by optimization.

The paper by Delleur and Rao is concerned with one of the most important problems related to parameter identification, namely how to overcome instability of the computations.

Considering a basin as a linear filter and making use of the Fourier transformation, Delleur and Rao have shown that the oscillations occurring in the process of the identification of the transformation function accomplished by various methods are attributed to the computational noise and the errors in the input data. The oscillation distinctly diminishes if a longer time step is adopted for the computations. The authors propose another method for noise reduction — smoothing the input function by means of a small filter. The moving average can be used as such a filter.

An analysis was made of noises due to errors in the input data. On the basis of the observation data, the probability distribution of the noise was developed which proves that noises are random variables.

It was shown that there are less noises in the streamflow data than in the precipitation data, and that, in the case of streamflow data, they are less intensive.

A numerical example is given based on the field measurements. On the basis of this example, the efficiency of diminishing oscillations of the kernel function with the increase of the time step and filtration was analysed. It is underlined that the best treatment of hydrological data of the linear systems can be achieved by the judicious combination of the time step selection and filtering.

Dujardin and Jouhet isolated two types of rainfall—runoff transformation models in their paper. The first type of models include those in which infiltration and evaporation are considered as aggregate losses taken into consideration by the incorporation of a runoff coefficient. The Previk model developed by SOGREAH in 1966 may serve as an example of this type of model. The second type of model which attempts to reproduce in detail all elements of the water balance, is illustrated by the Bilic model developed by SOGREAH in 1969. This model incorporates computations of evaporation, infiltration, underground runoff and the transformation. The model has seven parameters. Examples of the application of the Bilic model are presented, in particular the model for generating discharges of the Parana River.

The reviewed papers reflect the most up-to-date trends in the mathematical modelling of flood events. They suggest many ideas concerning further direction and perspective in the development of methods for the computation of the hydrographs. Some of these ideas are of not so typical character and they will not be presented in this general report.

There are, however, two basic questions on which depends the direction of further development of hydrology. I would like to concentrate the discussion on the two following points:

(1) The world literature concerned with computations and forecasting of flood events is characterized by a continuously growing number of various models without indication of the limits of their application. Such a situation is leading to an unlimited number of streamflow models and to duplication of individual works. There is a question about the possibilities of determining application limits for various models.

In my opinion, the ways leading to the solution of this problem are limited to the following possibilities:

(a) Development of some general streamflow models which would be relatively accurate and physically well justified.

(b) Broad utilization of physical and numerical experiments for testing all links of the model as well as the models on the whole.

(c) Testing the models with the sufficiently complete standard material (including special and standard observation data, field and laboratory experiments).

(2) Unfortunately, in many cases, the development of the model directly from physical analysis proves to be impossible. Is it advisable to use in such cases complex streamflow models with large numbers of empirical parameters? This question is specially related to the multiple input nonlinear models with complex networks.

DISCUSSION

J. W. Delleur:

I would like to express my thanks to the general reporter for his excellent review of the papers presented at this session.

I would like to make two comments about the paper by Rao and myself. In particular, I would like to say that this paper is not concerned with the identification of model parameters as are many of the other papers of this session, but rather it is concerned with the identification of the kernel function relating the input to the output in a linear hydrological system. It is also concerned with the effect of noise of this identification. As Professor Kalinin very well mentioned, his noise is made of two parts: that which is due to the computational procedure, and which is essentially related to the spacing of the time interval, and that which is due to the errors in the data. I should perhaps like to add a comment regarding the first kind of noise and to indicate that one way to eliminate it is by digital filtering and that the filters that were used were of the moving average type with trigonometric coefficients. It is perhaps also worthwhile noting that the noise attributable to the input was obtained and its spectrum was found to be essentially uniform and hence probably corresponding to a white noise. Also, the probability distribution of the input noise of the rainfall was found to be essentially uniform over a limited range.

O. Pfafstetter:

I would like to give some explanation about the review of my paper made by the general reporter. He questioned mainly the meaning of all those parameters used in the model. In fact, I used 32 parameters which seem to be too many to give a stable result. From the 32 parameters, only about one dozen independent values are left over. In many of the empirical relations used in the model, there appears three or four parameters which are not really completely independent because they have to describe a hydrological factor with a reasonable and continuous law. I think this will answer the very reasonable question why there are so many parameters in the model.

I would also like to point out that this model has a special feature because it was prepared for a very flat river reach with wide overbank flooded areas. There was a noticeable change in the hydrograph due to evaporation, precipitation and infiltration on the flooded area.

The last point I would like to call attention to is that the model tries to promote the reduction of time spent with processing on computers. Modelling is important to advance hydrology a little further than pure hydrometry. If we use complicated models which need much time or large computers for their processing, they will not be easily handled by every one. Large rivers which are normally broken down into 10-km reaches are in this model broken into 50-km. Using large river reaches and large time intervals the model becomes unstable. This instability has no physical meaning but is of a mathematical origin. In this case, the trial-and-error computation of flow in the river has to be improved with a forced trial which leads more quickly to the result. This instability is more pronounced in rivers with flat gradients and with a large backwater effect.

T. O'Donnell:

My first comment on the remarks of the last speaker concerns the large number (32) of model parameters that he said was forced on him. It was not made clear whether the optimization he carried out was an automatic technique or not. It seems to me that a number of his parameters were dependent on each other. This would make it very difficult to use an automatic optimization technique successfully. One faces the ever present difficulty that either one wants to use a large number of parameters, some of which are very likely to be dependent and so cannot be automatically optimized, or one uses a small number of independent parameters and has what may not appear to be a very satisfactory model.

My second comment concerns the remarks made by Professor Kalinin on the desirability of a classification of models. I would add that as well as a classification of models, we need a classification and proper use of fitting criteria for whatever models are used. This is a very important point which too many builders of hydrological models ignore. Papers are presented which show great complexity of hydrology in a model but which say very little, if anything at all, about the criterion of fitting used.

Either no algebraic form of criterion is given, or if an algebraic form is given no numerical values of the fitting criterion are reported. I would like to suggest that no organizer of an international meeting or editor of a journal should accept a model-fitting paper which does not give an algebraic statement of the fitting criterion used and, just as important, numerical values achieved in the fitting process. It is unscientific claiming that the fitting 'looks good' in some figure and leaving it at that.

C. Toebes:

Professor Kalinin raised the important point of classification and, of course, O'Donnell noted that one cannot divorce classification from the needs of having fitting criteria. I should like to mention that UNESCO, through its Working Group on Representative and Experimental Basins, is aiming at the preparation of a classification of conceptual models. The objective of such a classification is to promote or to arrange for the exchange of standard sets of data and for this purpose a paper has been prepared which proposes a classification and also standard fitting criteria. We intend to discuss this in more detail but I should like to raise one contentious question on this matter of classification.

This question is associated with the English language and deals with the difference between deterministic and stochastic models, but I believe this question also arises in French and German.

We consider that a differentiation between deterministic and stochastic models can be assisted by relating them to the concepts of forecasting and prediction, respectively. A forecast implies that, no matter how many times an event is processed under a given set of invariant conditions, the same outcome will always result. On the other hand, a prediction implies that every time an event is produced it will be different. The fundamental difference between the two concepts is one of probability. Forecasts are certainties while predictions contain an element of uncertainty or randomness.

However, the concept of prediction is more than one of pure randomness since the sequence of occurrence of variation is also taken into account. Stochastic models make predictions. They do this by generating sequences of events. The generating process itself takes into account the order of preceding events in the sequence.

Provided the basic process has been correctly described, the average characteristics of the generated sequence of events will match the average values of those recorded. The sequence of events could be in time or space.

Deterministic models make forecasts. To do this they transform one sequence of events into a second sequence.

Seen in this light, the two types of models are complementary rather than competitive.

It may be argued what would happen if one has stochastically generated input sequences. There is no problem however if one uses this particular type of classification. Because, if one has a stochastically generated input and one puts this input in a deterministic model (e.g. conceptual), then a forecast is made of what will happen, given what could happen. The outcome, of course, is a prediction and is stochastic, because it contains a stochastic element.

G. A. Alexeev:

I should like to make some general fundamental critical comments on mathematical models in hydrology.

(1) During simulation and evaluation of models it is often neglected that initial data obtained as a result of measurements and observations have random errors. It particularly refers to hydrological events where measurements and observations are not accurate and to short-term series. Under such conditions it is not always possible to use one-valued functional analysis as is often the case during hydrological events simulation without using the theory of probabilities and mathematical statistics. It is known that according to relatively short series of hydrological observations available we may determine more or less reliably sampling (empirical) statistic moments of the first and the second order. For the determination of sampling moments of the third order and the moments of higher order it is necessary to have very long observational series. This well known statement is often neglected.

For instance, in one of the papers here the influence function was written down as a power series — polynomial of the sixteenth order (as was pointed out by the general reporter), and the unknown coefficients of the polynomial are found from the condition of the minimum of the sum of square deviations. This means that selective statistical moments up to the 32nd order were used, which are statistically unstable and therefore are absolutely useless for the description and forecasting of the event despite the ideal formal description — approximation of the initial short observation series.

In other words, it is reasonable to intentionally use linear simulation and estimate random model errors resulting from the errors of measurements and from limited observational series.

A second example: while estimating the space homogeneity of time series and of their parameters Kritsky and Menkel in their paper (and other published works) did not take into consideration evident random errors of the initial hydrological data and the errors caused by series limitation. When they are not evident these errors considerably increase the space—time variability of the initial information and form false impressions about extremely great 'natural geographical' variability of the hydrological parameters which is in fact caused by initial data errors and series limitation.

(2) While estimating optical model parameters by the method of the least squares it is often forgotten that in the case of a non-equal variability (heteroscedasticity) of deviations of empirical points from the approximating regression function of the given model it is not correct to summarise and average all square deviations both large and small. Moreover, in some papers (and publications) neglecting any analysis of deviations they introduce a priori different weight coefficients for square deviations which intensify even more the non-equal variability of deviations and as a result the optimal parameters of the approximating model are actually determined from the condition of the minimum sums only of some greatest square random deviations.

To correctly estimate optimal parameters it is necessary first to normalize all empirical deviations by dividing them by corresponding conventional mean square deviations and then apply the least square method to normalized (equally variable) deviations.

If the task were greater, i.e. to give a general mathematic description of any stochastic phenomenon as a multidimensional distribution function, it would be necessary to transform non-equally variable (heteroscedastic) correlation into equi-variable (homoscedastic) correlation, and to substitute nonlinear correlation by linear correlation by normalizing the distribution laws of each initial variable (Alexeev, 1971).

Recently, specialists more and more often use factorial or component analysis which allows a shorter form of multiple regression equation. In this case the probability distribution laws of the initial values and of their parameters are not studied and one may say that factorial and component analysis is far from the probability theory and this analysis is similar to the functional one-valued analysis.

(3) At present very little attention is being paid to the methods and criteria for estimating the accuracy of mathematical models although assessments are often made about the relative advantages and efficiencies of different models.

Usually the mean square deviation of estimated values from the observed, as well as for individual independent short sample series are estimated for each model. Such estimates are of random character.

Therefore for each model it is also necessary to make an objective theoretical estimate of the total mean square root error which is caused both by the limitations of the series and by the errors of the initial data.

It is expedient to use as an accuracy criterion the size of the total error, i.e. the ratio of the total mean square root error to the mean square deviation of the value under study. This non-dimensional characteristic of random error means that it is possible to compare different models of one and the same phenomenon, and of the models of different physical phenomena.

REFERENCE

Alexeev, G. A. (1971) Objective methods of correlation smoothing and normalizing. *Hydrometeoizdat.*

L. S. Kuchment:

I think that the discussion concerning the models classification should not be continued since in my opinion it is not so important whether we have any classification. The problems and their solutions are important regardless of whether they are stochastic or deterministic. Of course, depending on the fact which of them dominates we can call the method a stochastic or a deterministic one. But I think that the precise classification of the models does not deserve our efforts.

I would like to return to the paper by Delleur and Rao who have carried out some very interesting and useful work concerned with a very interesting problem. In my opinion, however, the attempts at the elimination of noises with the help of some filters or by changing the computational time step are in some way subjective. Instead of eliminating the noises already recorded in the observations and which generally cannot be avoided, more important is the utilization of a proper method for the determination of the kernel function or for the solution of some other problems subject to analysis. As we know, there are very many methods for the identification of the kernel function for linear models with lumped parameters, but most of them show instability. At the present time, we use the method developed on the basis of the theory of incorrect problems. This method, which we consider very appropriate, is based on the concept that, whilst searching for the solution, we can use some complementary information. In most of the cases we search for the solution in a very wide range and this is the main reason of the instability. Very often, physical situations provide us with complementary criteria and information which allow the determination of this range. In hydrology, there are no pure black boxes — we always know something about the processes we are interested in. However, in the identification problems, such problems as those discussed in the paper by Delleur and Rao as well as in the first paper by Johnson and Clements, it seems that a pure black box method is applied. This leads to some difficulties with the instability of the model.

For these reasons, I would like all who work on such problems to acquaint themselves with the identification methods based on the theory of incorrect problems.

J. Amorocho:

I will address myself to the remarks of the general reporter, and then to the paper by Delleur and Rao. Professor Kalinin's comments are particularly appropriate: they rightly condemn the attempts of some hydrologists to extract from a given set of data more information than is actually present, by the simple device of constructing hydrological models of high complexity. Clearly, the structure of the model should be consistent with the quality and quantity of the input data, since the model cannot by itself compensate for the lack of knowledge on physical facts concerning the process to be simulated. On the other hand, a tendency to oversimplify the mathematical description of a hydrological system is equally dangerous. The results of an excessively coarse model may be inexpensive but of poor quality; the model may betray also a naive view of nature and fail to extract from the data much information which is effectively present and potentially usable.

To be objective on the relative merits of various types of models, I should like to suggest a simple technique consisting of comparing the results of their use in the simulation of the performance of the same hydrological system under the same set of conditions. For example, operating on identical data, the prediction of runoff from rainfall for a catchment can be made either with a linear model based on the convolution integral alone (unit hydrograph method) or with a more complex and sophisticated nonlinear analysis model, at a slightly higher cost. This is illustrated in Figs.1 and 2 showing the results of the prediction of streamflow in a basin in California, for the same period, by both methods. Both models were calibrated on the same rainfall and

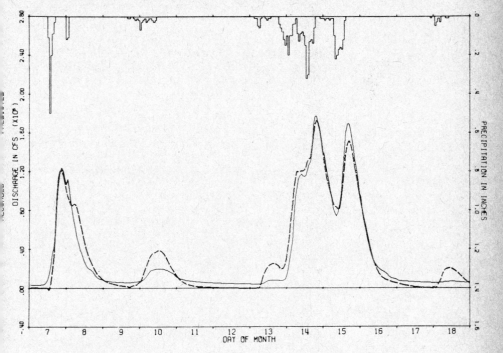

FIGURE 1. Comparison between recorded runoff and a prediction made by linear analysis for a basin in California (Amorocho and Brandstetter, 1971).

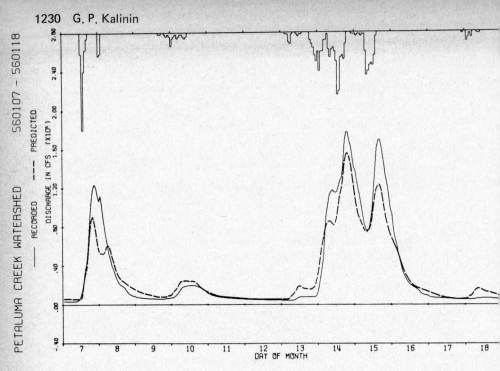

FIGURE 2. Comparison between recorded runoff and a prediction made by nonlinear analysis for a basin in California (Amorocho and Brandstetter, 1971).

runoff data, which anteceded the period of prediction by about one month. It can be seen that the nonlinear prediction is clearly superior, in the sense that it approximates better to the measured behaviour of the basin. The degree of fit was tested not only graphically, as shown, but also by the least squares criterion. The conclusion is that the nonlinear model made a better and more complete use of the same set of data.

In summary, I believe that while a general warning against unwarranted model complexity is sound, each case should be judged primarily on the basis of the performance of the model and on its cost of operation.

Now, with respect to the paper by Delleur and Rao, I wish to suggest that, although it is quite correct, as they have shown, that oscillatory linear kernels often result from the inversion of the convolution integral when data noise is present, this result may depend upon the method of computation. Methods based on the generalized harmonic analysis, such as those studied by the authors (ordinary harmonic analysis, Fourier, Laplace, Fourier-gamma transforms, etc.) are indeed very prone to reflect the data noise as amplified oscillations. In contrast, some other types of inversion, such as those based on the use of Laguerre function expansion are less likely to yield oscillatory kernel configurations, although the general shapes of the kernels themselves may indeed be affected by the data noise in other ways. A comparison of the costs of various inversion schemes, together with an evaluation of the adverse results of kernel deformations and instabilities should provide some criteria between advantages and disadvantages of the various procedures.

REFERENCE

Amorocho, J. and Brandstetter, A. (1971) *Wat. Resour. Res.* 7, No.5.

K. M. O'Connor:

The paper by Johnson and Clements proposes yet another method of instantaneous unit hydrograph identification 'capable of yielding realistic results with an acceptable degree of fit'. The method is analogous to the orthogonal function techniques of O'Donnell (1960) and Dooge (1965) in that the effective rainfall and storm runoff functions are expressed as series, and substitution into the convolution integral

$$q(t) = \int_0^t i(\tau) . u(0, \tau - r) d\tau$$

yields the linkage equations from which the coefficients of the series for the unknown $u(0, t)$ function can be derived. The effective upper limit of the integral in the above equation is t, and not T as in the paper, although, for the purpose of obtaining transforms, we would write ∞ for the upper limit as in the superposition integral.

The originality of their method lies in the choice of series, namely, the finite least square polynomial, and on the derivation of the corresponding linkage equations.

The choice of series is an unusual one as, in contrast to the straightforward evaluation of the coefficients in a series of orthogonal functions, it requires the solution of the set of $n + 1$ linear algebraic equations

$$mf_k = \sum_{j=0}^n p_j . A_{kj} \qquad \text{for } k = 0, 1, \ldots, n$$

to evaluate the coefficients p_j of the polynomial. It also requires the evaluation of the moments mf_k (referred to as Mellin transforms by the authors) up to order n. For large n, the higher-order empirical moments are unreliable and the inversion problem becomes a considerable one. The same problem arises in the solution of the $n + 1$ linkage equations where the corresponding moments of $i(t)$ are required.

The degree of polynomial n for the $q(t)$, $i(t)$, and $n(0, t)$ functions is not a parameter to be optimized in the usual sense, but must be an arbitrary *a priori* choice. A new set of coefficients must be solved for each time we try a new table of n. Table 1 (vol.1, p.490) shows the residual variance between computed and actual storm hydrographs for n varying between 4 and 24 inclusive. The authors note that high values of residual variance occur for polynomials with degrees 4, 9, 13, 18, 22, 23, but especially for degree 9. They attribute this to 'a cycle of potential inaccuracy related to polynomial degree'.

It is not quite clear from the paper if Table 1 is based on the residual variance calculated at the 3-h data intervals only or if it is based on the integral square difference between the continuous polynomial approximation and a continuous approximation in the form of a histogram or straight line segment diagram. The latter is the more likely as the moments were probably calculated using such a bar diagram or histogram. The erratic behaviour of the residual variance can possibly be explained for this case, although it must be remembered that the variance given is the variance between two approximations and not between the 'computed' and 'actual'. For the former case of variance only at the data points the whole pattern of Table 1 would be unreasonable.

It is doubtful that nature displays a particular aversion to polynomials of the 9th degree. I would suggest that these peculiarities are an expected outcome of approximating the $q(t)$ and $i(t)$ function by polynomials.

Theoretically, the approximation of a function $f(x)$ by a polynomial is sound. All we need is an adequate specification of the function in terms of moments mf_k, derivatives, or, better still, an analytical expression of the function. What we have, in effect, is discrete equidistant data. Linkage equations derived from the convolution integral do not apply. Fourier series, Laguerre functions, Laplace transforms, and all of the other continuous system techniques do not apply to this case unless we can adequately approximate the $q(t)$ function. All of these techniques, however, have their discrete counterparts in harmonic analysis, Meixner functions, Z- transforms, etc. The transition from the continuous to the discrete case has given rise to some misunderstanding among hydrologists. For example, the instantaneous unit hydrograph of O'Donnell's harmonic analysis method (1960) was in fact the instantaneous unit hydrograph specified at equidistant intervals of time. The authors are apparently prepared to accept the polynomial interpolation as a good approximation to the continuous function. This is indeed the basic assumption of their method.

The dangers of equidistant polynomial interpolation were discovered independently by C. Runge (1901) and E. Borel (1903). Lanczos (1956) reviews their work and gives this problem a lucid treatment. We must distinguish immediately between the case of polynomial interpolation, with just a few points, often termed local interpolation, and the case of interpolating the complete set of points, known as global interpolation. It is only in the latter case that the Runge phenomenon occurs. It can be shown that even for simple analytical functions we can obtain quite ridiculous results by equidistant interpolation with least squares polynomials. Increasing the density of points will not necessarily improve the overall convergence. The authors' result for the 9th degree polynomial would seem to be a typical example of the Runge phenomenon. We could improve matters by using high order difference approximations but this is not possible if there is noise in the data. The equidistant nature of the data is at the root of the problem. In effect, our demand on the function to guarantee success by polynomial interpolation with equidistant data is too rigid. This is partly reflected in the authors' method by the demand for the first n moments of the $q(t)$ and $i(t)$ functions.

It may well be pointed out that the authors' method showed considerable success for particular degrees of polynomials. A final value of n chosen from these results may, however, prove very unsuccessful in approximating another set of data from the same system for which, perhaps, a 9th degree polynomial might give satisfactory results.

Even assuming that the nth degree polynomial interpolation of the $q(t)$ function was adequate, and assuming the first n moments of $q(t)$ and $i(t)$ were available for n up to 24 (as in the paper), it would be far simpler to use these moments alone in the Nash moment-linkage equation

$$Q_S = (1 + U)^s$$

to solve for the first n moments of the instantaneous unit hydrograph directly. If desired, these moments could be fitted by a polynomial of any degree m, $m \leqslant n$, by the solution of an equation of the same form as the authors' equation (2a) (vol.2, p.485).

The weakness in the application of the moment linkage equation for high order moments has long been recognized. The authors' method compounds this weakness with the dangers of equidistant interpolation by polynomials.

To summarize:

(1) The degree of polynomial to be tested is an arbitrary a priori choice.

(2) Evaluation of the coefficients q_k, u_k, for $k = 0$, to n, each require the solution of a set of $n + 1$ linear algebraic equations, a procedure which must be repeated for each value of n tested.

(3) There is no objective criterion given for a single optimum n valid for the

three functions. The present writer doubts that this approach will yield a unique optimum value of n to characterize the operation of the system.

(4) The demand of the method on both the input and output functions, namely the first n moments, is too severe and can not be met using empirical data.

(5) Due to the special nature of hydrological data, interpolation by harmonic analysis rather than by polynomials is more suitable as it avoids the difficulties of the Runge phenomenon and operates in the global rather than in the local sense.

(6) Comparison of the authors' method and the method of moments shows that the latter is more direct if the high order moments are available, but that both methods are doomed to failure in their application to hydrological data because only the first few moments are generally considered to be reliable.

(7) It would greatly clarify the situation regarding the polynomial interpolation of the discrete runoff function if the authors would describe precisely how they evaluated the 24 moments for that function.

REFERENCES

Dooge, J. C. I. (1965) Analysis of linear systems by means of Laguerre functions. *J. SIAM Control, ser A*, **2**, No.3, 396–408.

Lanczos, C. (1956) *Applied Analysis;* Prentice-Hall.

Nash, J. E. (1959) Systematic determination of unit hydrograph parameters. *J. geophys. Res.* **64**, No.1, 111–115.

O'Donnell, T. (1960) Instantaneous unit hydrograph derivation by harmonic analysis. *Eaux des surface – General Assembly of Helsinki*, pp.546–557: IAHS Publ. No.51.

P. Johnson

I would like to make the following points in reply to O'Connor's comments:

(1) The method as developed seeks to combine the merits of good fitting as obtained by normal least squares methods, Snyder (1955), Eagleson *et al.* (1966), and realistic shapes of the instantaneous unit hydrograph as imposed by parametric relationships of a series, O'Donnell (1960), Dooge (1965). In this it partly fulfils its aim.

(2) The choice of polynomial degree n is arbitrary. Repeated computation for different polynomial degrees to determine a 'best' fit is envisaged; with a computer programme and a modern machine this is not necessarily a serious drawback. A polynomial of degree between 8 to 16 is indicated for a good fit.

(3) The comments on the Runge instability phenomenon are gratefully acknowledged. This problem needs a more exhaustive study in connection with convolution. From a practical viewpoint it has been found that, provided the data sets are 'suitable' (a subjective criterion), there is at least one degree n which will produce acceptable results.

(4) A few professional mathematicians expressed the same concern over the use of high order moment transforms. Not withstanding these expected problems, the method, when applied to natural data, does give useful results. The fact that the expected problems do not arise is certainly worth studying.

(5) The residual variance is calculated based on the sum of the squares of difference between the convolved and data hydrographs at discrete time intervals. Since the paper was written, subsequent computation has shown marginal relative improvement in the residual variance for a much increased number of data. In contrast, volumes of runoff computed from both hydrographs are much more comparable when more data are used to define the runoff hydrograph.

(6) The method does *not* require the fitting of a polynomial to the hyetograph $i(t)$. Since this is usually represented as a simple bar diagram, not having to fit a polynomial is considered to be a potential advantage. Provided the rainfall data are accurate and representative (this of course is open to question), the only numerical manipulation of them, the Mellin transforms, are precise and easily determined.

Polynomials are only fitted to the runoff hydrograph and to the unit hydrograph. Mellin transforms of the runoff hydrograph are calculated based on a histogram approximation. Numerical inaccuracy is of course possible in the process and was the reason for carrying out further study referred to in (5) before.

(7) Reference to the Nash moment-linkage equation by which transforms of the I.U.H. may be calculated knowing transform values of runoff and rainfall is very relevant. This point was overlooked but will be looked into as an alternative procedure.

(8) In subsequent work, two significant features have been confirmed. The first of these was already noted in the paper.

(a) The major source of residual variance lies with the initial lack of fit of a polynomial to the runoff hydrograph. This is particularly true when the event is double or multi-peaked which the fitting process seems only capable of smoothing into a general shape. Should the initial fitting of a storm hydrograph be improved, the derivation of a realistic unit hydrograph will be assured.

(b) The derived unit hydrograph is very sensitive to the relative phasing of the hyetograph although the finally convolved shape seems little influenced. This feature accounts for potentially high non-zero initial values of the instantaneous unit hydrograph (positive or negative). Such values are simply obviated by a phase shift of the rainfall; at the moment the amount of shift is subjectively judged until a near zero initial value is obtained at the origin of the instantaneous unit hydrograph.

REFERENCES

Dooge, J. C. I. (1965) Analysis of linear systems by means of Laguerre functions. *J. SIAM Control, ser.A*, **2**, No.3, 396–408.
Eagleson, D. S., Meijia, R. and March, F. (1966) The computation of optimum realisable unit hydrographs. *Wat. Resour. Res.* **2**, No.4, 755–764. (Discussion **4**, No.1, 1968, 213–217.)
O'Donnell, T. (1960) Instantaneous unit hydrographs derivation by harmonic analysis. *Eaux de Surface*, pp.546–557. *IAHS Publ. No.51.*
Snyder, W. H. (1955) Hydrograph analysis by the method of least squares. *Proc. Amer. Soc. civ. Engrs, Hydraul. Div.* **81**, No.793.

G. de Marsily:

I was very interested in the method presented by P. Johnson and M. Clements, and would like to have additional information on several points:

(1) Is there a limit for the amount of data that can be handled by this technique?

(2) When increasing the degree of the polynomial, is there not a tendency to unsmoothness (or oscillation) of the calculated unit hydrograph, and is there any smoothing procedure adaptable?

(3) Can constraints be imposed to the unit hydrograph, such as to be positive, which, in certain cases, may prove useful with actual data containing errors?

On this subject of the influence of errors on the unit hydrograph, I would also like to ask Delleur and Rao if the very efficient procedure of filtering that they use in their paper (appropriate moving average on the original data) can be applied to the calculated kernel function itself, without any modification of the original data?

Our experience is that oscillations in the kernel function (or unsmoothness sometimes) which can be attributed to the existence of errors in the data, can be controlled by moving average procedure on the kernel function and reiteration of the computation (*cf.* our paper 'Déconvolution et identification automatique des paramètres en hydrologie' vol.2, pp.709–735). There is a compromise to find between an acceptable shape for the kernel function and the agreement between actual output and reconstituted output.

J. W. Delleur:

I would like to answer some of the questions asked, in the reverse order in which they were posed.

First, in answering the last discusser, I would like to say that it is correct that one can filter the kernel function directly without filtering the input and output. This, in fact, has some advantages. Filtering with moving average filters results in some loss of information and usually the number of points available in the rainfall hyetograph is small. Consequently, the method of filtering the kernel function may be useful. This filtering actually can be done either in the time domain or in the frequency domain. In the frequency domain, the convolution sum is transformed into a product and naturally results in a lesser loss of information than in the time domain and can be useful in particular in the case of cascades of linear models.

Let me come next to the point raised by Amorocho in which he said that the instability was method dependent. I do not agree entirely with this. There are instabilities which are related to the time interval. Numerical experiments which we have made show that the same instabilities are found with Laplace and Fourier transforms and are also found in a numerical inversion of the convolution or Duhamel integral which does not involve trigonometric functions.

I would also like to point out at this time that the calculation of the linear kernel is of engineering interest because of the simplicity of its application. Coming to the first comment, it was said that the filtering was an attempt to avoid the analysis. This is really not what we were trying to do. We were trying to obtain the kernel function by eliminating the effect of noise which is present in the data.

P. Johnson:

The following comments are made in answer to questions put by de Marsily.

Firstly, the amount of data which can be handled by the method would seem to be limited mainly by the storage capacity of the computer being used. From the viewpoint of degree of fit, it has been discovered that, both in the method described in the paper and in other techniques used by the authors for deriving unit hydrographs, the fit does improve if the number of data representing the rising limb of the runoff hydrograph is increased. Difficulty is sometimes encountered when the runoff hydrograph exhibits more than one peak discharge, having been caused by complex storm patterns. Work continues to overcome this difficulty; increasing the amount of data in this case gives only limited improvement.

With reference to the sinuosity exhibited in the tail of the derived unit hydrographs, no special attempt has yet been made to reduce this. It is certainly caused by the presence of the higher order polynomial terms. On occasions these waves become greatly amplified and result in a high residual variance or lack of fit. Inspection of Table 1 (vol.2 p.490) reveals this feature; a particularly bad fit occurred using a polynomial of degree 9. Good fits (a low residual variance) occurred with polynomials of degree 12 and 16. The reasons for this phenomenon are not known and since experience indicates that there are ususlly polynomial degrees which give good fit for most data, the phenomenon although interesting, is considered not to be a drawback.

Apart from sinuosity in the tail and the occurrence of non-zero initial values of derived unit hydrographs, constraints on the unit hydrograph shape have not been found to be necessary. Non-zero initial values of the unit hydrographs have been found to be the result of ill-conditioned phasing of precipitation with the runoff. Subjective adjustment of the time of precipitation is noted to give sensitive control over initial ordinate values of the unit hydrograph.

Several pure and applied mathematicians were consulted on the derived method and on the occurrences of occasional extreme lack of fit at certain degrees of polynomial. Generally they considered that the method could lead to great inaccuracy due to inversion of a matrix whose elements consisted of high order Mellin transforms. So far their fears have not been confirmed. Tail sinuosity was predictable.

Work is continuing on the method.

General Reporter   Mr. David R. Dawdy, Research Hydrologist, US Geological Survey, Fort Collins, Colorado, USA

Chairman   Professor Emanuele Guggino, Instituto di Idraulica Agraria, Universita di Catania, Catania, Italy

The fitting of conceptual catchment models — II   Parametric hydrology

David R. Dawdy

The six papers for this session can be divided into two classifications. Three papers are concerned primarily with building models (system identification) and three are concerned primarily with optimization techniques (system specification). However, there are elements of both identification and specification in most of the six papers. The papers (all published in volume 2 of the Symposium Proceedings) are as follows.

1. *Gerhard Glugla*   A multi-layer model for gravity drainage of unsaturated soil
2. *J. Amorocho and Don Morgan*   Convective storm field simulation for distributed catchment models
3. *S. I. Solomon, A. S. Qureshi and U. Korngold*   Use of a parametric model as a tool for hydrometric network planning
4. *Uwe Grünewald and Siegfried Dyck*   On the application of optimization techniques to conceptual catchment models
5. *J. P. Fortin, R. Charbonneau, J. Lefevre et G. Girard*   Proposition et analyse de quelques critères adimensionnels d'optimisation
6. *Richard P. Ibbitt and Terence O'Donnell*   Designing conceptual catchment models for automatic fitting methods

The papers by Glugla and by Amorocho and Morgan are almost wholly concerned with model building. Glugla presents a multilayer model for simulating the movement of moisture into and through the soil column. The justification for such a model is three-fold. First, there is a time lag between infiltration (input) and drainage (output). Second, evapotranspiration occurs from different depths at different rates. Third, soil properties vary in the soil column. The third justification is ignored by the author, because λ, the retention parameter of the soil, is assumed to be the same for all layers, although it is considered to vary during the year. The time lag, therefore, is introduced into the model by means of a series of nonlinear reservoirs (this is quite similar to Nash, 1958). Glugla's model may be compared with the standard linear reservoir formulation as follows:

Linear reservoir model
$$I - O = \mathrm{d}S/\mathrm{d}t$$

$$S = KO$$

$$\frac{\mathrm{d}S}{\mathrm{d}t} = K \frac{\mathrm{d}O}{\mathrm{d}t}$$

Glugla model

infiltration $- (ET + \text{drainage}) = \mathrm{d}W/\mathrm{d}t$

$$\frac{\mathrm{d}W}{\mathrm{d}t} = O = -W^2 \text{ (for drainage only)}$$

where I is an input, O is an output, S is storage, K is a coefficient, t is time, and W is the moisture storage in the soil column in Glugla's model.

It would be interesting to compare the results of prediction with the nonlinear model and with a linear alternative in order to measure the relative influence of the nonlinearity. Both the number of layers (10) and the evapotranspiration function are arbitrarily chosen. The sensitivity of model results to these assumptions would also be of interest. Is evaporation from layer l in the model always at the potential rate, and from other layers at a proportional rate, or does soil moisture deficit influence ET? 'The accuracy of calculated rates of water content and gravity drainage is greatly determined by the accuracy of precipitation and by the accuracy of evapotranspiration and its distribution within the soil profile.' The only data on soil moisture were for the total soil column. Therefore, there is no way to determine whether or not the resulting model is more physically meaningful than simpler alternative models.

Both the multi-layer and the nonlinear assumptions are reasonable. A set of data on the internal state of the system would be helpful for formulating models. In addition, such data could help in determining whether a model which better predicts the internal states of the soil moisture column improves the prediction for a basin rainfall–runoff model.

Amorocho and Morgan present a model for the generation of the rainfall input to a distributed parameter rainfall–runoff model. The precipitation field for convective storms in time and space is considered. Two alternative models are discussed, one of which is white box and the other of which is black box. The work is still in progress, so that complete details are not given for either model. However, one might ask whether adequate data will be available to either calibrate or simulate with the model, other than for areas with dense research networks. In addition, the belief that a distributed input field coupled with a distributed basin model will improve prediction of output streamflow is an article of faith at this time. How can an underdetermined system be calibrated, in general? How many data are needed? How many constraints must be placed on the system in order to calibrate such a model with available data? What will such constraints do to the physical meaning of the model parameters? The attack on and, perhaps, the answer to these questions will be interesting to observe over the next few years.

The paper by Solomon *et al.*, an interesting extension of the grid-point method (Solomon *et al.*, 1968), forms a bridge between the previous two papers and the three remaining, and also attempts, in part, to answer some questions analogous to those asked above. A deterministic model is used to generate streamflow sequences from precipitation sequences, and vice versa. The purpose of the model is to supplement the data base. The ultimate goal is to determine data needs in terms of time (longer records) and space (more records). Both the time–space tradeoff and the cost–accuracy tradeoffs are mentioned. The intercorrelation among stations (and presumably the serial correlation at a station) is 'assured through the intercorrelation of the climatic data.'

One of the major problems in streamflow synthesis is the preservation of the statistical properties of time series. In multivariate synthesis, this may include the simultaneous preservation of serial correlation and intercorrelation (Matalas, 1967). Any deterministic model has a regression effect, and that regression effect will reduce the preserved variances, serial correlations, and intercorrelations for synthesized records unless special measures are taken. These problems may not be of major interest to the authors because their synthesized records are for purposes of extension of the data base rather than for project planning and design. However, the information content of the synthesized sequence is of considerable importance (Matalas and Langbein, 1962).

Flows were synthesized using the optimized and regionally derived parameters, and the results were compared. In addition, results were compared with the more standard regional hydrological analysis which correlates the hydrological parameters directly with physiographical and meteorological characteristics (Benson, 1962). Split-sample testing was used throughout so that the prediction ability for each model could be judged on the basis of data not used in the analysis. Most analyses of this sort present only the standard error of the fitted data, which can be quite misleading.

The most amazing conclusion of the authors is that 'about the same level of relative error of estimate can be obtained from the (authors') parametric model using 20 stations as with the method based on the correlation of hydrological and physiographical characteristics (standard method) using 200 stations.' There certainly is a compelling argument for the joint design of meteorological and hydrological networks. These indicated results would make such an argument even stronger. Any final report on this study should be of great interest to all network planning agencies.

The final three papers are concerned primarily with the optimization of parametric rainfall–runoff models. Grünewald and Dyck consider a spectrum of problems from systems identification through optimization techniques to optimization criteria. They compare three conceptual models for the instantaneous unit hydrograph: (1) n identical reservoirs in series without translation, (2) the same system with a Poisson-shaped translation. Two optimization criteria were also compared. Each criterion was based on a least squares formulation adjusted to an equivalent sample size of 100. The first criterion gives equal weight to all values of the instantaneous unit hydrograph. The second has a weighting function, with larger values receiving larger weights. The fitting results were compared to a black-box instantaneous unit hydrograph. The method of fitting was by a linearization of the objective function by Taylor series expansion, solving for an optimum, and, by iteration, finding the true optimum. There apparently was rapid convergence for these simple models. The sensitivity of the value of the objective function to changes in parameter values is shown only for values less than the optimum for one model and one criterion (Fig.2 vol.2, p.499). The degree of symmetry of response of all the models to both objective functions used would be of interest. The response (or non-response) of the Poisson translation parameter is rather puzzling, and should be explained. The major difference between model 1 and models 2 and 3 is the use of a translation hydrograph, and that certainly is the reason why models 2 and 3 were found to be superior. The superiority of the non-equally weighted criterion function over the equally weighted one is based mainly upon results with model 1, which gives a gross underestimation of peak values. Therefore, assigning greater value to the peak flows could not but help. Other studies on similar problems which might bear on the results of this paper are the weighting criterion problems discussed by Dawdy and Thompson (1967) and the parameter sensitivity studies on a similar model by Betson and Green (1968).

Fortin *et al.* are concerned only with optimization criteria. They consider both bias errors (in amplitude and phase) and random errors in the study of response of seven different dimensionless criteria for daily streamflow sequences. Actually, several of their criteria are not dimensionless in terms of placing equal weights on large and small flows. For instance, C_1, C_3, and C_6, all have a constant value as a denominator, so that although the results are dimensionless, each error term is not. Thus high flows are inclined to have greater weights than low flows.

On the other hand, C_2, C_4, and C_7, all have each error term given weight as a ratio to the observed value. Several variations of least squares and of mean absolute deviation criteria are investigated and compared to C_1, a criterion previously proposed by Nash:

$$C_1 = -\frac{\sum\limits^{N} (Q_0 - Q_c)^2}{\sum\limits^{N} (Q_0 - \overline{Q}_0)^2}$$

The purpose of the Nash criterion would be to make different periods or different stations comparable, but it does not actually include a weighting function among flows in a single record. A very simple dimensionless criterion which gives equal weight to the proportional errors is

$$C_8 = \frac{1}{N} \sum\limits^{N} (\log Q_0 - \log Q_c)^2$$

and it would have been worthwhile to compare this with the other seven. Studies such as these help the modeller gain insight into the meaning of model response and of the fitting process.

Ibbitt and O'Donnell consider optimization techniques and general problems encountered in the objective fitting of parameters of rainfall–runoff models. An idealized case with perfect information is constructed. The parameter values for the model are perturbed, and the operation of the optimization method as a result of the perturbation is described. Also, the objective function response surface is mapped. The operation of the optimization method and the shape of the response surface are then compared, and, where possible, related. Threshold functions in a model are shown to produce plateaus in the response surface which may remove parameters from the optimization scheme. The major problem of local optima in general is discussed, and local optima are shown to be related to model errors and to data errors. The case usually encountered in rainfall–runoff modelling is one for where there are many small local optima in the vicinity of the global optimum. Ibbitt and O'Donnell term these local optima 'potholes'. In practice one usually finds such potholes associated with one or more runoff events. Parameter interaction such as occurs in hydrological models is shown to create valleys in the response surface which are not parallel to the parameter axes. This particularly creates problems for optimization techniques which move only along the parameter axes.

Strategies for overcoming these various problems are discussed. These include recommendations concerning model component structures. Thus, the consideration of the optimization problem alone may class some seemingly physically meaningful formulations as indeterminate in terms of system specification. This paper will be intensively studied and extensively quoted by many researchers concerned with objective fitting of parameter values for hydrological models.

These papers as a group show the trends in parametric hydrology today. More models for more uses, more detailed models, and better understanding of the meaning of models and of the fitting of parameter values to models are all major areas of research. The papers raise some interesting questions which can be discussed. How much detail should a model contain? Should an intuitively correct formulation be constructed for each component of an overall model? What criterion should be used to judge if a change in a model (usually adding more detail and more parameters) improves a model? Are hydrological models for prediction or for understanding? If models are for the understanding of hydrology, how can the model formulation be judged other than by real-world data? Are distributed-parameter models actually better than lumped-parameter models? In what way are they better? Lacking data, how can they be calibrated? What are the uncertainties introduced into models as a result of uncertainties in data? These six papers raise all these questions and, to an extent, provide tentative answers to some.

REFERENCES

Benson, M. A. (1962) Factors influencing the occurrence of floods in a humid region of diverse terrain: *US Geol. Survey Water-Supply Paper 1580-B.*

Betson, R. P. and Green, R. F. (1968) Analytically derived unit graph and runoff. *Proc. Amer. Soc. Civ. Engrs.* **HY6**, (paper 6256), 1489–1505.

Dawdy, D. R. and Thompson, T. H. (1967) Computer simulation in hydrology. *Amer. Water Works Assoc. J.* **59**, No.6, 685–688.

Matalas, N. C. (1967) Mathematical assessment of synthetic hydrology. *Wat. Resour. Res.* **3**, No.4, 937–945.

Matalas, N. C. and Langbein, W. B. (1962) Information content of the mean. *J. geophys. Res.* **67**, No.9, 3441–3448.

Nash, J. E. (1958) The form of the instantaneous unit hydrograph. *Eaux des surface – Evaporation (Proceedings of the General Assembly of Toronto 1957)*, pp.114–121: IAHS Publ. No.45.

Solomon, S. I., Denouvilliez, J. P., Chart, E. J., Woolley, J. A. and Cadou, C. (1968) The use of a square grid system for computer estimation of precipitation, temperature, and runoff. *Wat. Resour. Res.* **4**, No.5, 919–929.

DISCUSSION

J. Amorocho:

There are some remarks I wish to make on Mr. Dawdy's comments on our paper. The model proposed deals with the simulation of rainfall intensity fields produced at ground level by convective storms. Convective precipitation, as is well known, is characterized by very localized activity, which is manifested at ground level in the form of rainfall intensity fields having limited areal development and rapid change. Figure 1 shows an oblique areal photograph of a convective storm in the United States Midwest, in which the presence of individual storm cells can be noticed. The event illustrated is

FIGURE 1. Oblique aerial photograph showing new cells developing in growing cumulonimbus associated with a complex convective storm. (USAF photo, published in *Weatherwise* vol.18, No.2, 1965.)

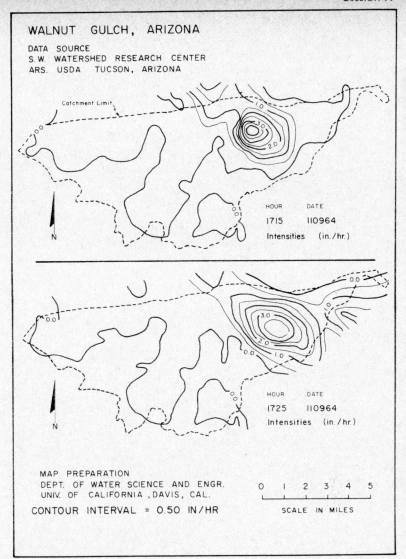

FIGURE 2. Instantaneous isohyetal maps of a thunderstorm in Arizona, showing storm intensity patterns at two times, 10 min apart. The basin drains towards the left.

fairly large as convective storms go, having a multiplicity of cells; many other storms in arid and semiarid regions possess a single cell and, while they only cover a very small area, are nevertheless responsible for often destructive flash floods. Studies in Arizona and New Mexico in the United States have shown that typical summer thundershowers are associated with ground level rainfields covering areas with maximum mean diameters of the order of 3 to 5 miles only (approximately 5 to 8 km). The precipitation rates at the storm centre can be very intense and the position of the storm within a basin is crucial in determining the runoff at the basin outlet. Figure 2 shows two instantaneous isohyetal maps for a storm recorded over an area in Arizona; as indicated above it covers a very limited area. It appears obvious, even intuitively, that a generalized storm model that may be located randomly over a catchment should be very useful in connection with distributed rainfall–runoff models, for maximizing flood events for the design of engineering structures.

A. Becker:

I would like to make some comments on the paper of Glugla and the questions raised by the general reporter on this paper.

To understand the structure of Glugla's model it is very useful to compare it with the well-known Nash model as the general reporter has done. There is, however, a certain obscurity in the basic equation which I would like to clarify. Glugla's model is a cascade of n nonlinear reservoirs of the type

$$O = \lambda W^a \tag{1}$$

where (in the symbols used by the general reporter) O = outflow (gravity drainage), W = gravity water storage, n, λ and a = parameters.

According to the physically based assumptions of Gardner, $a = 2$. After this we get, from the continuity equation, $(dW/dt = I - O)$ by inserting (1)

$$dW/dt + W^2 = I \tag{2}$$

where I may be called the 'effective input rate', in Glugla's case the difference between infiltration and evapotranspiration rate, which can be positive or negative. Equation (2) can be integrated for the case of constant input rate. This solution has been taken by Glugla for calculating the moisture regime of a soil profile with a calculation time step of one day.

Glugla's comments on the question raised by the general reporter may be summarized as follows:

A major task of the model is to be physically meaningful and to give insight into the moisture regime of the soil in different depths. For this purpose, it has been proved to be better to take $a = 2$ or greater (instead of $a = 1$) and to take more layers into account (the more the better). Those n-values which are obtained by subdividing the soil profile considered into layers of about 20 cm thickness seem to optimum.

With regard to the evapotranspiration function, it may be said that Glugla knows the total daily evapotranspiration EP from direct measurements or calculations using climatological data. A linear increase of the evapotranspiration EP_i of the single layers i from zero at the bottom of the rooted zone up to a maximum value at the surface layer is assumed where ΣEP_i must equal EP.

S. Dyck:

According to the discussion of our paper by the general reporter, I would like to make some further comments.

The aim of this research was to test some powerful automatic fitting methods in applying them in a first step to simple conceptual models of catchment behaviour. In a second step they are applied to more complicated models.

We compared three conceptual models for the instantaneous unit hydrographs. Three methods of fitting were applied:

(1) The method of linearization of the normal equations (Gauss–Newton method).

(2) The statistical optimization. A special optimizing method for the second order regression polynomial derived from a central composite design in the 'ridge-line-analysis'.

(3) The method of the nonlinear adjustment. This is a special interpolation between the gradient method by Rosenbrode and the Gauss–Newton method.

The paper reports only on the results obtained with the first and second fitting method. Two optimization criteria were also compared.

Figure 1 of the paper (vol.2, p.495) shows the black-box instantaneous unit hydrographs of the four basins used as an exact representation of the basin behaviour.

Figure 3 (vol.2, p.499) represents the optimal impulse response functions (IUH) according to the method of linearization of the normal equations for the catchment of the Ammelsdorf station.

Figure 2 (vol.2, p.499) shows a graphical representation of the relation between the objective function and the values of the parameters for model (2). The fitting method was the ridge line analysis. A surprising result is that the Poisson translation parameter a is nearly constant. A similar result was obtained for model (3).

However, this is only true for the smallest of the four basins. We have a noticeable response of the Poisson translation parameters for the other three basins.

This is demonstrated for model (2) in the graphical representation for the basin of the Sichtenwalde station. Here the parameter a is no more constant.

The same applies to model (3) for this basin and the same applies to the other two basins.

It appears to me that the considerably uniform behaviour or reaction of the small Innelsdorf basin can be described with only one constant Poisson shaped translation.

On the other hand, we have a noticeable response of this parameter for larger catchments.

T. O'Donnell:

The paper presented by myself and Dr. Ibbitt gives an account of only a part of a much larger study of conceptual models.

In another paper based on that study, it is shown that the Rosenbrock method is the 'optimum' automatic optimization technique for the catchment models used.

I should like to emphasize again that our present paper concentrates on two aspects of catchment model automatic optimization. Firstly, it describes the sorts of problems that may arise; secondly, it suggests ways of avoiding those problems, or at least of reducing their effect.

The paper offers warnings to the builders of conceptual models, warnings which may possibly be summarized by saying that what may be good hydrology in a model may be bad for automatic optimization methods.

As an example, Fig.6 in the paper (vol.2, p.468) shows the possible problem arising from a threshold mechanism in a model. Contours of the objective function, F, the sum of squares of the differences between computed and observed runoff ordinates are plotted for various values of two of the parameters in a model, the remaining parameters being held fixed. If the interception storage threshold parameter, R^*, exceeds a value of about 3.2, the F value is constant for all values of both R^* and the other parameter, K_s. Thus, if the optimization search enters such a region, it will wander about uselessly and declare convergence in all probability. It cannot get off the 'plateau' presented by the response surface.

One way of avoiding such undesirable threshold effects is illustrated in Fig.10 (vol.2, p.474). The discontinuity in the upper diagram arises from a threshold parameter, a, on the axis. The operation of this gross to net precipitation mechanism is seemingly good simple hydrological model construction. The lower diagram shows a continuously curved approximation to the straight lines of the upper diagram. This approximation alters the model operation by only a small amount but it removes the threshold parameter, a, and the risk of incorrect convergence somewhere on a plateau. The continuous curved characteristic can be made to lie very close to the straight line asymptotes with simple equations as given in the paper. A difficult optimization problem can thus be avoided by appropriate simple changes in the mathematical formulation of a conceptual hydrological process.

A second example of how apparently good hydrology may cause optimization problems is demonstrated in Fig.9 (vol.2, p.472). Here we have the famous parameter inter-action problem shown in a reponse surface plot for two of the parameters in the Horton equation used in an infilatration component of a model. Many optimization techniques would converge at points on the inclined ridge which are far away from the true opti-mum point yet which have almost the same F value as that optimum. If we merely wished to obtain satisfactory optimization results, we would be satisfied, but the general aim of modellers to relate optimized parameter values to measurements of catch-ment properties in the field would not be achieved very satisfactorily. The paper sug-gests ways of lessening or even avoiding the parameter interaction problem.

To conclude, I should like to comment on two of the questions raised by the gene-ral reporter. One question asked was: How much detail should a model contain? I would suggest that it all depends on what the model is to be used for, i.e. what is the aim that is to be fulfilled with the model study. One should include only that amount of detail necessary to meet a specified aim, and need not include details that meet other aims. For example, I am presently involved with a study of isolated flood events. The model postulated therefore makes no attempt to reproduce low flow behaviour or dry period characteristics. The redistribution of soil moisture either by evaporation or by deep drainage is not included in the model. Slow processes are not of interest in the modelling of fast flood events. Of course, an index of soil moisture at the start of a large event is needed and this is done via an antecedent precipitation measure. However, soil moisture changes throughout the event are not modelled. Just enough detail exists in the model to reproduce satisfactorily large flood events, which is the aim of the study.

The general reporter asked a second question: Should an intuitively correct formula-tion be constructed for each component of an overall model? Bearing in mind that intuition can be a dangerous thing, we must recognize as hydrologists that we are still in a state of considerable ignorance and so our necessarily approximate intuitive for-mulations must be tempered with practicality. It may be sterile for two hydrologists to argue that one's formulation is better hydrology than the other's when both formula-tions may present threshold problems or parameter interaction problems to fitting techniques. The two hydrologists would be better employed in settling their differen-ces and finding a formulation which avoids fitting problems. Intuitive correctness in constructing model components must always make due allowance for trouble-free fitting.

A. S. Qureshi:

My remarks are in regard to the paper by myself and co-authors Solomon and Korn-gold. The study reported is part of the efforts to determine the optimum future expansion of the data-collection system in Canada. As remarked by Mr. Dawdy, the paper basically deals with the development of a parametric that is, deterministic, model for generating streamflow sequences from precipitation sequences, and subse-quently discusses the possible use of the model as a means of estimating optimum future expansion of a streamgauging system in terms of additional stations and addi-tional sampling periods. The latter was achieved, as explained in the paper, by inter-facing the model with appropriate cost factors concerning the operation of the hydro-metric network. Finally, the general reporter has raised the point as to whether the intercorrelations among stations and other similar conditions are assured through the climatic data. In this regard, mainly for the sake of simplicity and to keep the size of the study down, this point was not investigated during the course of the study.

Closing remarks by the general reporter

I have only two particular comments to make. One is to Amorocho, and this is really a question which, I think, is of interest to all of us. It is not a question that can be answered but a problem which must be considered. This is the problem of the data needs for distributed models. How much data are available? How much data are necessary? And how much gain does one make in distributed modelling in the face of the present extent of our data base? Concerning O'Donnell's remarks, I am glad that the Rosenbrock method is still best. That saves me having to re-programme anything. I must say that, as usual, I agree almost completely with O'Donnell's remarks. Many of my questions are not particular, they are general questions. And I think those questions will always be with us. They are not really meant to be answered, they are meant to be posed. We must develop techniques for answering them, each and everyone of us. But I feel that actually, included in those answers, really, is one's philosophy concerning hydrology.

General Reporter   Professor T. O'Donnell, Department of Civil Engineering, Imperial College, London, UK

Chairman   Professor D. A. Kraijenhoff van de Leur, State Agricultural University, Wageningen, The Netherlands

The fitting of conceptual catchment models — III

T. O'Donnell

The following papers are reviewed and discussed in this general report. (All are published in volume 2 of the Symposium Proceedings.)

1. *M. Brajković, S. Jovanović and R. Dakkak*   A mathematical model for flood wave simulation using synthetic rainfall data for a vast watershed
2. *V. Jelen, J. Němec and J. Zezulák*   Behaviour and physical stability of sheet flow in surface runoff models
3. *Gábor Kienitz*   Problem of applying the conceptual model developed for the flat catchments of Hungary
4. *V. C. Kulandaiswamy and T. Babu Rao*   Digital simulation of a drainage basin
5. *Aleš Svoboda*   Linear and nonlinear models for peak flow forecasting
6. *W. C. Visser*   The model of the water balance and nutrient uptake as a basis for hydrological, agro-hydrological and other projects

Of the six papers reviewed, only two give explicit attention to techniques of fitting the parameter values of a conceptual catchment model. Kienitz attempts to impose parameter values based on field measurements of catchment characteristics for all but two parameters. Visser uses an automatic parameter optimization method to fit 15 parameters. Three of the other four papers are based on conceptual models but make little or no mention of their fitting, while the remaining paper is concerned with a hydrodynamic model rather than a conceptual model, and does not discuss the fitting question at all. It is therefore somewhat difficult to use the material presented in the six papers to draw together a unified report on the topic 4 theme. Instead, some general comments on this theme are offered as possible subjects for discussion.

As a first general comment, the obvious point must be made that no conceptual model will ever reproduce catchment behaviour perfectly. Some sort of compromise has to be made and the compromise must depend to a large extent on what the designer of a catchment model aims to achieve, i.e. what characteristics of catchment behaviour it is important to preserve as faithfully as possible, at the expense of poor preservation of less important characteristics.

Not only will the *structure* of a model be oriented towards specific preservation objectives but also the choice of a *fitting criterion* must reflect required objectives. A given model could be used with different fitting criteria to suit different objectives, and different sets of 'best fit' parameter values would emerge (i.e. best for any particular objective sought).

In order that studies using different models and different fitting criteria can be compared with each other, it is vital that both the *form* of the fitting criterion and its *numerical value* at the end of the fitting procedure should be stated. There is a further need for standard measures of fitting to be adopted and values quoted wherever relevant. Such standard measures should be made non-dimensional and independent of the length of fitted record. Fitting a model over part of a record followed by examining the goodness of fit with the rest of the record should be standard practice. Again, numerical values of the fitting criteria should always be quoted for both the fitting and the proving runs.

A sound principle to adopt in conceptual modelling is to exercise the utmost economy in the number of parameters used. This not only greatly eases the parameter fitting effort but also it reduces the possibility of parameter interaction. In view of the gross lumping built into conceptual models when the natural processes are highly distributed, the temptation to construct over-elaborate models must be resisted. If automatic optimization methods are to be used, care should be taken to avoid certain features in the structure of a model that cause great difficulties to such methods.

Brajković *et al.* describe the use of a deterministic river basin model to generate runoff hydrographs from synthetically generated daily rainfall data. The rainfall is first transformed to storm runoff via regression equations involving API and week number in year, without details being given, the storm runoff is then presumably passed through some form of storage and routing model. As regards fitting procedures, details are given of how the rainfall to storm runoff regressions are found from historical data, but no details are given as to how the deterministic model parameters are fitted, or what objective criterion of fit, if any, is used.

The authors would seem to have adapted a standard linear regression approach for *total* storm runoff volume regressed on *total* storm rainfall, API, storm duration and time of year. Their regressions omit storm duration and, though not explicitly stated, presumably are based on *daily* rainfalls and runoff volumes, in order to provide daily inputs for the basin model. If so, one might expect a lag one or lag two regression model involving rainfalls on earlier days as well as that on the concurrent day to be superior. Inadequacies in their regressions are indeed pointed out by the authors. A further point to add is that only one value of the API decay rate ($K = 0.9$) was used; other values may have improved the results.

In order to reconstruct historical flood events whilst proving the procedures, it was found necessary to insert a random element in the storm runoff regressions based on the distribution of residuals. Somewhat surprisingly, such random elements were omitted during the synthesis of non-historical flood events based on synthetically generated daily rainfalls. The authors rightly conclude that their preliminary results need further investigation.

The paper by Jelen *et al.* is only marginally concerned with conceptual catchment models and their fitting. It concerns a study of sheet flow caused by a distributed rainfall on a plane sloping surface and the conditions leading to wave formation in the flow. Firstly, a numerical solution obtained with an analogue computer is described and this solution is then examined with an analytical treatment.

The work is preparatory to an experimental programme using a laboratory catchment and will provide essential information for the design of adequate depth-sensing instruments, in particular their response times. The study will be of considerable use to experimenters operating laboratory catchments.

The study described by Kienitz is again exploratory rather than fully developed. It is concerned with a conceptual model of very flat catchments in Hungary and the difficult problems presented by the large surface depression storages in such areas. A sensitivity analysis using total runoff volumes and waterlogged area as the response criteria showed up two critical components in the model *viz.* S_0, the excess precipitation gathering in the depressions, and k_1, a loss-factor allowing for evaporation and infiltration losses in simulating the runoff process.

The main attention in the paper is given to S_0. The author bravely attempts to specify the values of the model parameters as far as possible from measured catchment characteristics, an ideal sought after by all conceptual catchment modellers. A critical factor in fixing both the evaporation and infiltration amounts from the depression storage is M, the current moisture stored in the upper soil layers. The author does not report how M_{max} and M_{min} were found from field data. These two constraints on M, and a vegetation factor B (also not adequately discussed), critically control the evaporation procedure used in finding the S_0 values.

M is also used in determining infiltration amounts via a complex process based on the Horton equation and Boughton's method for avoiding sharp breaks in processes at threshold values. The author uses field data to specify f_c in Horton's formula, but employs an optimization routine to fit f_0 and c in that formula.

No details of the optimization routine are given, nor is the form of the objective criterion used stated, in contrast to the great detail given on the working of the model components. It is known that strong interaction exists between the Horton parameters f_0 and c making them very difficult to fit by automatic optimization methods. A Philip-type equation is far superior in this respect.

The starting value, M_0, for the soil moisture may have a strong influence on the model behaviour. The 'fair guess' for this parameter made by the author could be rather critical.

To conclude, the early results presented by the author in Fig.2 (vol.2, p.509), in which the simulated total S_0 value is only some 60 per cent of the observed, might be greatly improved by automatically optimizing M_0, M_{max}, M_{min} and B. At the very least, a sensitivity analysis on those parameters would provide useful guidance.

Kulandaiswamy and Rao present a further account of work already fairly well-known without adding any major advances in technique as far as can be seen from their condensed exposition. An elaborate numerical analysis treatment of a nonlinear relationship between rainfall excess and surface runoff is used after losses have been removed from total rainfall by a ϕ-index and varying loss-rate procedure. The rather coarse empiricism of the latter would not appear to be a good foundation for the refined mathematical treatment of the former as a general proposition. Undoubtedly, for large flood producing events, where the nonlinearity of the surface processes dominates, the technique should be at its best, as is demonstrated in the one example presented. Even so, several such examples of reconstruction along with a comparison with the results obtained by the standard unit hydrograph procedures would have made a more convincing case.

The method involves a number of empirical curve-fitting processes and average parameter value selection procedures, based on historical data. It would probably be difficult to overcome the subjectivity in these processes and procedures to allow them to be objectively assessed via an automatic parameter optimization procedure. One iterative procedure for minimizing the difference between the volumes of excess rainfall and computed surface runoff is used, but it is not clear from the paper what factor in the procedure gets changed between iterations.

Svoboda examines both numerically and analytically the response of a series of equal nonlinear reservoirs (for which $Q = S^b$) to combinations of sets of input rates (P_k) and input durations (T). In particular, he studies the peak rate of outflow (Q_{max}) from such a series of reservoirs.

In the case of a single reservoir, the interesting result emerges that when the *volume* of input (i.e. $P_k \times T$) equals approximately 1.1 units, then Q_{max} is virtually constant for a given T for *any value of b* in the range from 0.6 to 2.0. As T increases and P_k decreases (such that their product remains at 1.1 units), then Q_{max} decreases. The range of T values used is from 0.2 to 1.0 units.

Similar results for 2, 3, and 5 equal nonlinear storages in series are reported, i.e. that there is a particular *volume* of input for each cascade such that whatever the duration of that input (up to 1.0 units), the peak runoff rate is constant (for that duration) whatever the value of b.

As the author comments, under these conditions the nonlinearity of the system does not play a significant role in forecasting a peak outflow and a simple linear scheme may be used. This may explain the success of linear models in certain cases.

No mention is made of any application of the conceptual reservoirs model to actual catchment records. Strictly, there needs to be a dimensional coefficient in the reservoir equation (e.g. $Q = aS^b$) and the input rates and durations would need to be expressed in specific units. A fitting problem would then arise involving the number of reservoirs in the series as well as the parameters a and b.

The very ambitious studies by Visser compound the difficulties of a water balance model with the complexities of a plant growth model. Parameter adjustment problems are discussed at some length. Without doubt, the author has attempted a very sophisticated modelling of catchment processes with the inevitable consequence of ending up with the problems arising from the large number of parameters that need to be fitted.

It is not easy to follow the detailed accounts of such complex models. As a result, it is not at once obvious why the full plant growth model of equation (1) (vol.2, p.582), and the ensuing term by term treatment of that equation, appears to be replaced by the much simpler model of equation (15) (vol.2, p.587) with only 4 parameters. The linking of the plant growth model to the water balance model is also difficult to ascertain. It is likely that amongst the 15 parameters of the water balance model there is appreciable interaction, and this would account for the difficulties experienced in the use of the Ibbitt automatic optimization method.

The rather elaborate step-size procedure used by the author in adjusting the model parameters is interesting. Without comparative testing, it is not possible to decide whether such refinement produces a net gain over the much simpler step-size procedures of optimization techniques like Rosenbrock's. With high speed computers, simple procedures used many times are often more efficient than complex ones used a few times.

The paper concludes with a general discussion of the philosophy of conceptual catchment modelling and their fitting. Some very sound observations are made.

DISCUSSION

S. Jovanović:

I would like to stress firstly that there is no misinterpretation in the general report, but that all questions posed by the general reporter are a consequence of a very condensed text of the paper and of a wish of the authors to avoid redundancy. Figure 1 (vol.2, p.625) represents the scheme of the Zapadna Morava basin and, at the same time, the computation scheme. Here we cannot speak of a strictly conceptual model. It is rather a routing model. This model was verified by the use of historical data. The procedure for flood wave simulation entails the following steps: (1) Simulation of daily rainfall series (by means of a strictly stochastic model). The simulation was performed for 22 raingauging stations within the basin. (2) Calculation of the mean precipitation depths for each sub-basin for the rainy spell concerned, as well as for the previous days. (3) Computation of effective rainfall (runoff) using a model giving the relation between gross and net rainfall. Net rainfall represents the input data of the system (sub-basin). (4) Determination of output hydrographs for each sub-basin, representing the input data (input hydrograph) for the fundamental model (Fig.1). (5) Using the fundamental model, the computation of corresponding hydrographs on key sections on the main water courses is performed. According to the procedure described, 29 flood occurrences, giving the highest peak discharge in the corresponding season, were simulated, and a frequency analysis of these peak discharges was made. The results of analyses for 8 sections are shown in Fig.8 (vol.2, p.632).

In Fig.7 (vol.2, p.631), an example of the model verification procedure is shown. Here, step (1) of the simulation procedure described above is omitted, since the historical rainfall data were used. Rainfall depths on the upper part of Fig.7 represent the mean basin precipitation, but for each sub-basin in the calculation procedure, a different amount of recorded precipitation was used.

A. Svoboda:

My results indicate the relation of input magnitude and its duration for which input we get the same output, whatever be the nonlinearity of a system.

The shape of this relation is of a hyperbolic nature, which is fortunately the nature of the relation between the intensity and duration of storm rains. This is, or at least seems to be, the reason why a linear conceptual model yields fair results when used for peak flow forecast.

My second comment concerns the problem of fitting the scheme I have used in the paper, i.e. a series of equal storage elements, in general, nonlinear. I agree completely with the general reporter that this problem has not been solved in the paper and I am aware of the difficulties which would arise in fitting this 3 parameter model to a particular catchment. As a matter of fact, I have used the model with the aim of expressing the nonlinearity of a system by a unique parameter (exponent b in storage function). The problem of fitting has been solved here only by giving the answer to the questions: (1) linear or nonlinear models in peak flow forecasting? and (2) if linear models, for what type of input?

To conclude, I think that no difficulties will arise in the application of the results with respect to units, this application being done for a particular catchment. The dimensional coefficient (U) in the reservoir equation I have used is equal to unity and it is always possible to transform this equation of a specific case to that unit form by using proper units for the outflow rate and corresponding storage. The same could be said about the use of time units. Those used in the paper are equal to the time constant of a system and the input durations have been expressed by these units.

W. C. Visser:

The remarks of the general reporter concerning my paper seem to consist of four different points, namely:

(1) The degree of sophistication of the model influences the result of adjustment.
(2) The adjustment of the biological model is carried out with 4 of the available 7 unknowns.
(3) The adjustment can be improved by decreasing the number of parameters.
(4) Mathematical troubles arise from too large a number of parameters.

(1) With respect to the first remark, it is good to bear in mind that the model represents the practical problem posed by the technicians, which is solved by generally accepted formulae. If, in the model, formulae would be used which are not generally accepted by hydrologists, biologists and soil physicists, the possibilities to obtain a correct mathematical solution would be small. The word 'sophisticated' is acceptable if it stands for complicated. If it would mean that a simplication might lead more quickly to a better solution, it should be remarked that in that case one would try to solve a problem differing from the practical design problem with formulae which are not normally used in the more basic sciences. Such formulae have not been subjected to a long process of checking.

(2) The model is adjusted to find the parameter values. This is an indirect method which, for some of the parameters, can be taken over by laboratory or field determina-

tion methods. The agricultural model with 4 parameters, mentioned in the second remark, can always be adjusted because data, obtained by repeated determination of the yield as a function of time or temperature, can be easily made available. The other parameters are only to be adjusted with the aid of results of drainage or sprinkling experiments, which are rare and may not give reliable results if the climate is adverse to the purpose of the experiment. These data have to be derived from accidentally available experimental results or from subjective evaluation. The adjustment is only advocated for the parameters for which the data can be easily made available.

(3) By describing the number of parameters as mentioned in the third remark, a solution for the variate used to adjust the model to the observations can be improved. The question is to be decided, however, whether a smaller number of parameters will still give the solution of the same problem. If not, the result is no improvement and a better remedy would be to increase the amount of data and to make use of those data which more forcefully solve the parameters otherwise giving trouble. A wet year will give better results with respect to drainage, and a dry year with respect to evaporation.

(4) Too many parameters may, according to the fourth remark, lead to mathematical difficulties, but this is probably described in more detail by the statement that parameters describing curvature properties will give more problems than parameters describing levels or tangents. By simplifying a formula, the number of curvature parameters will commonly be more reduced than the parameters for levels and tangents. Therefore, a reduction of parameters leading to linearization decreases the mathematical troubles. However, it also reduces the value of the set of parameters for calculations dealing with other and simpler problems than the adjustment of such complex properties as discharge or groundwater depth which was the task of the adjustment process.

The fitting can be done in order to get a reliable formula to calculate the adjusted value, be it discharge or groundwater depth. Of much more technical importance is the assessment of the parameters with which all kinds of related hydrological problems can be solved.

The adjustment should therefore deal with a formula with a wide field of application as, for example, the water balance equation, and should not only be limited to a smaller field of application as, for example, discharge. Further, the basic formulae should, for each calculated hydrological property, be present in a recognisable form, in accordance with generally accepted formulae.

J. C. I. Dooge:

I wish to make some comments on the paper by Svoboda. In it, he uses a conceptual model based on a series of equal nonlinear reservoirs and points out a special property of this particular conceptual model. The model he uses is a special case of the model which I have defined (Dooge, 1968) as being models of uniform nonlinearity. These models have special properties that preserve some of the advantages of linearity but which cannot, of course, preserve full superposition since they are nonlinear.

I would like to quote from my earlier paper (Dooge, 1968), in which I said:

'a uniform nonlinear time system may be defined as one whose actions can be simulated with sufficient accuracy by a model consisting of some arrangement of equal nonlinear storage elements.'

Since such a system is nonlinear, the properties of proportionality and superposition will not hold. It can, however, be shown that for a uniformly nonlinear system, the property of proportionality will hold provided that the time scale has been previously transformed in accordance with the intensity of input. This special property of these systems, I think, lies at the basis of the peculiar relationship which the author has found. I do not want to discuss this in detail in the limited time, but just to point out that it is part of a general approach.

I would like to go on to discuss the question of how the parameters of the author's model, or of a similar model, could be found, and to suggest that if we are attempting to apply this model to a nonlinear catchment we can determine the value of the power constant b by comparing a characteristic time for each storm with the intensity of that storm. In the cases that have been examined, this is found to be a power relationship and gives the proper power relationship for the model. So that parameter b is then determined.

It is then necessary to find the two remaining parameters: the scale parameter, which the author has not made explicit in his paper, and the parameter of the number of non-linear reservoirs that are cascaded. This can be done by means of techniques which have already been developed in relation to linear systems for the determination of the optimum time scale for the Laguerre representation of such systems. If one choses a length of cascade, it is possible to determine, by a routine based on Clowes theorem, the optimum value of the scale constant. So one could take a number of values of the parameter N for the number of elements of the cascade, determine the optimum scale for each of them, and then determine which of these pairs of values gave the best fit to the original data.

Given a number of storms of sufficiently different intensity to show up the non-linearity, we can, by a nonlinear model of a time constant carefully defined, determine the first parameter and then, by a method of Laguerre expansion time scaling, we can determine the other two parameters. In regard to the definition of this time constant, it must not prejudge any of the results. I feel the time constant used by Svoboda in his paper does this by being defined only for a value of $b = 1$. To my mind, the proper time-scale to use in such studies is one of the following types. We could define the time scale of any given storm event which had gone to equilibrium as the storage at equilibrium divided by the outflow at equilibrium. This would be an operationally defined time-scale, independent of the parameters that we seek. There are other definitions which can be used. We can take the time at which the cumulative outflow is 50 per cent of the cumulative inflow. There are many more which are operational and objective. But I do suggest that this model is a useful model; it is part of a general class and its parameters can be objectively determined by techniques that are known.

REFERENCE

Dooge, J. C. I. (1968) Hydrological systems with uniform nonlinearity. *Hydrological Aspects of the Utilization of Water* (Proceedings of the General Assembly of Bern, 1967), pp.409–413: IAHS Publ. No.76.

Closing remarks by the general reporter
Very briefly, I should like to thank the authors for their comments on my report and Professor Dooge for his contribution. I greatly sympathize with Jovanović who is frustrated in his optimization work by the cost of his computing time. I myself am in the fortunate position of not having to pay for my computing and fully appreciate how restrictive a financial constraint would be. Dooge is to be complimented for his demonstration that a particular treatment is only a special case of a general theory.

General Reporter   Dr. G. Kienitz, Research Institute for Water Resources Development, Budapest 1428, Hungary. Because of Dr. Kienitz's absence, the report was presented by Dr. I. Bogárdi

Chairman   Dr. C. Toebes, National Water and Soil Conservation Organization, Ministry of Works, Wellington North, New Zealand

Distributed parameter catchment models and input fields — I Parametric hydrology

G. Kienitz

The papers reviewed and discussed in this general report are the following. (All are published in volume 2 of the Symposium Proceedings.)

1. *U. B. Vinogradov*   Deterministic and stochastic modelling of floods in mountain regions
2. *A. P. Zhidikov, A. G. Levin, N. S. Nechaeva and E. G. Popov*   Mathematical model of spring runoff formation
3. *L. N. Borovikova and Yu. M. Denisov*   Mathematical modelling of the formation of river runoff in mountain drainage basins
4. *V. G. Andreyanov and V. I. Babkin*   Mathematical model of hydrometeorological regime formation of a river basin
5. *D. E. Overton*   Simulating overland flow on hillslopes with a kinematic cascade
6. *K-R. Nippes*   A new method of computation of the suspended sediment load

Within the concept of deterministic — also called parametric — hydrological models, there are essentially two categories, namely the synthesis and analysis type of models. Both can consider the catchment as a lumped-parameter transformer of precipitation into runoff. Synthesis-type models do this by constructing a conceptual model describing how this happens, while analysis-type models are not in need of that, because they use the black-box method. But the synthesis-type models are capable of performing more than that: they can consider the catchment not only as a lumped-parameter system, but also as a distributed-parameter one. The basic difference between the two approaches is that when lumping the parameters, detailed information about the internal distribution of the physical characteristics are lacking, and the calculation of results related to sites within the system are of no interest. As a contrast to this, in a system with distributed parameters, both detailed internal information is available and internal site results may be sought.

Although well known to those who work within its narrower field, I felt it necessary to define the subject of this topic. A distributed parameter catchment model is characterized by *regional knowledge,* within the catchment, of the parameters concerned. Analogous to this, a distributed input field means *regional knowledge* about the input within the catchment. These conditions naturally imply that, in such models, calculations are performed for *catchment elements* which are ultimately integrated in some way to produce the response of the catchment as a whole.

The road to distributed parameter models, which of course are more informative and therefore more attractive than lumped parameter models leads through tedious detailed researches which provide the means for building up such models. These researches, having as their aim the clarification of the behaviour of the *components* of hydrological phenomena, involve *component modelling,* when not the whole catchment, but only some hydrological processes within the same are examined. This aspect brings, to a certain extent, component modelling within the range of this topic.

The six papers presented within this topic should be judged according to what has been said. Some of them present approaches to catchment response from a more or less distributed parameter model starting point, while others are rather presentations of results in component modelling. They have, however, one merit in common, namely *clarifying new aspects of this field of hydrological research.*

The study of Vinogradov corresponds directly to the concept of distributed parameter models. His model considers storages and runoff in separate 'micro-watersheds', each of which is homogeneous in itself concerning meteorological, hydrological, geobotanical, soil and other aspects. He rightly alludes to the fact that a lumped approach to an inhomogeneous basin produces unstable parameters which will change from rainfall to rainfall, from flood to flood, owing to the different reactions of the composing elementary runoff-producing areas to changes in the hydrometeorological situation. In this model, the runoff hydrograph at the catchment outlet is gained therefore as a summation of all the elementary hydrographs coming from the 'microwatersheds', taking into account the time of channel concentration. This latter solution – favouring a pure *translation role* of the channel, instead of embarking upon more sophisticated methods which approach channel flow with more or less approximations as an *unsteady* one – may be justified if the time-factor can be considered as being of only minor importance in the calculations. Distributed parameter models, owing to their very nature, usually involve great amounts of calculations, and sophisticated calculation of channel flow would add to their difficulties. It needs consideration however, whether, in a given case, a translation approach of channel flow is permissible or not.

As far as the calculation of catchment runoff is concerned, the Vinogradov model is deterministic (parametric). At that stage, however, when stochastically generated inputs are fed into it, it looses this character and its outputs also become stochastic ones.

In many areas of the world a great proportion of precipitation appears in the form of snow. The calculation of runoff originating from snow melt is therefore of major interest for researchers in many countries. As the water equivalent of the snowpack is a locally changing factor within a catchment, *snow melt is a typical example for the distributed input field.* Two studies have this problem as their main concern.

Zhidikov *et al.,* though using a lumped system model, elaborated a very complex method to account for input originating from melting snow. Space non-uniformity of the distribution of water stored in the snowpack, of snow melt intensity and of water yield coming from snow, of absorption of water by the soil, of capacity of closed concaves, of snow melt runoff temporary control, etc. were considered as obeying statistical regularities, and were introduced into the calculations as such. Use was made of physical as well as empirically established equations in order to determine inputs resulting from snow melt.

A notable aspect of the same study is the splitting of the convolution integral into two parts, one for open and one for forested parts of the basin. This, unless there are in effect two sub-basins supplying water separately to the same outlet, may be disputed on theoretical grounds.

Borovikova and Denisov also present elaborate calculation methods for determining input fields of melting snow and spacially varying rainfall. Their model of catchment behaviour is however a lumped parameter one.

The idea that there is non-conformity between distributed input fields and lumped parameter models, is quite obvious. Researchers should aim to combine their tediously determined distributed inputs with equally distributed parameter models, because in a lumped model much of their efforts are wasted. The prerequisites for such approaches, even if only the less sophisticated distributed parameter models are made use of, lie at hand.

The study of Andreyanov and Babkin is based on data collected in an experimental basin. In this case, the *physical characteristics* were known in what can be considered a distributed pattern, which helped to calculate reliable averages of input to the introduced model. This model covered the hydrometeorological regime of the basin, comprising a system of equations of water and heat balances and interrelationships between elements of the same balances. Such models prove to be useful when the water resources of a basin are forecasted.

One of those well based studies in component modelling which are necessary to approach the ideal of distributed parameter models, is presented by Overton. In the path of surface water flow towards the outlet of the basin, the least clarified section is that of *overland flow*. This study establishes with a practical engineering approach the limits within which approximations are permissible, and then introduces such approximations in the form of very useful new concepts. Ultimately, hillslopes with uniform shape are substituted by equivalent planes whose handling in calculations proves to be very comfortable. The demonstration of the limits within which committed errors are still reasonably low, promises a good change for the application of this method in catchment modelling.

The paper by Nippes is concerned with another problem in hydrological modelling, namely with that of *suspended sediment load*. Basing its calculations of suspended sediment yield on the rate of increase in river discharge, this study is a valuable contribution to modelling in this field.

To conclude the report the following can be stated. Progress in this field is characterized by researchers laying emphasis either on component modelling as separate studies, or over-emphasizing some parts of the overall catchment model at the expense of making great simplifications concerning others. Input fields, flow towards the channels and channel flow seem to be competitors in gaining the researchers' attention in favouring them with a distributed type approach, or leaving them at the lumped stage.

Although such choices are in most cases determined by the special conditions of the catchment serving as a basis for the investigations of the particular researcher, the general intention should be to bring into conformity with each other all parts of the elaborated catchment models. Component modelling, as has been emphasized, provides the means for creating such models. The aim should be to construct models with physically meaningful parameters which can be interpreted in the possession of input fields reflecting changes of physical characteristics in both time and over the investigated basin. The idea of distributed parameter models is unseparable from this goal. Subjective determination of bulk parameters, or even optimization of the same are concepts related to lumped parameter models. The advent of distributed parameter models should mean that parameter values become more and more defined functions of physical characteristics determined with the required precision all over the investigated catchment. This, at least, should be set as an ultimate objective of our endeavours.

DISCUSSION

M. S. Gruszewskij:

I would like to comment briefly on the paper by Andreyanov and Babkin from our Institute. These authors have developed a mathematical model of the hydrometeorological regime of the river basin with monthly time steps, in compliance with the conditions of the Walday laboratory of TTI which is located in an excessively humid zone.

Andreyanov has asked me to convey the information that he has presented the fundamental methodological assumptions underlying this model in a recent research paper (Andreyanov, 1970). This information is not included in the paper presented at this Symposium.

In short, these assumptions are as follows:

(1) Elementary processes generating precipitation, evaporation, filtration, flow, etc. are in fact elements of one complex and self-controlling process. The analysis of this process calls for a genetic approach which reduces to the establishment and simultaneous solution of a number of equations. The set of such equation is the Andreyanov and Babkin model. Such an approach to the problem calls for detailed observations of the basic elements of the balances. It means that multi-annual investigations to be carried out by special expeditions are required, and that this approach can be applied just for the basins with discharge gauging stations. However, this is the only way which allows complete tracing of the process of river flow generation and to analyse this process in detail.

The objective evaluation of the accuracy of determination of all factors influencing both balances is necessary – both evaluation of the instrumentation errors and even more important errors due to the spatial interpretation of the measurements. The only objective criterion is the check on the accuracy of the balances on the whole and of their elements. It is worthwhile to underline that not all the information shall be used for this purpose. Only information from one year or from its part shall be used.

(2) After the completion of such investigations, the statistical methods must be applied for the generalization of the obtained characteristics in the area under consideration.

The real characteristics should be constructed on this basis, in accordance with the mean regional values of the basic parameters. Finally, they must be adjusted in accordance with the surface area of the individual regions. I would like to add also that the methodology proposed by Andreyanov and Babkin makes use of both the deterministic and statistical approach.

REFERENCE

Andreyanov, V. G. (1970) Development trends in the investigations and computations of the river flow (in Russian). *Hydrology Papers No.10.*

N. S. Nechaeva:
In our paper presented at this Symposium, we have described our work on the model intended for the computation of a spring flood hydrograph on the basis of some investigations concerning the snow melt process. The proposed model is based on the following general assumptions.

The river basin is considered to be a certain dynamic system with lumped parameters being fed by water inflow from the snow melt; the reaction is a hydrograph in the last (closing) profile of the river. Regional non-uniformities of the distribution of water resources in the snow cover, infiltration, time rate of snow melt and the surface runoff are subordinated to certain laws described by statistical relationships, distribution curves, etc. Intensity of snow melt, water capacity of the snow cover, temporary storage of water on the river basin surface, water inflow to the primary hydrographic network – all are considered to be time functions. The equations used for the determination of the individual components of the water balance and for describing basic processes of the hydrograph formation have been formulated on the basis of some theoretical considerations and more general ideas resulting from the analysis of empirical data and the schematization of some processes. The parameters of these equations are computed or fitted with due consideration given to the specific water storage conditions in the basin. The standard observation data on snow cover, air temperature, precipitation, cloudiness and winds make the initial material for computation of the hydrograph. In all cases where the river basin could be divided into two parts with and without the forest cover, the components of water balance were computed separately for each part. The river basin was considered to be open or forest covered depending on the percentage of the forest covered area. For the evaluation of the computation results, the criterion of squared differences between the computed and natural flows was used. The results of hydrograph computation at a given set of parameters were considered optimal when this criterion was taking its minimum value.

The proposed model was used for the computation of an inflow to the Gorky Reservoir. The basin was divided into 16 elementary basins with flows determined in the profiles closing each of them. Next, the 16 hydrographs were transformed into total inflow to the reservoir.

U. B. Vinogradov:

I would like to present some comments concerning my paper and to discuss simultaneously some general problems concerning the modelling of the hydrological processes.

First of all let me underline three points concerning deterministic models.

(1) Very often we are dealing with various approximations and the number of such approximations is unlimited. For this reason, we should use those which lead us to the most reasonable solutions. For instance, instead of the typical parabolic relationship between water storage and its release, we can use the exponential function. This allows for the analytical solution and consequently increases the accuracy of the computations and saves computer time.

(2) In the models describing river flow processes, the parameters should be standardized. The elimination of the influence of the river basin surface area and its slope are of primary importance. The standardized parameters can be compared, systematized according to the types of the basin surface and transferred to some other basins which have not yet been investigated.

(3) The evaluation of the parameters of the flood hydrograph models should be carried out only on the basis of observation data from small homogenous basins. Only in this case can we count on the reliability of the evaluated parameters. I would like to stress again that this problem is well reflected in the general report.

All these points are in some way or other reflected in my paper. I have presented the procedure for the computation of flood flows for small mountanous river basins. Taking into account that the procedure is applicable for small (tens or hundreds square/kilometres) and mountainous basins, I have decided that the incorporation of a complex scheme allowing for the consideration of the river channel regulating effect shall be advisable. It turned out that the concentration time τ can be taken into account with the help of the moving average operation along the value of τ (in case of necessity the weighing function can also be used).

One of the basic hydrological problems is the determination of flow characteristics, especially the maximum flood flows with a given probability of recurrence.

This problem was not sufficiently reflected at our Symposium. I would like to mention, therefore, that the deterministic models should be constructed with due consideration given to their possible applications in stochastic modelling. The purpose of such modelling is the determination of the computational distribution of probabilities of the hydrological characteristics which we are interested in.

The analysis of a broad spectrum of different questions is required for the solution of this problem. Within the framework of the physical description of the problem, it is necessary to construct the models of infiltration, runoff, evaporation and transformation of runoff into flow hydrograph. In this phase it is advisable to use the theoretical methods which have sufficient physical justification. The correlation methods are not advisable in this phase.

As far as the probabilistic tasks are concerned, it is necessary to describe the random process of precipitation, its transformation into the process of flow formation, integration of this process in the time domain, computation of the maximum flows, description of the random field of precipitation depths, description of the distribution of the computed process parameters (depth and duration of precipitation, duration of time intervals without precipitation, etc.).

All these problems are in some way reflected in the paper, of course not all of them with the same precision.

L. N. Borovikova:

The lumped parameter model of rainfall transformation into discharge presented in our paper, has been developed for the mountainous basins located at an altitude varying from 3 to 4 km a.s.l. Similar to many other models of this type, it has been structured under the assumption that the basin can be represented by a number of storage capacities. It should be mentioned that the analysis of the sequentially and parallel connected storage capacities means that the model consisting of a large number of elements seems to be more justified for mountainous rivers. The river flow in the closing profile (Q) originates from the releases from various storage capacities (Q_1). The model presented in our paper has no allowance for the river channel transformation, but in the case of mountainous rivers with large longitudinal slopes, short inflow time, and small capacities of the river channel, this transformation does not play any important role.

The input function Q_n includes, therefore, the volume of water originating from the whole basin surface area caused by snow and ice melt as well as by occasional precipitation.

In such a case, the problem of the regional distribution of precipitation and snow melt had to be solved for the computation of the input function. The sub-model of snowcover formation was structured as a component of the main model. I would like to underline again that the mountainous rivers in Central Asia are located at an altitude of 3—4 km, but in all cases there is a distinct relationship between various characteristics considered in the computations and the altitude of these characteristics points. For this reason the whole model is built in the system of altitude and time coordinates.

The division of computations into two stages allows for the evaluation of the results not only on the basis of the final result i.e. the flow hydrograph, but also according to the intermediate results of computations based on the model of snowcover formation.

In this way, the model on the whole represents the lumped parameters flow model with distributed input. However, this solution can be considered only as the initial one, as it was reasonably stressed by the general reporter. Nevertheless the model is very promising.

D. E. Overton:

I want to make a comment concerning my paper about the kinematic cascades, and then secondly allude to one of the questions that were raised by the general reporter. I find the structure of many of the component elements of basin runoff models to be simplifications of the flow equations of the various processes. Therefore, it has been my intention to work with one component, namely overland flow, whereby I could derive lumped parameter models in terms of the flow equations which, therefore, would be a criterion for their use. I feel the key to success in this area, if success it is to be. is the use of a characteristic time, which I call lag time. In overland flow, as in many processes, equilibrium is a matter of degree, and I found that subjectivity in determining time to equilibrium from overland flow produces great sensitivity in testing models. The lag time that I proposed, I think, has hydrological significance. It is a more stable parameter, can be computed from the equilibrium storage at infinity and the input intensity. Depending on the model you are working with, you can predict lag time if you can compute equilibrium storage. For example, in the Nash model I found that equilibrium storage was equal to input intensity times lag time. Therefore, for the Nash model, the lag time is equal to $N.K$. I use this as a constraint which reduces the zone of optimality. Thus, I am promoting the use of lag time. I think that it is a very important time, it has made life very easy for me, and I think that others working in this field may examine this as a possible parameter. Finally, I would like to address an answer to the first question that was posed by the general reporter as to how we can develop methods which can be used by designers. I run across this problem often in the United States and I find generally a large gap, in terms of vernacular. between operations people and research people. From my point of view, I think that if we would attempt to popularize our research results more, if we would try to tone down the vernacular when we are speaking to operations and design people. I believe that we can close the gap. A good example of this is the 'universal erosion equation' which was developed by statisticians and engineers in the US Agricultural Research Service. The way that they presented their universal equation permitted immediate implementation by operations personnel.

J. Amorocho:

I do not know why hydro-biological models were not presented at this Symposium. It is certain that a considerable amount of work has been conducted in the recent past in this area. The importance of these models is very great, in the United States it has been found that, in situations in which water is not scarce, the biological constraints, insofar as they affect water quality, are often much more important in water resources planning than strict considerations of water abundance.

I join you in lamenting that biological models were not presented here.

M. Roche:

Je ne peux qu'être d'accord avec le principe de ce qui vient d'être dit, et en particlier les allusions de M. Toebes à l'importance de l'intervention d'autres spécialités dans les problèmes concernant les modèles mathématiques et l'aménagement des eaux. Mais il faut bien dire qu'un symposium, forcément limité dans le temps, doit limiter son sujet. Je pense quant à moi que l'évantail des questions qui ont été traitées ici est déjà suffisamment large. J'avais cru comprendre, et je pense que c'était effectivement dans l'esprit des organisateurs, qu'il s'agissait d'un symposium méthodologique; de ce point de vue. l'aspect sous lequel les questions ont été traitées me semble parfaitement conforme à ce qu'on pouvait en attendre.

Je ne dis pas par là qu'on n'aurait pas pu s'orienter vers un côté plus pratique et, en particulier, essayer de parler non plus d'un point de vue un peu universitaire, comme l'a fait remarquer un commentateur, mais d'un point de vue plus opérationnel. C'est une idée qui a commencé à se développer; en Amérique d'abord, puis en Europe et en Extrême-Orient, on a crée. avec le concours de l'OMM, des commissions d'experts pour les modèles opérationnels.

Les modèles peuvent être imparfaits, mais cette imperfection n'entraîne souvent qu'une imprécision modérée en regard de celle qui résulte des données servant à les construire et à les régler.

On a fait allusion également, et on a même fait plus qu'allusion puisque la question vient d'être mise explicitement sur le tapis par M. Toebes, puis par M. Amorocho, aux questions concernant les aménagements et la qualité des eaux. Ce point de vue est extrêmement préoccupant; en ce qui concerne la France, on peut même dire que c'est un problème qui passe au tout premier plan, puisque l'action de l'organisme national le plus officiel qu'est le Secrétariat Permanent pour l'Etude des Problèmes de l'Eau, est essentiellement axée sur les questions de qualité des eaux, de pollution et de traitement. Récemment, nous avons eu à Paris une réunion du Comité Technique de la Société Hydrotechnique de France entièrement consacrée aux problèmes de pollution et en particulier aux modèles mathématiques qui s'y rapportent. C'est donc une question que nous n'avons donc pas du tout perdue de vue.

Dans un symposium, il faut choisir. Si on veut faire un colloque concernant finalement l'utilisation des eaux, on peut traiter à la fois des problèmes quantitatifs (volume des ressources) et des problèmes de qualité en y incluant le traitement des eaux, c'est-à-dire la modification de leur qualité en vue de leur consommation, et l'épuration, c'est-à-dire la modification de cette qualité après usage et avant rejet dans les circuits naturels. Mais alors, on laissera de côté les discussions un peu académiques telles que celles auxquelles nous avons participé depuis le début de ce symposium.

Ma conclusion est qu'on ne peut pas à la fois faire un symposium méthodologique sur des questions de modèles mathématiques, avec tout l'arsenal mathématique qu'on nous a exposé, et traiter des problèmes généraux et fondamentaux d'utilisation et de qualité des eaux.

G. F. Pinder:

I would like to discuss an idea that is just becoming important in the United States and which is in a rather embryonic form because of the dependence of it upon the size and speed of digital computers. In this concept we look at the hydrological system, including the transport of chemical constituents as a problem in continuum mechanics. I think that it is universally recognized that the deterministic equations describing the movement of fluids and the transport of dissolved solids are based on continuum mechanics. However, there has been minimal use of this concept because the coupling between the resulting equations makes their solution difficult. For example, if we look at the transport equations which describe the movement of dissolved solids in porous media. or in surface flow, these equations are dependent on the unknown variables in the surface water or groundwater flow equations. And it is only by coupling these equations, or, more specifically, solving the equations of mass, momentum, and energy transport simultaneously, that we can in fact describe totally any hydrological system. What I am proposing is that some of the problems that we are considering, we are looking at from a very narrow point of view. We attempt to solve a particular problem, failing to recognize that it is a small part of a far larger and perhaps more important problem.

H. W. Anderson:

I would like to respond to the question about the biological aspects of hydrology and in particular, of course, to its effects on forest and mountain area hydrological modelling. The question has been asked: Why is the biological phase not more closely examined in this Symposium? Of course, modelling has created a tremendous data acquisiton problem over many years. Now, I believe, tools are becoming available to help with some solutions; satellite and improve photographic methods and computer systems which can document the biological conditions of the various parts of basins. We must know how these biological conditions are associated with the geology, the topography and, ultimately, with the soils formed and how the basins perform. We may now better appraise this biological, or if you wish, ecological system in our hydrological analyses. One of the big problems that has always presented itself in the biological phase of hydrology is how do we give dimensions to biological entities? We start out with, as it has been pointed out by many people here, a variety of products: flood generation, total water yield, water quality, or the timing of water yield, that is, when is water delivered. Each product must be separately dealt with. And so we have to determine a variety of models, depending on which of these characteristics we are particularly interested in. So we have had great problems in dimensioning, in particular in our forests. Concepts have included such entities as the total biomass. Is biomass a useful concept and is its measurement useful in evaluating a hydrological product? Well, it might be useful in one aspect, such as the total water yield where evapotranspiration is involved, but it may not be so useful when we are considering snowmelt when arrangement or patterns rather than total biomass fits the hydrological model in which we have particular interest.

I should also like to stress that we are not dealing with minor variations in our outputs as a result of these biological factors. We are dealing with variations which have been going on in time, have been affecting both deterministic and stochastic hydrology. Take for example, forest fires which I have examined. These may cause changes in floods for example, that do not vary by a few percentage points but by as much as five times, depending on the fires history in a particular basin. More recently we have examined the effects of logging of forests, that is the harvest of the forest trees, on one hydrological aspect, the discharge of sediment. This suspended sediment problem came up in one of the papers. What was the effect when we have a combination of a major flood in areas which have been recently logged? After such a combination, as much as five times as much sediment occurred in the years after the flood for the same streamflow distribution. So we are dealing with major changes in a basin, associated both with the biological phase and the interaction of the biological phase and the hydrological phase.

Session XIII   29 July 1971

General Reporter   Dr. J. Amorocho, Professor of Water Science and Civil Engineering, University of California, Davis, USA

Chairman   Dr. Frederick V. Appleby, Department of Civil Engineering, Imperial College of Science and Technology, London, UK

Distributed parameter catchment models and input fields — II

J. Amorocho

The papers reviewed and discussed in this general report are all published in volume 2 of the Symposium Proceedings. They are listed below.

1. *Alfred Becker*   Applied principles of catchment simulation
2. *V. I. Koren and L. S. Kuchment*   Physio-statistical model of rainfall flood formation and determination of its parameters
3. *Arie Ben-Zvi*   The velocity assumption behind linear invariant watershed response models
4. *G. Girard, J. P. Fortin and R. Charbonneau*   Un modèle hydrométéorologique simplifie et quelques applications régionales
5. *F. V. Appleby*   A determinate model for runoff as a nonlinear system
6. *Y. Emsellem, G. de Marsily, D. Poitrinal and M. Ratsimiebo*   Déconvolution et identification automatique de paramètres en hydrologie
7. *Ven Te Chow*   Role of WES in the development of hydrodynamic watershed models

Introduction

One of the fundamental question that remains largely unsettled in hydrology is to what extent one must go in the faithfulness of a mathematical description of fields of input and motion to obtain models having an acceptable predictive power. This problem is a very difficult one because of the large uncertainties involved in its formulation in the first place. Not only are the spatial and temporal distributions of hydrological variables enormously varied and difficult to measure, but the standards for acceptability of estimates resulting from hydrological modelling tend to be as flexible as the proponents of the models, as well as the users wish or are forced to make them. Therefore, a direct answer to the question may remain elusive for some time. On the other hand, it is certainly worthwhile to persist in the effort of improving hydrological models by the application of new techniques and by more thorough analysis of the phenomena involved, through what amounts to computational experimentation. This process in the past has involved essentially a comparison between results obtained by old methods and by new methods; almost invariably the new methods turned out to be better, at least in the view of their proponent. The group of papers summarized here may be said to follow this scheme at least implicitly. None of the authors has attempted to answer the larger question mentioned above.

Summary of papers

The paper by Chow follows more closely the theme of this session and will be discussed first. The author gives a summary description of the Watershed Experimentation System (WES) operated at the University of Illinois, and lists bibliographical data on more complete discussions given elsewhere on this laboratory device. It consists of an installation which permits the application of controlled fields of water spray over various surfaces, and the measurement of the rate of the water outflow after it has been routed through the catchments being tested. The size of the installation is such that true overall similitude between model catchments and natural catchments is not obtained due to scale effects; however, it permits the measurement of a number of local variables in prototype scale, and thus may lead to a better understanding of the basic components of the flow mechanisms operating in nature.

Chow also introduces two types of hydrodynamic field models based on certain geometrical and analytic simplifications. The models represent the flow field generated over impervious, inclined surfaces by uniformly distributed rain. The WES installation is used for the physical simulation of these same conditions, and some of the constants and parameters to be entered in the mathematical model are determined from the laboratory results. The WES is also employed for the verification of the predictions of the mathematical models.

The paper reports on the results of tests for a catchment consisting of an inclined trough formed by two equal rectangular planes joining along one of their sides. The model designated as IHW model II treats the problem by approximating the flow system with two distinct elements: the first is an overland flow component, treated as one-dimensional flow in the direction of the transversal slope of the two intersecting planes forming the through, and the second is a stream component, also treated as one-dimensional flow with lateral inflow. This second element represents the mean flow along the line of the trough. The results of the application of hydrodynamic equations for the two elements in cascade were compared with the observed results of applying simulated rain to a trough in the WES. The author reports that, although the rising limb of a simple simulated hydrograph agrees well between the IHW model II and the WES, some differences are apparent during the recession.

The second mathematical model, designated by the author as the IHW model III, is a hydrodynamic two-dimensional representation of the flow field over the trough. In the sense that this model allows for variation in the flow directions over the catchment surface, it is more nearly a true field model. The formulation, still in the process of modification and refinement, involves interesting algorithms for the normalization of the continuity and momentum equations and permits the preparation of multi-dimensional computer displays of the flow field.

Chow's analysis is one of the first in which the full hydrodynamic equations have been applied to overland flow. Most of the previous work in this area has concentrated on the use of the kinematic wave approximation. While the latter may be more economical in computational time, the more rigorous treatment given by Chow is likely to shed considerable light on a number of interesting aspects of 'watershed hydraulics'. Among these are such phenomena as the change in the value of the friction factor caused by impacting raindrops, the formation of transient hydraulic jumps and the effects of bed geometries on the flow fields. The expectation certainly appears promising that this line of investigation will be very fruitful.

The paper by Becker reports on the concepts embodied in a conceptual catchment model of some complexity, which is the fruit of work performed over a period of several years. Some of the ideas involved in this model were incorporated in an earlier paper (Becker, 1967). The model, which is described with great lucidity and compactness, involves as the author remarks, two basic types of distribution: (1) a 'real-position-distribution', or areal arrangement of the input and flow elements in the catchment, and (2) a 'statistical distribution', describing the characteristic variability of parameter values within the catchment. In the author's words '. . . the first type of distribution was used for modelling the surface flow system of the catchment. The second type was used in different forms for the interception [sub] model and the lower soil layer [sub] model – capillary water and gravity water storage . . .' Although the model permits the calculation of flow components for discrete portions of a catchment, the use of these 'distributions' does not make it a truly *distributed parameter* or *field* model, in the sense that a continuous variation in time *and* space is not explicitly recognized. This, however, should not detract from the usefulness and acceptability of the scheme.

The principal concepts involved in the structure of the model are the following:

(1) A linear distribution of the interception storage capacity within the catchment.

(2) A distribution defining the heterogeneity coefficient for lower capillary water.

(3) A separation of the infiltration process into several phases, in such a way that the computational time increments can be varied depending on the rate of change of infiltration.

(4) An area-variant concept, which permits the calculation of overland flow from partial areas of the catchment where this type of flow can physically occur.

(5) An adjusted time–area diagram, used as response function for the channel system.

(6) Algorithms permitting a modulation of the computational time increments depending on the demand for detail definition as the calculation proceeds.

The model has been used successfully in applications on a small (15.7 km^2) mountain basin. As indicated by the author, although the array of linked elements may become complex depending on the complexity of the basin itself and on the degree of subdivision used, the basic components are rather simple, but they involve consideration of the physical processes at play.

In contrast with the comparative rigour observed in the conception of the various catchment sub-systems of the work by Becker, the model by Girard *et al.* utilizes a much simpler set of assumptions in postulating the components of the runoff cycle and in linking them together. It is to be noted that, although both the Becker and the Girard *et al.* models perform lumping, or discretization throughout. the latter involves a considerably coarser vertical structure.

In general terms, the application of the model is based on subdividing the catchment into partial areas as determined by Thiessen's polygons drawn around the raingauges available. Further subdivision is performed when advisable to account for physiographical or cultural differences. The runoff produced by each partial area is routed following a set of isochrones which are superimposed on the subdivided basin map.

To compute the runoff generated at each partial area, a simple mass balance is performed for each time interval, wherein the water content of the element is modified by subtracting the estimated evapotranspiration, adding the precipitation and snowmelt and deducting the outflow. The latter is estimated by two simple functions of the water content of the element (expressed as an equivalent depth), which are entered in the computation depending on a fixed threshold value of this water content.

Snow melt is estimated as a function of ambient temperature and apportioned among the snow-covered areas.

The estimation of evapotranspiration is again simplified, and based on available record of pan evaporation, modified to account for water available in the soil at each computational step.

The authors recognize the severe simplifications and blanket assumptions made in their model, but remark that the closeness of the predictions to observed runoff hydrographs of a number of basins with large contrasts in area, climate and location justifies the procedure. The interesting fact brought out by these experiences, as well as by those of other hydrologists throughout the world, is that, not only do various types of models seem to perform 'satisfactorily' although their theoretical fundaments may be of different quality, but that the quality of the *input data* is about the same in all cases. This might suggest the intriguing possibility that beyond a certain level of model complexity, the quality of the prediction may be determined primarily by the data and only to a lesser degree by the refinement of the models.

Koren and Kuchment present a model which, like the previous two discussed, is lumped but with the lumping extending to the entire catchment. However, although the model is not distributed in a strict sense, a consideration of areal variability is implicit in the form of analysis used by the authors.

The basic relationship used for the rainfall excess $q_i(t)$ of an element of area i, is

$$q_i(t) = P_i(t) - E_i(t) - I_i(t) \tag{1}$$

where $P_i(t)$, $E_i(t)$ and $I_i(t)$ are the local precipitation, evaporation and infiltration rates.

The rates of evaporation and infiltration are given by empirical equations involving wind velocity, humidity contrast across the air—water interface, and soil moisture state.

Averaging equation (1) over the catchment, the expected value of the rainfall excess is given by the expression.

$$M(q) = M(P) - M(E) - M(I) \tag{2}$$

in which the symbol $M(\ )$ stands for the expected values of the variables in parentheses.

The authors then proceed to write the expressions for the various terms of equation (2) based on their empirical formulations and finally combine the surface and sub-surface runoff by superposition through the expression:

$$Q(t) = \int_0^t P_1(t-\tau)q_s(\tau)\mathrm{d}\tau + \int_0^t P_2(t-\tau)q_I(\tau)\mathrm{d}\tau \tag{3}$$

where $Q(t)$ is the basin outflow; $q_s(t)$ and $q_I(t)$ are the overall surface and sub-surface flow input rates; and $P_1(t)$ and $P_2(t)$ are kernels or response functions. The latter are assumed to have the form of the gamma distributions. $q_s(t)$ and $q_I(t)$ are found as empirical functions of the mean basin rainfall excess and the mean basin soil moisture deficit.

In the process of computation, the various parameters are determined by systematic trials, wherein the parameters, and the expressions where they appear, are first neglected and then introduced successively until the prediction by the model satisfies the criteria for fit established by the authors.

The special merit of this paper is that the authors have combined the method of direct linear analysis involved in the well-known linear convolution approach, with a broad physical consideration of the main processes operating in the catchment, taken in a statistical sense. Although the empirical equations adopted to define the main variables are admittedly rather coarse, the approach would seem to be amenable to the use of more rigorous formulations in the future.

A great deal of erudition is incorporated in the paper by Appleby. Its merits are notable not only as they pertain to the conception of the hydrological model proposed, but also regarding the copious scientific and technical background and reference material presented.

Again in this case, the model lumps the storm and catchment field properties, so that functions of time only are considered. Performing a drastic abbreviation of this very detailed contribution, at the risk of omitting some significant details, it can be stated that the main concept is the consideration of two principal components of catchment flow. The first, termed by the author a 'carrier field', represents elements involved ultimately in producing base flow. This is a slowly varying and stable field upon which is superimposed the second, more rapidly fluctuating field, representing elements associated directly with surface flow.

The input, or rainfall, is considered 'compact and uniform' and is assumed to be characterized by certain mean parameters of a time function.

Two kernel functions are proposed to represent the overall operation of the component fields in producing runoff at the catchment outlet. The parameters of these functions are determined by numerical analysis of recorded events and have led the author to perform successful hydrograph decomposition (base flow separation) in a number of catchments.

Emsellem *et al.* present a method for performing the inversion of the convolution integral which avoids the numerical instabilities and oscillatory behaviour often observed in other procedures in current use.

This contribution is strictly in the realm of linear theory with lumped parameters. However, as is well known, it is possible to use the convolution integral for the approximate solution of field problems by consideration of finite elements. Therefore, in the sense that the method proposed furnishes a valuable tool for performing economically many stable inversions, it can be considered pertinent to the topic of this session.

The procedure consists of constructing the kernel function from recorded input and output sequences by successive approximations, each of which is a correction of the previous approximation, having zero mean. The explanations given in the paper about the theory and application of the procedure are so precise and lucid that further abbreviation within the limits of this report would detract from their excellent presentation. Suffice it to say that the method provides a very compact form for determining linear response functions, that it is easily adaptable to any digital computer system and that the degree of approximation desired can be controlled at will.

The final paper, by Ben-Zvi, is devoted to an attempt to prove the proposition that the unit hydrograph and the time–area concentration curve are identical. The subject of the paper is again only loosely connected with the topic of distributed models, since the unit hydrograph is strictly a lumped parameter conception.

The argument presented can be summarized roughly as follows. As long as the time–area concentration curve for a catchment is assumed to be invariant and unique, it also satisfies the conditions imposed by the assumptions for existence of the unit hydrograph. A single time–area concentration curve can exist if the velocity vector field at every point on the catchment is invariant. However, a single curve may also exist '. . . if there are properly balanced variations in the velocity vector at different points of the watershed . . .' The author feels that since the latter process cannot conceivably exist, it must be concluded that only the invariant velocity vector field is plausible. By a simple secondary argument, the identity between the unit hydrograph and the time–area concentration curve is deduced.

The time–area concentration curve has been used in connection with unit hydrograph applications, either implicitly or explicitly, by a number of workers before (e.g. Dooge, 1959), and the assumption has generally been made that this curve exists. A proof such as that suggested by Ben-Zvi was not known to the writer, who must also confess that he does not find it entirely convincing. The entire argument is somewhat heuristic, and it has a particularly weak point, to wit: if it is regarded as inconceivable that the velocity vector field in the catchment could have properly balanced variations to give rise to an invariant and unique time—area concentration curve, why is it more conceivable that this field should remain invariant for all conditions of flow? It appears equally problematical that either situation may really prevail in nature, and if the existence of an invariant unit hydrograph is dependent on such conditions, it would appear that *it* also is problematical. Many hydrologists throughout the world have in fact found this to be the case.

To conclude this report, it is of interest to remark that none of the authors dealt in any specific manner with the input fields. It seems that in the examples of application of their proposed models, the storm fields were indeed very uniform and hence could be lumped, or the space variations of the input became so attenuated by the catchment that the lumped values used in the models led to acceptable results. In the experience of the writer, this is a more likely occurrence when dealing with cyclonic storms of relatively large areal development. It certainly is not if the main sources of runoff are local convective storms.

REFERENCES

Becker, A. (1967) Threshold considerations and their general importance for hydrologic systems investigations. *Proceedings of the International Hydrology Symposium*: Fort Collins, Colorado, USA, September 1967.

Dooge, J. C. I. (1959) A general theory of the unit hydrograph. *J. geophys. Res.* **64**, No.2, 241–256.

DISCUSSION

Ven Te Chow:
It is not the purpose of the WES project to develop physical models of watersheds. It is hoped that the conceptual hydrodynamic models to be developed in this project will be applied to sub-areas of a watershed in the practical application for the purpose of calculating runoff from rainfall which may vary substantially from sub-area to sub-area. This will be a part of future research of the project.

A. Becker:
There is a topic of special interest in this session which I would like to bring into discussion, namely the question of the classification of distributed parameter catchment models.

The most complete description of the hydrological processes taking place within a catchment would be that all the interesting variables – the input field p, the vertical flow field f (infiltration, percolation), the lateral flow field q, the storage field S – and all the parameters k_N (storage and flow parameters with $N = 1, 2, ...$) could be expressed as functions of space (x, y, z) and time (t) as shown in Table 1. Such a system model could be called completely distributed. The opposite is the lumped system model (no distribution), where all the variables mentioned above are only functions of time, and the parameters k_N are lumped parameters (e.g. areal averages). Most of the current research in catchment modelling is at this time between these two extremes and we should discuss what intermediate groups of distribution principles are in application. According to my paper, I propose an intermediate group of so-called 'statistical distribution' which may be defined as follows: The variables or parameters are expressed in dependence of the percentage $a = A_p/A$ of the total catchment area A under study, where A_p represents parts of A ranging from zero up to A.

This distribution principle was introduced for the infiltration capacity distribution in the Stanford model and for the interception storage capacity distribution in my paper being presented here. The basic idea of this concept is to respect 'statistically' the areal differences of a definite capacity value within a basin and so their effect, e.g. in the form of a linear distribution around the areal average of the capacity considered.

Regarding the papers being presented at this session. that of Chow would fall into the group of so-called 'real-position-distribution', though not of the complete type but as a partial real-position-distribution. My own model and that of Koren and Kutchment would belong into the group of 'partial statistical distribution' and that of Girard *et al.* into another intermediate group which may be called 'lumped partial areas distribution' (see Table 1).

There is no doubt that the main development of current research is from right to left in Table 1. Every model builder has to decide for himself what type of model he should take. In this connection, however, he should respect the really essential remarks of O'Donnell which he makes in his general report on this topic: 'Some sort of compromise has to be made and the compromise must depend to a large extent on what the designer of a catchment model aims to achieve', and later: 'A sound principle to adopt in conceptual modelling is to exercise the *utmost* economy in the number of parameters used. This not only greatly eases the parameter fitting effort but also it reduces the possibility of parameter interaction'. I agree completely with this statement and that is the reason for using extensively only the statistical distribution principle in our modelling work. It is an unquestionable advantage of the linear statistical distribution principle that, e.g. in the case of accepting the existence of zero-capacities within a catchment, no more parameters are necessary that under lumped conditions (see Fig.2 of my paper: vol.2, p.766). Only the mathematical scheme for calculating the time dependent variables is more complicated as is simply seen from the interception sub-model flow chart which I presented in Fig.3 (vol.2, p.766). Recognize, please, that, contrary to this, already in the case of 'lumped partial area distribution' the number of the parameters to be distributed increases as a product with the number of the

TABLE 1. Classification of distribution principles for catchment modelling

Variables, parameters	Real-position -distribution	Lumped partial areas distribution	Statistical distribution	No distribution (lumped conditions)
Input	$p(x, y, t)$	$p_j(t)$	$p(a, t)$	$p(t)$
Storage	$S(x, y, z, t)$	$S_{ji}(t)$	$S_i(a, t)$	$S_i(t)$
Vertical flow (infiltration. percolation)	$f(x, y, z, t)$			
Lateral flow	$q(x, y, z, t)$	$q_{ji}(t)$	$q_i(a\ t)$	$q_i(t)$
Parameters	$k_N(x, y, z)(t)$	$k_{Nji}(t)$	$k_{Ni}(a)(t)$	$k_{Ni}(t)$

N = number of parameters considered
i = number of layers considered
(t) = time dependence respected if necessary (time variant systems)
j = number of lumped partial areas considered
a = A_p/A = percentage of the total catchment area A
A_p = part of A being considered

Note: Above listed are the cases of complete distribution in applying the distribution principles mentioned in the table (all parameters and variables distributed). Mostly only partial distributions are applied (only one or some parameters and/or variables distributed, the others lumped).

1 2 3

transfer function and actual output (intermediary step)	agreement between actual and computed output (intermediary step)	transfer function (final step) actual and computed output are identical

Example on theoretical data

FIGURE 1. Slides taken on a cathodic screen display unit.

partial areas introduced. And in the case of real position distribution this increase may be a power one. Here really the essential question of data needs and availability as pointed out by Dawdy is of major importance.

The second reason for preferring the 'statistical distribution' is that, according to many other authors, a more complex vertical and a nonlinear input separation (as described in my paper) seem to be more important. Still more important for catchment modelling seems to be the introduction of the concept of saturated area overland flow. I think this concept is new in catchment modelling, and it has been proved during our work as very essential because this flow component occurs far more in nature than the usually assumed 'catchment-wide' (Hortonian) overland flow. This was clear not only from our model tests but also from the valuable experimental results of Dunne and Black as mentioned in my paper.

However, despite my defence of the effectiveness and usefulness of 'statistical distribution principles', I would like to say that we, at the same instant, should appreciate thankfully that some researchers in deterministic hydrology are working ahead within the complicated field of 'real position distribution' preparing a basis for possible future improvements in catchment modelling.

G. de Marsily:

I would like to thank Professor Amorocho for his excellent report on our paper (Déconvolution et identification automatique de paramètres en hydrologie), especially for his broadening of the theme of todays' discussion, and for his comment on the use of simple linear models with cheap cost of application, versus sophisticated ones, depending on the quality of the available data.

I will add three brief comments:

Firstly, this method presented here was originated on a more general identification problem, namely the determination of the transmissivities of an aquifer from piezometric data. Secondly, the successive approximation procedure that we use with increasing precision on the computed kernel function, is particularly well adapted for use with a cathodic screen display unit connected with computers, which enables us to follow the agreement between observed and computed output, and also the shape of the kernel function. This allows us to stop the procedure when sufficient precision is obtained, or to skip values of the data used which evidently contain systematic errors, or even to control, by appropriate smoothing procedure on the kernel function itself, the compromise between the smooth shape of this function, giving it a physical meaning, and good agreement between observed and computed output (e.g. Fig.1 as a theoretical example).

My last point is that this technique can be applied to various sorts of problems in hydrology, besides rainfall—runoff response. Relations between movement in a piezometer and in a river have already been commented during session VIII. I shall only mention another example, along the line of pre-occupation of Toebes during session

XII. Deconvolution, associated with a series of tracings along a river, has been used to try to localize an unknown point of reject of pollutants in the river, once a wave of pollution has been detected and measured several miles downstream.

G. Girard:

La présentation du modèle à paramètres distribués ayant été faite par M. le rapporteur général, je désirerais apporter les récents résultats obtenus avec ce modèle et indiquer que le but de celui-ci est de fournir les caractéristiques hydrologiques d'un bassin à partir des données météorologiques connues pour une période de 30 années et des mesures hydrométéorologiques complètes systématiques faites sur ce bassin au cours de deux ou trois années.

Notre méthode consiste:

(1) à recherches la période d'observation maximale commune des débits, des précipitations et des températures, à 4 ou 5 stations au moins, quitte à ne pas tenir compte des données incomplètes;

(2) *à caler notre modèle sur une période de 3 années;*

(3) à exploiter le modèle sur la période globale de 30 années;

(4) à traiter par la statistique la série simulée, puis la série observée;

(5) et, à comparer les lois de distribution des étiages annuels de 7 jours et des débits maxima annuels de ces deux séries.

Les valeurs des étiages de fréquence (0.1, 0.01, 0.9, 0.99) déduites de l'adjustement d'une loi log Pearson III sur les deux séries (simulées et observées) ne different pas de 2, 5, 10, 20 pour cent pour les quatre bassins étudiés de la Province du Québec dont les superficies sont comprises entre 1000 et 6000 km^2.

Les valeurs des débits de crue déduites de la méme facon sont plus précises et ne different que de 2, 5, 8 et 10 pour cent pour ces mêmes fréquences.

Par suite du nombre restreint de stations météorologiques utilisées, le modèle ne simule pas exactement les chroniques de débits.

Cependant, en injectant à nos postes, des précipitations représentatives de celles tombées effectivement sur le bassin, les crues même catastrophiques, telles celles produites par un orage de 325 mm en 6 h, sont parfaitement reproduites par le modèle.

Pour conclure, les modèle actuel est susceptible de nous fournir les caractéristiques hydrologiques essentielles pour les aménagements de bassins, dans les conditions données.

Compte tenu du nombre minima de paramètres utilisés, nous espérons pouvoir relier chacun d'eux aux caractéristiques physiques du bassin (pente, pourcentage de lacs, superficie).

J. C. I. Dooge:

I want to make some comments on the paper by Ben-Zvi. I particularly want to make these comments because in the acknowledgements in this paper, he expresses his gratitude to myself, among others, for comments on an earlier draft of the paper. I want to put the record straight. The earlier comments of mine, made on a draft of this paper, were to the effect that the conclusion of the author was wrong, and that also he had failed to acknowledge the earlier work in the one special case in which his proposition held.

The author has now included reference to these earlier papers, but he still, in this paper, comes to the conclusion that, 'the instantaneous unit hydrograph is identical to the time—area concentration curve.' Quite flatly, this is not true. A necessary condition for the existence of a time—area concentration curve is that the system be linear and time invariant.

A necessary condition for the existence of a unit hydrograph is that the system be linear and time invariant. However, the curves are not identical.

To say that the time–area concentration curve and the unit hydrograph are identical is to confuse the velocity of a particle of water and the celerity of a wave. It is the same error that is made in hydraulics, when somebody does not distinguish between the fact that a particle of water will travel at a mean velocity Q/A, whereas a wave will travel approximately at a Kleitz–Seddon velocity of dQ/dA. This confusion between the speed of particle motion and the celerity of wave motion is inherent in this paper.

The supposed proof, which the author gives, arises from his writing down of equation (5), which is not, in general, correct. In equation (5) the author states that the runoff due to an amount of rain occurring during an infinitesimal time ϵ is

$$Q(t) = u(t)I\epsilon \tag{5a}$$

where $u(t)$ is the unit hydrograph, I the intensity, and ϵ the infinitesimal duration. In fact the correct equation is

$$Q(t) = u(t - \tau)I\epsilon \tag{5b}$$

for a single input of rain of infinitesimal duration and

$$Q(t) = \int u(t - \tau)I(\tau)d\tau \tag{5c}$$

for an input of finite duration. In these equations τ is the time of input and t the time of output measurement. Equations 5(a) and 5(b) are not the same, unless the instantaneous unit hydrograph happens to have the shape of a rectangle. This fact was pointed out by Nash thirteen years ago. Only for that special case is the author's conclusion true, that 'the instantaneous unit hydrograph is identical to the time–area concentration curve'.

J. W. Delleur:

Je voudrais me référer au papier de Emsellem *et al.* et je voudrais les féliciter d'avoir présenté une méthode nouvelle et intéressante pour la détermination de la réponse percutionnelle d'un système linéaire. Je suis également très content de voir que les auteurs ont pu utiliser à bon effet des données qui ont été obtenues par un de mes élèves et moi-même. Je voudrais néan-moins demander à ces Messieurs de donner la référence correcte et complète dans l'édition finale des comptes rendus étant donné qu'en lisant la version actuelle on croirait que le bassin se trouve aux Indes, mais en effet le bassin en quesiton se trouve dans l'Etat d'Indiana aux Etats-Unis à 200 km environ au Sud-Est de Chicago. En ce qui concerne la Fig.3 (tome 2, p.713) je voudrais peut-être leur faire remarquer que les entrées et les débits sont très lisses et, par conséquent pouraient être représentés par les séries de Fourier avec dix ou peut être moins de termes. A en juger par la Fig.21 (tome 2, p.728) il parafrait que les auteurs ont employé au delà de 100 termes, l'instabilité serait doncdueici à un excès d'information. En ce qui concerne la Fig.14 (tome 2, p.721), il eut été également intéressant de comparer le résultat obtenu par les auteurs et le résultat obtenu par M. Blank et moi-même par la méthode de la transformée de Fourier. Je n'ai pas pu faire la comparaison directe étant donné que les unités dans le papier que je discute ne sont pas données sur la figure. En résumé, je voudrais donc souligner que la méthode qui est proposée parait être très intéressante et il serait désirable de la poursuivre, pour autant que le temps employé par les ordinateurs soit petit.

D. Poitrinal:

I would like to mention another application of the deconvolution procedure that we have presented here, it is dealing with the loss of information that is usually the fate of all direct measures of the repartition of a physical parameter. Speaking in terms of deconvolution, the input function would be here the actual repartition of the physical parameter considered, and the output would be the set of measures given by the measuring device.

In the first step, using experimental conditions where the actual input is known, deconvolution of the measured output by this input gives the unit response of the device.

The second step, corresponding to the normal use of the device, is to deconvolute the measured output by this unit response to get the input, i.e. the actual repartition of the parameter. By this technique, the quality of the information given by the measuring device is quite improved, without any other assumption on the characteristics of the device, but linearity.

We have applied this to the measurement of the activity of each layer of an aquifer in a well, during the injection of a radioactive tracer, with the final aim of getting permeability and dispersion coefficients.

F. V. Appleby:

It is necessary to correct and to comment on the reporter's summary as follows;

(1) *that 'functions of time only are considered'.* The mathematical discussion on the space of runoff sources (defined in relation to the rainfield space distribution) forms a central theme of my paper. It is demonstrated in the outcome that the nature of the P_2 runoff fluctuation model may be considered to be homogeneous as to E_n space and isotropic as to its phenomenological properties in the n directions of that space. Thus, a physical interpretation and a precise meaning is established for the 'lumped' space distribution parameters. It was Einstein (1956) who showed that a Brownian motion (P_2) system was, in effect, a physical realization of the gaussian in which the variance of displacements in space depended on the product of a 'lumped' isotropic diffusivity with the sampling time t. Thus, the P_2 model defines a stochastic process as to space migrations of runoff from distributed sources. The success of the application of this model to a wide range of conditions offers a clear demonstration 'that large-scale random phenomena in their collective action create strict non-random regularity (Gnedenko and Kolmogorov, 1954, 1968).

(2) *that 'the input, or rainfall, is considered 'compact and uniform'... characterized by ... a time function'.* My use of the terms 'compact and uniform' (vol.2, p.673) refers to rain-*fields*, i.e. the runoff source space. Considered to be compact in the sense that it is not disjointed in that space, and uniform in the sense that the layer dimensionality, n_i, may be approximated (without loss of generality in the argument) by means of a single number, n, for all m layers. The input, as rainfall measured at some representative point, is not restricted as to variability in time. In general the m layers may all be different as to their rainfall content. See Figs. 4. 5 and 7 (vol.2, pp.679, 680).

(3) *The reporter passes over the claim of 'nonlinearity' in the title.* An appreciation of the nonlinear association of the runoff fluctuations with the carrier field and with its own developing 'layers' is essential for a proper understanding of the principles of construction of the model. The linear superposition principle was abandoned at the outset, and thus presented at once a challenge to a long established school of thought in this field. This nonlinear development was made possible by means of a direct and general attack on the nature of the function space for runoff fluctuating models. It is seen that the 'canonical' model (6) is not an integral equation, so that the 'black-box' (input system function) inversion problem does not arise. The input is transferred to

the layer space development and the fluctuation system *as a whole* is seen to possess, as to its collective action in space and as to its layer development in time, the properties of an influence function or Green's function which is not dependent on a linear convolution operation *in time* for its definition. The zero of this collective influence function is located near to the mass centre of the effective input (as a time sequence) so that the rising limb may be filled out in detail, by a linear convolution (composition) operation (in time) with the collective P_2 Green's function.

REFERENCES

Einstein, A. (1956) *Brownian Movement*: Dover Publications.
Gnedenko, B. V. and Kolmogorov, A. N. (1954, 1968) *Limit Distributions for Sums of Independent Variables*: Addison-Wesley.

Closing remarks by the general reporter

I congratulate the speakers for their excellent presentations. In particular, I wish to refer to Becker's remarks on the meaning of the distribution of parameters in hydrological models. The tabulation he has presented is particularly useful because it shows, in a very orderly and logical fashion, the transition between the mathematical treatment of variables with continuous space and time distributions (scalar or vector fields), through discretized approximations of space—time variability, to lumped representations (time functions only).

I consider myself very fortunate in having been exposed to the varied approaches given by the authors to these problems. These presentations cover the entire spectrum of modes of analysis which Becker has to elegantly summarized.

General Reporter   Dr. Boleslaw Kordas, Associate Professor of Hydraulics and
Hydrology, Cracow Polytechnic University, Cracow, Poland

Chairman   Dr. Slavoljub Jovanović, Professor at the 'J.Cerny' Institute for Water
Resources Development, Belgrade, Yugoslavia

Modèles hydrodynamiques pour les systèmes fluviaux, les lacs et les estuaires

Boleslaw Kordas

Depuis quelque temps, nous observons le renouvellement d'une liaison intime entre les
deux branches de la science de l'eau: l'hydrologie et l'hydraulique.

Je crois que le pas décisif dans ce processus de rapprochement a été franchi au cours
du Symposium de Fort Collins (1967) où une large représentation d'hydrauliciens.
groupés dans l'Association Internationale de Recherches Hydrauliques. a participé à un
symposium purement hydrologique. A cette occasion. ils ont pu constater que les deux
domaines scientifiques en question deviennent de plus en plus complémentaires et que
les méthodes développées dans l'un peuvent être fructueusement appliquées dans
l'autre.

Qu'il me soit donc permis de vous présenter les sept rapports dont les auteurs traitent des divers problèmes qui intéressent le thème ci-dessus mentionné:

1. *Robert J. C. Burnash and R. Larry Ferral*   A generalized streamflow simulation
   system
2. *I. I. Krashin and D. I. Peresunjko*   Modelling groundwater resources replenish-
   ment due to river flow in the flood period during the operation of infiltration water
   intake
3. *J. Geringer*   Etude de l'écoulement nonpermanent dans un canal découvert
4. *Hanna Witkowska*   Mathematical model of the river-bed erosion below a dam
5. *G. F. Pinder and J. D. Bredehoeft*   Groundwater chemistry and the transport
   equations
6. *L. A. Roesner, W. R. Norton and G. T. Orlob*   A mathematical model for simulating the temperature structure of stratified reservoirs and its use in reservoir outlet design
7. *J. P. J. O'Kane*   A kinematic reference frame for estuaries of one dimension
   Les deux premiers de ces rapports (par Burnash et Ferral, et Krashin et Peresunjko)
   concernent l'influence du bassin versant (y compris sa partie souterraine) sur le
   régime d'un cours d'eau. Les deux autres rapports (par Geringer et Mme Witkowska)
   concernent les phénomènes hydrodynamiques qui se produisent dans le lit même
   d'un cours d'eau.

Les trois rapports qui restent (par Pinder et Bredehoeft, Roesner *et al.*, et O'Kane)
concernent des sujets particuliers.

En poursuivant le même ordre, je voudrais vous présenter le plus rapidement possible les sujets particuliers, les méthodes de solutions adoptées et les résultats atteints par les auteurs des rapports.

Le rapport de Burnash et Ferral représente un intérêt particulier par sa conception rationnelle et logique d'un modèle mathématique d'un système riverain.

Le but défini par les auteurs consiste à créer un modèle tel, qu'il permette de faire une bonne prévision du régime des cours d'eau. Pour atteindre ce but, ils analysent successivement les facteurs qui influencent le débit du fleuve.

Ils supposent que la partie souterraine du bassin versant est constituée par deux couches superposées qui communiquent entre elles. Ces deux couches assurent lacréation du débit de base dans le fleuve. L'écoulement direct provient d'une pluie qui tombe sur les parties imperméables de la surface du terrain. L'écoulement superficiel est créé par l'excédent de l'eau de précipitation qui dépasse la capacité de rétention de la couche supérieure. Cette couche s'épuise, d'une part par évapotranspiration (la zone supérieure non saturée), d'autre part par drainage latéral de la zone saturée. dont les eaux s'écoulent vers le récipient, et enfin par la percolation verticale vers la couche inférieure. Cette dernière, dans la zone non-saturée, s'épuise par l'évapotranspiration tandis que les ressources emmagasinées dans la zone saturée alimentent le débit de base du fleuve.

Les résultats définitifs de l'analyse sont donnés sous la forme d'un bloc-diagramme qui représente le schéma du fonctionnement d'un bassin versant permettant de transformer la pluie en débit et donne les formules qui permettent de calculer successivement les éléments de cette transformation.

Les auteurs annoncent que le modèle proposé a été appliqué à plusieurs bassins versants en donnant d'excellents résultats. Dans leur rapport, ils citent les deux exemples des bassins versants situés dans la partie ouest des Etats-Unis, qui se caractérisent par un régime intermittent. Les hydrogrammes inclus dans le rapport démontrent une excellente concordance de la prévision et de la réalisation.

Je crois donc que, grâce à la méthode proposée par Burnash et Ferral, le service des Prévisions Hydrologiques des Etats-Unis dispose actuellement d'une arme scientifique de haute qualité.

Le rapport de Krashin et Peresunjko présente un problème complexe de l'hydrodynamique souterraine. En se basant sur un cas réel, concernant la vallée du fleuve Serga (Oural), les auteurs analysent le fonctionnement d'un système hydraulique constitué par:

(1) le fleuve lui-même,

(2) la nappe alluviale voisine, constituée par deux couches anisotropes et non homogènes, alimentées par le fleuve au cours d'une crue,

(3) le captage, constitué par un système de puits, placés dans la nappe à proximité du fleuve.

Le captage fonctionne continuellement avec un débit constant, dépassant deux fois le débit minimal du fleuve. Ceci provoque des variations périodiques des ressources des eaux souterraines qui s'épuisent au cours de la saison sèche et se réalimentent au cours de la crue.

Pour résoudre le problème ainsi posé, les auteurs formulent tout d'abord un système d'équations différentielles aux dérivées partielles qui décrit le phénomène en question.

La solution de ce système d'équations est obtenue avec un modèle électrique type R. C. constitué par les deux réseaux résistances et capacités, dont le premier représente la nappe alluviale elle-même et le deuxième la nappe qui l'entoure.

Les conditions aux limites adoptées pour le modèle correspondent aux résultats de l'analyse hydrologique du système.

Les résultats définitifs des recherches sont présentés graphiquement sur un dessin donnant les variations du niveau de l'eau souterraine dans certains puits en fonction du temps.

Dans leur conclusion finale, les auteurs soulignent la grande valeur de la méthode choisie pour la prévision du régime des eaux souterraines influencées par le fleuve et perturbées par le fonctionnement simultané d'un captage d'infiltration.

Geringer, dans son étude, s'attaque à un problème classique de l'hydrodynamique des cours d'eau. En supposant un canal linéaire à pente fixe, il décrit la propagation d'une onde de crue dans un tel canal avec le système des équations de St. Venant. La solution des équations est obtenue par la voie numérique. En particulier, l'auteur utilise ici la méthode des différences finies avec un schéma de calcul dit explicite proposé par Lax.

Mais, ce qui attire surtout l'attention du lecteur, ce n'est pas tellement la solution elle-même que l'analyse de l'influence des paramètres physiques du cours d'eau sur la transformation de l'onde de crue. En particulier, l'auteur présente graphiquement les résultats des calculs concernant l'influence exercée sur l'onde par les variations du coefficient de rugosité du lit et de la profondeur initiale de l'eau.

Le rapport de Mme Witkowska représente un modèle mathématique de l'érosion du fond d'un cours d'eau en aval d'un barrage. Le problème est important parce qu'une telle érosion dégrade considérablement le lit déjà régularisé.

En se basant sur l'hypothèse que la construction d'un barrage perturbe l'équilibre de transport des matériaux, l'auteur analyse tout d'abord les différentes schémas hydrauliques correspondants du phénomène étudié. Ensuite, compte tenu d'un cas réel qui l'intéresse, elle décide de choisir un modèle d'un écoulement graduellement varié. Dans ce modèle, le phénomène étudié est décrit par un système de trois équations dont la première est l'équation même de l'écoulement graduellement varié, et les deux autres concernent le transport du matériel. Comme équation de transport, l'auteur a choisi la formule de Meyer-Peter.

La combinaison des trois équations mentionnées a permis à l'auteur d'arriver à une équation différentielle décrivant la position instantanée du lit en fonction du temps et des paramètres physiques tels que les caractéristiques géométriques du lit et les caractéristiques du matériel transporté.

L'intégration de cette équation a été effectuée par une méthode numérique et les résultats des calculs ont été comparés aux résultats des mesures réalisées sur le terrain pour un cas réel d'érosion du lit de la Vistule Supérieure à l'aval d'un barrage situé à Przewoz (région de Cracovie).

Cette comparaison a montré une bonne concordance, ce qui prouve que la méthode proposée par l'auteur peut être fructueusement appliquée comme méthode de prévision dans les cas analogues.

Le rapport de Pinder et Bredenhoeft constitue une sorte de revue des possibilités offertes par l'application des méthodes numériques et des ordinateurs de la troisième génération dans la solution des problèmes classiques de l'hydrologie.

Bien que la majeure partie du rapport soit consacrée aux problèmes de l'hydrodynamique souterraine, les auteurs commencent leurs considérations par l'analyse du problème de l'écoulement dans les canaux découverts. A ce sujet, ils citent les équations du bilan de masse et de moments proposées par Frederickson et Bird (1961) et constatent que ce couple d'équations peut trouver une excellente application dans la solution des modèles undimensionnels de la propagation de l'onde de crue et des estuaires.

Dans le domaine des écoulements des eaux souterraines, les auteurs considèrent tout d'abord le problème général du transport dans un milieu poreux. Ils citent à ce propos les équations correspondantes proposées par Raats et Klute (1968) et constatent que de nombreux problèmes quantitatifs de l'hydrologie souterraine peuvent être fructueusement décrits avec les formes simplifiées de ces équations. Par la suite, ils analysent successivement les possibilités offertes dans ce domaine par les méthodes numériques.

Ils commencent par le problème de l'écoulement bi- et tri-dimensionnel des eaux souterraines dans la zone saturée, en soulignant la valeur de la méthode des différences finies (méthode itérative avec l'accélération du processus de relaxation) et la méthode des éléments finis basée sur le principe variationnel proposé par Galerkin.

Ils passent ensuite au problème du transport où, dans le cas le plus général, certaines phases peuvent changer ces caractéristiques au cours de l'écoulement (cas de la dissolution chimique).

En citant des équations du transport, les auteurs soulignent qu'il s'agit ici de la superposition de deux (ou plusieurs) champs physiques accouplés, dont l'un est le champ de filtration et l'autre le champ du facteur physique étudié. Ils constatent que dans la plupart des cas, où le transport par convection est prédominant, le problème se réduit à un problème hyperbolique où la méthode de caractéristiques peut être appliquée avec profit.

Le rapport renferme aussi des considérations générales sur l'influence réciproque des eaux souterraines et des eaux de surface, ainsi que les possibilités offertes par l'application de l'équation du bilan d'énergie.

Dans leur conclusion finale, les auteurs soulignent la valeur des modèles basés sur la théorie de la continuité et constatent qu'il reste toujours à résoudre le problème le plus général, celui du transport avec les réactions chimiques simultanées. Dans la solution de ce problème, les méthodes numériques peuvent démontrer toute leur valeur.

Roesner *et al.* présentent dans leur rapport un modèle mathématique d'un bassin de retenue thermiquement stratifié. Ce modèle suppose la division de la profondeur du bassin par les tranches horizontales d'une hauteur uniforme. Les auteurs se basent sur l'hypothèse que l'écoulement de l'eau dans le réservoir se réalise horizontalement, mais que l'échange thermique s'effectue uniquement dans le sens vertical. Ils donnent ensuite les équations du bilan de masse et du bilan thermique.

Dans ce dernier, ils tiennent compte de la radiation solaire (ondes courtes), de l'advection verticale, de la diffusion verticale, ainsi que de l'échange thermique avec le contour du réservoir. Finalement, ils donnent une équation de l'énergie thermique pour une tranche quelconque du réservoir sous la forme des différences finies.

La solution du problème consiste en la solution simultanée du système d'équations correspondantes à toutes les tranches compte tenu des conditions de la concordance sur les limites des tranches. Cette solution est obtenue par l'application de l'algorithme de Thomas.

Pour les cas concrets, il est nécessaire d'introduire les caractéristiques géométriques du réservoir, des données météorologiques (température de l'air, vitesse du vent, couverture glaciaire, pression atmosphérique) ainsi que les données hydrologiques telles que le débit et la température à l'entrée du réservoir et la façon de sa gestion.

Le modèle a été fructueusement appliqué pour le calcul du régime thermique des nombreux bassins de retenues situés aux Etats-Unis. Dans leur rapport, les auteurs citent l'exemple du réservoir de Dworshak.

Le dernier rapport que j'ai le plaisir de vous présenter est celui de O'Kane sur le cadre de référence cinématique pour les estuaires undimensionnels. Le rapport est basé sur une idée originale concernant le remplacement de la rive de l'estuaire, considérée habituellement comme un cadre de référence, par un cadre cinématique.

Ce cadre oscille tout en maintenant un volume d'eau constant entre un observateur situé sur lui-même et la tête de l'estuaire. L'auteur présente les équations de transformation d'un cadre traditionnel en cinématique et examine l'équation de la diffusion.

Le rapport contient des considérations analytiques fort intéressantes concernant l'analyse harmonique de la houle. Ses conclusions peuvent être appliquées avec profit dans les calculs hydrodynamiques des estuaires undimensionnels.

Ainsi, j'en ai terminé avec la présentation générale des rapports appartenant à notre groupe. Je crois que ces sept rapports sont bien représentatifs de toute la diversité des sujets et des méthodes appliquées et qu'ils permettent de se faire une idée sur l'utilité d'application des modèles hydrodynamiques en hydrologie.

Mais je crois que nous sommes parvenus à un niveau si élevé de la liaison des deux branches en question que nous pouvons nous pencher sur une synthèse des méthodes appliquées par elles.

Pour être plus précis, je vais vous présenter un exemple et, puisqu'il s'agit ici des modèles riverains, je vais choisir mon exemple dans ce domaine.

Supposons une onde de crue. Pour un hydrologue moderne, une onde de crue dans une section donnée constitue un phénomène stochastique dont les caractéristiques peuvent être calculées sur la base des données historiques existantes.

Pour un hydraulicien étudiant la transformation de l'onde de crue de long d'un tronçon de fleuve, le problème se ramène à la connaissance de l'opérateur bien déterminé (par exemple les équations de St. Venant) et les caractéristiques de l'onde à la section initiale du tronçon.

Si nous supposons maintenant que ces caractéristiques portent un caractère stochastique, nous arrivons à la solution d'un problème de la transformation d'une 'onde de crue stochastique' par un opérateur déterministe.

En résultat, nous obtenons à la sortie du tronçon étudié du fleuve, une nouvelle 'onde de crue stochastique' transformée par rapport à sa forme initiale, par l'intermédiaire de l'opérateur déterministe en question.

Nous avons donc un problème classique de l'analyse des systèmes avec l'entrée stochastique donnée, l'opérateur de transformation bien connue et où l'on cherche les caractéristiques stochastiques de la sortie.

Je crois qu'une telle façon de poser le problème synthétise les points de vue des hydrologues et des hydrauliciens et permet de créer toute une gamme de solutions nouvelles dont l'utilité pour la prévision hydrologique me semble incontestable.

Qu'il me soit donc permis de proposer qu'un tel sujet figure au programme de notre prochain Symposium consacré aux modèles mathématiques en hydrologie.

BIBLIOGRAPHIE

Frederickson, A. G. and Bird, R. B. (1961) Transport phenomena in multicomponent systems. In *Handbook of Fluid Dynamics* (edited by V. L. Streeter), 56 pp.: McCraw-Hill. New York.
Raats, P. A. C. and Klute, A. (1968) Transport in soils: the balance of mass. *Soil Sci. Soc. of Am. Proc.* **32**, No.2, 161–166; Transport in soils: the balance of momentum. *Soil Sci. Soc. of Am. Proc.* **32**, No.4, 452–456.

DISCUSSION

R. Larry Ferral:
I wish to thank Professor Kordas for his clear and very generous discussion of our paper.

I would like to elaborate briefly on his report. The model we propose is a deterministic lumped parameter model with a flexible and varying time scale. Its primary purpose is for use in short-term river forecasting, but we hope also to use it with stochastic inputs for seasonal or water-year predictions. We developed it because we needed a continuous flow forecasting model to replace the antecedent precipitation index system that we are now using, and we were not satisfied with any existing model of which we were aware. We wanted a model that was simple but accurate in reproducing the total hydrograph. We wanted it to be physically realistic, so that we could be confident that it would not display some strange quirk during a flood forecasting emergency. We wanted a system in which most of the parameters could be estimated analytically before beginning an optimization routine.

Some small changes have been made in the model since the paper was written. The percent of impervious area now consists of a constant term until upper-zone tension water storage has been satisfied, then a term which varies with the square of the relative wetness of lower-zone tension water storage. This rather strange relationship may actually represent area variations in moisture storage capacities in the basin, the sealing of wetted clay sub-soils, or the development of seepage outflow areas which thus become impervious.

The model has been fitted to seven widely varying basins in California near the Pacific Coast, one in Maryland near Washington, DC and one in the Mississippi Valley in the state of Oklahoma. The Office of Hydrology of the National Weather Service in Washington, DC is now testing this model in comparison with a version of the Stanford model and two other National Weather Service models in a programme designed to determine which model should be recommended for river forecasting. Testing has been completed on two out of six test basins using a hill climbing optimization procedure with standard error of mean daily flow as the objective function. This model has performed better than any of the others in these two basins.

George F. Pinder:

Earlier this afternoon I proposed looking at the problems of hydrological simulation within the framework of continuum mechanics. In the discussion of my paper I would like to clarify this rather nebulous concept with an example.

This field problem involves the movement of contaminated ground water from two source areas toward the water supply wells of Brunswick, Georgia. The contaminating water enters the aquifer with a concentration of approximately 2000 mg/l. of chloride. The resulting concentration distribution of chlorides in the aquifer is the result of convective and dispersive transport.

In the most general case the simulation of a hydrological system requires the simultaneous solution of the equations of mass, energy, and momentum transport. In this case, however, the problem could be simplified by assuming an isothermal system and by considering only the two-dimensional aerial plane.

As can be seen from the figures, we were able to simulate the history of contamination and to predict the movement of the contaminated water by solving the equations of groundwater flow and mass transport. It was also possible to use the mathematical model to test the effectiveness of various remedial schemes. As shown in the figures, the introduction in 1972 of an interceptive pumping well in the aquifer between the source of contamination and the pumping centre significantly improves the quality of the pumped water by the year 1975.

H. Witkowska:

I should like to express my thanks to the general reporter for the excellent presentation of my work.

I shall make only two comments.

(1) I want to stress that in my work I did not claim to build a universal model which, I believe, cannot be constructed for the moveable river bed problems. I tried only to discuss all the parameters which have to be taken into account before choosing the model and I prepared the model for the conditions of the Upper Vistula River. I think that such a model in every case has to be based on carefully prepared field data. In my paper it was maybe not too clearly shown that in our work we checked the few diameters concepts, flows equations, different bed-load formulae and some other para-

meters. The wrong choice of only one parameter involved gave at least a 100 per cent error.

(2) The second comment concerns the general reporter's suggestion on the possibility of stochastic input to the deterministic operator. I think that in fact all models concerning moveable river beds should be stochastic but, as far as I know, stochastic bed-load theories ready for application to other sediment problems do not exist. So it rests meanwhile with the possibility of stochastic input concerning flow.

J. P. J. O'Kane:

Two further properties of kinematic reference frames are

$$\text{(a)} \frac{\dfrac{\partial}{\partial z'}\left[\dfrac{\partial z'}{\partial t}\right] = \dfrac{\partial}{\partial z}\left[\dfrac{\partial z'}{\partial t}\right] \cdot \dfrac{\partial z}{\partial z'} = \dfrac{\partial}{\partial t}\left[\dfrac{\partial z'}{\partial z}\right]}{\dfrac{\partial z'}{\partial z}}$$

This says that the rate at which the velocity of the bank varies along the kinematic frame measures the percentage change in the coefficient of distortion per unit time or the percentage rate of change of length per unit length of the moving bank. A similar result holds for an observer on the bank and is due to Kaplan (1962)

$$\text{(b)} \frac{\partial z}{\partial t} = - \frac{\partial z'/\partial t}{\partial z'/\partial z}$$

This follows from result (7). If a kinematic frame has unit coefficient of distortion then the velocities of the bank and the frame are equal and opposite and are functions only of time [from property (a)]. Hence the tidal excursions, which may be regarded as integrals of the velocities starting from the half-tide position, are also equal, opposite in sign and invariant with position. A particular case of this is the exponential estuary dealt with in the body of the paper. The Thames estuary appears to be of this kind.

The application of deterministic mathematical models to environmental systems is strengthened when a definite methodology is stated at the start. The following steps might form such an approach and may be taken as comment on the papers of topic 6.

(1) Statement of the criterion of fit as dictated by model use.
(2) Application of a split record test to validate the model.
(3) The analysis of the residuals between the predicted and observed response of the system using the techniques of time series analysis. (This suggests using white noise residuals as a criterion of fit.)
(4) Modification of the model to improve the structure of the residuals.
(5) So you understand the model, now do you understand nature?

W. Parzonka:

Je voudrais aborder le problème des modèles, tenant compte des transports solides. J'ai feuilleté soigneusement les Comptes Rendus du Symposium et j'en ai trouvé deux seulement: celui de M. Nippes concernant le transport de sédiments en suspension et celui de Mme Witkowska concernant l'erosion du lit en aval du barrage, d'un tronçon de la Vistule. Alors on peut constater qu il y a un certain manque di pourrait être éliminé dans le Symposium suivant. On le sait pourtant bien que les problèmes concernant l'hydrologie fluviale devraient prendre en consideration le transport solide — aussi bien celui de l'erosion que celui de l'accumulation. Naturellement ce problème est beaucoup plus complexe, beaucoup plus difficile que celui qui ne tient pas compte des transport solides, mais la réalité est telle que parfois le transport solide peut d'une façon visible déformer les modèles admis ne tenant pas compte de ce phénomène.

De ce point de vue, je voudrais souligner que les deux travaux cités sont très nécessaires. Je voudrais aussi propose que le symposium suivant de l'Association prenne en consideration peut-être au moins un terme consacrés au modèles mathématiques avec transport solide.

En ce qui concerne les détails je voudrais souligner surtout le rôle énorme du travail de Mme Witkowska, qui est arrivée à vérifier quelles sont les formules de transport solides qui sont applicables pour la Vistule. On sait bien que dans la plupart de cas on admet l'application des formules telles ou autres, mais on n'essays pas de le vérifier. Mais ici, sur la base des trauvaux soigneusement menés et sur la base de vérification sur modèle on a obtenu le résultat intéressant que pour la Vistule (en tout cas pour le cas étudié du tronçon Przewsóz-Nawady) la formula de Meyer—Peter—Müller est la meilleure. On peut donc tirer une conclusion pratique très intéressante pour les ingénieurs et pour d'autres spécialistes, par ex pour ceux qui élaborent le plan de la Vistule dont on parlera samedi. On sait maintenant que pour la Vistule supérieure la formule de Gontcharov de vas pas du tout et que au contraire la formule de Meyer—Peter—Müller décrit bien la realite (si on peut décrire bien un problème si compliqué comme celui des transports solides . . .) D'Autre part je voudrais souligner, que c'est aussi un des premiers travaux qui a introduit la variation du diamètre des grains avec la longeur du tronçon étudié. Actuellement dans la plupart des travaux on admet que le diamètre des grains d'est constant. Du resultat de Mme Witkowska on peut donc tirer aussi une conclusion pratique intéressante comment varie d'en aval des barrages. Ce sera entre autre un problème pour les autres parties de la Vistule, ou on prévoit la construction des barrages. Ici je démanderai l'avis de Mme Witkowska est ce qu'on peut estimer sur la base de ses études la variation de la profondeur du lit et celle de diamétre des grains dans d'autres tronçon de la Vistule que le tronçon étudié.

Les remarques finales par rapporteur général

Je voudrais ajouter quelques petites remarques à ce qui j'ai dit dans mon rapport général. Tout d'abord je suis très reconnaissant à Ferral, O'Kane ainsi qu'à Mme Witkowska, il aurait peut être fallu que je la cite la première, pour les remarques qui enrichissent considérablement le rapport général. Comme a dit M. le Président les slides que vous avez vus au cours de la discussion étaient l'arme secrète de nos auteurs. Ce choses ne se trouvent pas dans leurs rapports, mais je crois que leur présentation ici a permit d'eclaircir des problèmes étudiés. Je voudrais tout de même ajouter quelques remarques concernant tous les rapports ici présentés. La première concerne l'analyse très soigneuse des paramètres physiques qui sont considérés comme éléments du modèle. Les hydrauliciens sont bien habitués à l'analyse des paramètres, donc ce n'est pas quelque chose d'inattendue, néanmoins c'est l'effet qui doit être souligné ici. Ce ne sont pas n'importe quels modèles, ce sont des modèles qui sont basés sur une étude qui précède leur construction. Et c'est cette calibration et cette identification des paramètres qui constitue un problème aussi important que la construction elle-même. Je voudrais souligner que tous les auteurs des rapports ont consacré une bonne partie de leur travail pour cette analyse des paramètres. La deuxième question que je voudrais souligner est celle présentée à la fin de mon rapport. Moi je suis hydraulicien. Il faut vous dire que les hydrauliciens ne se posent pas le problème de la prévision. Les problèmes classiques de l'hydraulique ce sont certaines généralisations des résultats d'expériences ou un travail théorique qui don-

nent certaines formules qui permettent d'écrire ainsi bien que possible le phénomène étudié.   Mais la prévision dans le sens employé par les hydrologues ne se pose
pas en général en hydraulique, et si nous voulons la poser, il nous faut employer
l'entrée stochastique, il n'y a pas de doute.   Et pourtant on ne trouve pas jusqu'au
present un nombre énorme de solutions faites par les hydrauliciens qui utilisent une
telle entrée.   Et je crois que c'est un probléme très important pour le developpe-
ment future de l'hydraulique-même.   Je crois que c'est également un problème
important pour les hydrologues, parce que ceux-ci emploient dans leurs analyses les
opérateurs relativement simples, tandis que l'hydrauliques a développé déjà un certain nombre d'opérateurs qui sont très complexes et très riches en possibilités.   Je
crois donc qu'il s'ouvre ici un champ de travail en commun.   Et ensuite le dernière
remarque, plutôt personnelle.   Je ne sais pas certain si j'ai réussi ce que je voulais
effacer le rapporteur général dans mon rapport.   Je supposais que le rapport ce n'est
pas la présentation de points du vue du rapporteur général, mais que c'est la présentation la meilleure possible des idées, des méthodes et enfin des résultats atteint par
les auteurs.   La deuxième chose que je voulais faire, et je ne sais pas non plus dans
quelle mesure j'y ai réussi c'est de vous présenter non pas tellement une synthèse,
parce que c'est très difficile, sinon impossible, compte tenu du sujet qui m'a été
proposé de traiter, mais plutot une large gamme de possibilités, une énorme variétés
de possibilités, offertes aux hydrologues par les méthodes fructueusement appliquées
en hydraulique.   Et je crois qu'il serait très profitable, si notre Comité des Modèles
Mathématiques soutient cette proposition qui dit qu'à chaque Symposium d'hydrologues il y a un sujet où les hydrauliciens pourront présenter leurs propositions concernant le développement futur de l'hydrologie.

Session XV   29 July 1971

General Reporter   Dr. Chester C. Kisiel, Professor, Department of Hydrology
and Water Resources, University of Arizona, Tucson, Arizona, USA. This report is
being published posthumously, as, regrettably, Dr. Kisiel's sudden death occurred on
5 November 1973.

Chairman   Dr. Ulrich Maniak, Leichtweiss Institute for Water Research, Technical University of Brunswick, Brunswick, GDR

Objective function and constraints in water resources systems

Chester C. Kisiel

The papers reviewed and discussed in this general report are as follows. (All are
published in volume 2 of the Symposium Proceedings.)

1. *M. Domokos*   Indices of water restriction and water deficiency tolerance
2. *M. Sugawara*   Water resources and negentropy
3. *K. Tiemer*   On large-scale simulation of groundwater flow systems
4. *Lucien Duckstein and Chester C. Kisiel*   Collective utility: A systems
   approach to water pricing policy

Introduction
The careful examination of objective functions and constraints is at the heart of the
collective efforts to evolve a science of decision making about water resource systems
or any other system. To place in overall perspective the set of papers assigned to
topic 7, a critical survey of the subject is presented. Within the prescribed space
limitations, no such review can be all-encompassing or complete in its acknowledge-
ment of the vast literature on the subject.

System modelling and objective functions
General systems theory (Simpson *et al.*, 1975; Duckstein and Kisiel, 1968) provides
a useful framework in which to conceptualize the interrelation of objective functions
and constraints. The system model is conceived in terms of the usual input of $X(t)$,
state $S(t)$, and output $Z(t)$. Let time be discrete $t = 0, 1, 2, \ldots$, days (months or
years) depending on the problem. Let the input $X(t)$ be chosen out of an input
alphabet: for example

$$X(t) = \{Q'(t), C'(j, t), \theta'(j, t), G'(j, t), P'(j, t), \ldots\}, \qquad j = 1, 2, \ldots, J \tag{1}$$

in which j is the site index (J sites or numbers of points studied);

$Q'(j, t)$ = net water inflow ($\text{m}^3 \text{ day}^{-1}$);
$C'(j, t)$ = pollutant flux ($\text{g day}^{-1} \text{ m}^{-3}$);
$\theta'(j, t)$ = heat flux ($\text{cal day}^{-1} \text{ m}^{-3}$);

$G'(j, t) =$ geological, geomorphic or topographical information (type of rock, dispersion, storage coefficient, transmissivity, mean basin slope, roughness);

$P'(j, t) =$ pricing policy or interest rates, schedule of operations (power production, low flow releases, pumpage from an aquifer, recharge to an aquifer, . . .), air and water quality standards, forecasts of population growth, water demand, waste loads, technological change and changing social use of the water resource structure, aesthetics.

In other words, the input alphabet consists of controllable or uncontrollable inputs, natural or human inputs and other constraints. The input alphabet should include, for complete system definition, known system interfaces and socio-economic externalities. Let the state $S(t)$ be a set defined by

$$S(t) = \left\{V(j, t), Q(j, t), C(j, t), \theta(j, t), G(j, t), \ldots\right\}, \tag{2}$$

in which

$V(j, t) =$ velocity of flow (m day^{-1});
$Q(j, t) =$ volume of water in given material element (m^3);
$C(j, t) =$ pollutant concentration (g m^{-3});
$\theta(j, t) =$ temperature ($^\circ$C);
$G(j, t) =$ geological information.

Economical and social states may also be included. For example, interest rates may change as a consequence of a decision to conserve groundwater for future use based on prior results of system operation. The above variables in the state set (2) are updated as more geological, input, social or economic information on $X(t)$ are obtained.

In order to describe how input is utilized to update information on states, let the state transition function F (in discrete time) be defined as

$$S(t + 1) = F\left\{X(t), S(t)\right\} \tag{3}$$

Implied is the state at all J points in space. This deterministic relation F requires knowledge only of the present state $S(t)$ and present input $X(t)$; of course, to compute $S(t + 1)$ at the next discrete time $(t + 1)$ requires a well defined F, including where necessary, a coupling of the social, economic, biological, chemical, thermal, and hydraulic behaviour of the system. In equation (3), an example of one component of F is mass balance

$$Q(j + 1, t) = Q'(j, t) + Q(j, t) \tag{4}$$

or present volume = net inflow plus previous volume.

Continuing with the system model, let the output $Z(t)$ be chosen out of an output, such that

$$Z(t) = G\left\{X(t), S(t)\right\} \tag{5}$$

in which G is the output function. The output $Z(t)$ may be physical, economic, social or information (negentropy). Note that the name given to G varies with the writer and the orientation of the problem. Beard (1971) uses value function so as to include non-commensurate or non-quantifiable values. In decision theory the loss function defines the social or economic loss assignable to errors of overestimation and underestimation. Other names are error, penalty, criterion, objective, cost, ranking utility or goal function.

For example, a physical output at time (t) is the flow volume at one point j over $\Delta t = 1$

$$Z(t) = Q(j, t) \tag{6}$$

or at all joints j in J over the time interval Δt

$$Z(t) = \sum_{j \text{ in } J} Q(j, t) \tag{7}$$

Note that $Q(j, t)$ is an element of the state set; thus $Z(t)$ may be $S(t)$ if desired. The problem to be 'solved' determines the richness or complexity of F and the nature of $Z(t)$.

An economic output requires the specification of an objective function G for a management problem such as the cost of treating polluted river of aquifer water; for example,

$$Z(t) = p(1)C(t, 1) + p(2)C(t, 2) + \ldots = \sum_{j} p(j)C(t, j) \tag{8a}$$

in which $p(j)$ = cost of removing pollutant at jth location in units of \$/gram, or

$$Z(t) = q(1)C'(t, 1) + q(2)C'(t, 2) + \ldots = \sum_{j} q(j)C'(t, j) \tag{8b}$$

in which $q(j)$ = cost of removing pollutant flux from the jth input in units of (\$/gram) $\times$ unit time of \$-day per gram. Physically, the system outputs are concentrations of pollutants but economically the outputs must be expressed in a monetary metric if costs are to be minimized, or in a common value metric if aesthetic aspect of pollutants are to be assured.

Generally speaking, the economic output forms the most common class of objective functions used in water resources systems analysis. To transform equations (8a) and (8b) into objective functions, one need only specify $C(t, j)$ and $C'(t, j)$ in terms of a decision variable such as the size of treatment unit. Depending on the way the problem is structured, the objective function G may be scalar or a vector, linear or nonlinear, and subject to constraints that are linear or nonlinear, deterministic or stochastic, and convex or nonconvex. The state transition function F is one of the constraints on the optimization of G. F may be:

(a) uniquely defined (that is, it is deterministic and well posed in terms of the identification of the model's parameters),
(b) non-uniquely defined in terms of its parameters,
(c) defined only as a system simulation, for example, as a spatial configuration of existing hydraulic structures (Beard, 1971),
(d) stochastic, and expressible only as a transition probability matrix (for example, storage levels in a reservoir),
(e) uncertain and not expressible in any of the above forms.

The general systems model, if properly constructed, should clearly identify what is unknown about the inputs, and other externalities, state transitions, and outputs.

This includes what is not quantifiable but which may be quantifiable in the future.

The form of G and F determines the method of solution. These methods undoubtedly form the substance of many of the papers given at this Symposium and have been reviewed by Buras (1974) for the IUGG General Assembly. For deterministic F, both mathematical programming and simulation are employed (Beard, 1971) even with stochastic inputs. Simulation is the only apparent computationally feasible method if a stochastic F must be defined. The calculus of variations is of value to optimal control problems but is of limited value for large-scale and complex water resources problems. Simulation is the only apparent method for simultaneously coping with optimization in space and time or for handling the combined effects of horizontal and vertical externalities; too often, either space is 'lumped' or time is 'lumped' to permit solution. A major problem with simulation results is the lack of confidence in the optimality of the solution even if techniques of experimental design (analysis of variance) are employed, such techniques are expensive on the computer. An important challenge is the concurrent optimization of designs and operating rules; often, one or the other is fixed to facilitate the evaluation of the goal function. Generally speaking, analysts tend to structure the objective function according to the criterion of computational feasibility rather than to the criterion of correspondence to the real world problem. To the extent that many constraints, state transition functions and objective functions are well-defined or known, might our confidence in the optimal solutions be strong. On the other hand. if such prior knowledge is not available, then varying degrees of risk must be taken in the use of the results.

An information output, taken as Fisher's information measure and proportional to the reciprocal of the variance of the parameter of interest, is the variance var $[S(t)]$ or var $[X(t)]$. Thus, to minimize variance is presumed to maximize information content as used by Matalas and Langebein (1962). Because a flat frequency distribution (like the rectangular or uniform) has a high variance, its entropy is higher than a highly peaked frequency distribution. The information content of the latter is considered to be high, the variance is very small, the central tendency is close to certain, and the entropy is very small.

In the context of optimal control theory, objective functions may be formulated so as to achieve certain goals with minimum expenditure of time, energy, power, or money. Not all can be concurrently achieved.

The general systems model also provides a framework for considering that special class of objective functions arising in estimation theory. A substantial review of these 'loss' functions has been given elsewhere (Kisiel, 1971) in the context of errors in modelling and parameter estimation. Given the correct model or state transition function F, the problem is one of obtaining the best estimate of model parameters from a limited set of field data. Let Z_m be the model output and Z be the observed output, for example, of river discharge or of water levels in an aquifer. This is the so-called inverse or 'non well posed' problem. Two methods have been used for its solution. The classical method is simply a trial and error adjustment of guessed parameters until the observed state set $S(t)$ and estimated state set $S_m(t)$, that is, Z and Z_m, are 'perfectly' matched. This match is usually judged by eye. Of course, the solution is non-unique. More recently, hydrologists and water resource system analysts have formulated objective functions as a means of coping with the system indeterminacy. The most popular loss functions are exemplified by the integral squared error (ise), mean square error (mse), and integral of the absolute value of error (iae), respectively, as given below:

$$G_{ise}(Z, Z_m) = \sum_{k=1}^{K} \{Z(k) - Z_m(k, \beta)\}^2 \tag{9a}$$

$$G_{\text{mse}}(Z, Z_m) = \frac{1}{K} \sum_{k=0}^{K} \{Z(k) - Z_m(k, \beta)\}^2 \qquad (9b)$$

$$G_{\text{iae}}(Z, Z_m) = \sum_{k=0}^{K} |Z(k) - Z_m(k, \beta)| \qquad (9c)$$

The quantities to be summed are the errors $e(k)$. In equations (9), $T = K\Delta t =$ length of observation period used to calibrate model (that is, to estimate the parameter set β of the true parameter set b) and $t = k\Delta t$ is the continuous time equivalent of discrete time integer k. In each case, the loss function is symmetric, that is, all three criteria assign equal weight to errors of underestimation, $-e(k)$, and errors of overestimation, $+e(k)$. No economic interpretation has been given thus far to these errors but seems logical because of the nonsymmetric character of socioeconomic losses assignable to $-e(k)$ and $+e(k)$. However, such loss functions are difficult to estimate except for special cases. Some hydrologists choose to optimize with respect to deviations at one time point of the hydrograph, for example

$$G_i\{Z, Z(k), Z_m(k, \beta)\} = \{Z(k) - Z_m(k, \beta)\}^2 \qquad (10)$$

which is an instantaneous criterion function resulting in a static optimization problem. Each of the above objective functions has, as constraints, the state transition function F, physical bounds on the variability of model parameters, physical realizability, stability and continuity. Subsequent use of optimal parameter values in the same model to predict independent conditions will, in part, be a model validation step. The entire procedure depends on the chosen state transition function (it may be physically-based or formulated in the spirit of parametric hydrology), loss function and sample size K. The optimal parameters permit the use of F for subsequent management studies. Note that the estimation problem has been decoupled from the control or management problem.

Recent developments in Bayesian decision theory present the possibility of putting the parameter estimation problem in an economic decision or control framework. However, at this time, its application to water resources systems has been confined to uncertainty in the stochastic parameters of $X(t)$, for example, the mean, μ and variance, σ^2 of the log-normal distribution for the annual sequence of flood peaks. Davis and Dvoranchik (1971) have presented computer results for the design of bridge piers, and Davis (1971) has used the same approach for the design of flood levees. In the objective function $G = G\{d, b, X(t)\}$, d is the decision to be made and the possible alternatives (depth of bridge pier or height of flood levee), $b = \{\mu, \sigma^2\}$ is the true parameter set, and $X(t)$ is the design flood. The Bayes risk,

$$R(d, b) = E_b G(d, b, X) = \iint G(d, b, X) p(X/b) q(b) \mathrm{d}X \mathrm{d}b \qquad (11)$$

is the expected value of the objective function for each alternative d. $p(X/b)$ and $q(b)$ are probability density functions, respectively, of the flood data $X(t)$ given the true parameters b, and of b if the true parameters are uncertain. $q(b)$ is the classical prior probability and is shown to be the normal χ^2 distribution for $b = \{\mu, \sigma^2\}$ of the log-normal distribution. The optimal alternative, d^* to maximize the expected value of the objective function is obtained from the minimization of the Bayes risk:

$$R(d^*, \beta) = \min_{d} \ E_{b} \ G(d, b, X) \qquad (12)$$

The above formula also serves as the framework for judging the worth of additional data for a water resource project (Davis, 1971)

Objective functions in a philosophic framework

Objective functions and constraints are usually given formal meaning in the framework of mathematical models. A major advantage of models is their frame of reference for the consideration of a problem, for example, in suggesting informational gaps and fruitful lines for research. Major issues, however, in the use of management or socio-economic models of water resource systems are the adequacy of their testing (calibration) and subsequent validation. Little attention has been given to these issues on the management side in contrast to the physical sciences. Generally, the modeller appeals to the rational side of potential users or to a comparison of his modelling approach to traditional practice. Notwithstanding the difficulties in the calibration and validation of hydrological models (Kisiel, 1971), the same challenge is even more severe, if not impossible, with respect to the management models. Consistency and adequacy of objective functions and constraints with a given real world problem is unalterably a function of the validation problem. The calibration entails obtaining best estimates of parameters from currently available data, whereas validation involves a reasonable matching of model assumptions and real world conditions and comparison of model predictions and actual future events. Those who exude confidence about their models should be challenged to give formal justification for their subjective kind of validation.

According to Wilde and Beightler (1967), optimal design and decision require three steps:

   (a) knowledge of the system,
   (b) finding a measure of system effectiveness – often involving a value judgment,
   (c) optimization.

The first two steps, often lightly pursued, imply careful definition of the objective function and its constraints. Given the present state of development of operations research, optimization theory, or systems analysis, it is useful to recall that Leibniz, in coining the word 'optimum' in 1710, speculated on our living in the 'best of all possible worlds.'

Jantsch (1970) and Dror (1970) argue for the evolution of the policy sciences as a superdiscipline that would integrate the *policy, strategic* and *tactical* levels of the *process of rational creative action.* That process is defined as the horizontal sequence of *forecasting, planning, decision making* plus *action.* The 4 X 3 matrix of the model shows how the policy, stratigic, and tactical levels interact with the four elements of the sequence. The resultant *human action model* is structured about social values and the consequent norms that influence forecasting, planning and decision making. This construct is in contrast to the mechanistic model so common today wherein goals are given from the outside or internal organization is independent of purpose. The human action model selects values, invents objectives and defines goals; it is self-regulating and self-adaptive.

In the context of water resources systems, the model circumvents the optimization of such systems independent of other components of the social system (urban, transportation, agricultural, cultural); it argues against the common but fallacious assumption that the goals of the water system are independent of the social process which initially gave it birth. Dunn (1971), in agreeing with this view, states that 'in evolutionary experimentation it is the goals that form the test of the (water resources, developmental hypothesis, developmental hypothesis, and through its practice the goals themselves are brought under periodic review and modification.' Dunn points out (as does Jantsch) that physical scientists and economists (and other social scientists) have tended to believe strongly that the domain of values, norms and goals is off-limits to science. There has been a slavish emulation of the physical sciences in the construction of models in management science, Franksen (1969) has shown the remarkable correspondence between the classical Hamiltonian variational calculus, mathematical economics and mathematical programming. Mechanistic concepts have apparent value as approximations to social phenomena where the process of social learning is not dominant. In addition, the mechanistic model of the socio-economic process implies a deterministic view of the world in which all of the mechanistic equations are seemingly known and the entire process must evolve in the direction set by the decision-maker according to his probability forecasts of the future. In a highly constrained world or system, the optimal course for the future should be highly predictable. For example, the optimal design and/or operation of a single purpose reservoir should be easy provided no uncertainty in the Bayesian sense exists.

Jantsch (1970) shows that the management sciences, so vigorously pursued in water resources management, and administrative sciences, have value at the tactical (operational) level.

> Their concepts dominate the entire process of rational (but, in this case, non-creative) action to such an extent that linearity and sequentiality prevail, and that institutions and even instrumentalities (the latter particularly in the public domain) are perpetuated by this process. Disciplines and inter-disciplines providing aspects for this level, include, above all, 'correct economics' (with linear economic forecasting, econometrics, Keynesian growth economics, cost–benefit assessment, and discounted cash flow approaches), administration, operations research, resource allocation and deployment, and others. The emphasis is on *expansion*: expansion of actual activities, of system variables, of resources, markets and operations in general. The result is the conservation of systemic structures and increasing 'system clogging.'

Also quite relevant to our thinking on objective functions and constraints is the following outline of the inadequacies of contemporary systems analysis for treating complex social issues (Dror, 1970):

(a) It (systems analysis) focuses on proposing preferable policies, neglecting the institutional contexts, both of the problems and of the policy making and policy implementation process.

(b) It does not take into account political needs, such as consensus maintaining and coalition building.

(c) It has difficulties in dealing with 'irrational' phenomena, such as ideologies, charisma, high-risk commitments, martyr tendencies, and unconventional styles of life.

(d) It is unable to deal with basic value issues and often inadequately details the value assumptions of analysis.

(e) It deals with identifying preferable alternatives among available or easily synthesized ones.

(f) It requires some predictability in respect to alternatives. Situations of qualitative or primary uncertainty (in contrast to quantitative or secondary uncertainty as used in the Bayes risk formulation) cannot be handled.

(g) It requires significant quantification of main relevant variables.

(h) It does not explicitly face basic strategy choices such as attitudes to risk or time. Instead, it usually assumes maximum or minimax and discount of the future. At issue is the discounting of future generations to benefit the present.

The implication from this listing is that these deficiencies may, in general, mitigate the presumed benefits of systems analysis. In the context of water resources, these deficiencies tend to become important as the analysis proceeds from single-purpose use to multi-purpose use, as space–time scales of the analysis become larger, as more interfaces or externalities impinge on the system being analysed, as the environmental and ecological values compete in importance with the economic bene-fits accruing to water, as the biological and chemical constraints on water use and needs become tighter, and as more than one decision-maker is involved.

At the strategic level, the emphasis is on analysis (system, policy, need, institu-tional and market). The disciplines include sociology, technology, system simulation, decision theory, and the entire spectrum of integrative approaches (comprising, among other subjects, Planning-Programming-Budgeting, and generalized 'social cost-effectiveness' approach). The results at this level manifest themselves as social and technological innovations. A substantial outline of the cost-effectiveness approach from an engineering economics viewpoint is given by English (1968).

Relationship of papers to the survey

Three of the four papers assigned to topic 7 directly concern themselves with eco-nomic and social issues. Paper 3 by Tiemer, the only exception, is in the spirit of defining the state transition function F for a large-scale aquifer. His approach to aquifer management is classical in the sense that the aquifer model is not explicitly used in an economic objective function. A recent effort that couples the physics and economics is reported by Bredehoeft and Young (1970).

Tiemer derives a general differential equation for aquifers that allows for time variation of transmissibility T; but the storage coefficient is taken to be constant. He assumes mean values for the time-dependent values of T because of the small changes in water level. The resulting linear time-invariant differential equation is solved with the aid of an RC network for the case of a gallery of wells (for municipal water supply) coupled to facilities for artificial recharge. The objective was to establish the mathematical model so as to evaluate the effects of several operating policies for the aquifer. At the start of the simulation, the author recognized the incomplete-ness of his information on T and S and endeavoured to supplement this information by simulating the aquifer before construction and during test operation of the infil-tration basins. His results led him to conclude that the available data on aquifer inhomogeneities did not permit a unique detection of the hydrogeological structure. In this general report we earlier suggested that seldom should we expect such unique determination of parameters. Thus, the RC network model simply forces the real sys-tem to behave in a different manner. Tiemer gives in his Fig.1 (vol.3, p.890) what is called an effort at model validation but it is not clear whether the model predictions and actual observations were systematically adjusted in order to find T and S. If this were the case, then the author is simply calibrating his model. If, on the other hand, T and S were pre-determined, then Fig.1 is correctly termed a validation. However, to me the goodness of fit for wells V5 and V7 is not satisfactory. In fact, there is no clearly defined objective function. In the context of Bayesian decision theory, one may inquire about the relative economic importance of these errors which may be in the order of 2 to 3 m. It would be of interest to know how the author judges the fit to be satisfactory.

Sugawara introduces the intriguing idea of negentropy as a basis for qualitative rea-soning about water resources sytems. However, no equations or results are given on the estimation of negentropy, defined as a measure of order. This concept is in opposition to the classical entropy defined as a measure of disorder or maximum randomness. Sugawara argues that water use is mostly the consumption of the water's negentropy for example, in consuming the water's potential energy at a hydroelectric power station, in consuming the water's clarity, heat, healthfulness, or other quality properties through municipal, agricultural, or industrial use, in dispersing the water into space and time after use in irrigation systems or paddy fields. On the other hand, he points out that the entropy of rainfall in space—time is eventually transformed into a high level of negentropy in the form of baseflow from an aquifer; but this is gained in exchange for lost potential energy.

The statistical meaning of entropy leads Sugawara to show that information and the human will (to control pollution, for example) can prevent an increase of entropy in water resources systems.

I wonder if Sugawara implies that science and technology can reduce entropy as society wishes and that all randomness and disorder in nature can be overcome. Negentropy, be it information or highly controlled water resources systems, is bought at a price. His concept leads into the classical question about whether or not the entropy of the earth is increasing. Water resources systems do not evolve independently of other social systems. Their development arises from a continual competition for limited resources whose availability fluctuates in space and time. Seemingly, the objective function to be minimized is the sum of the entropies of all variables impinging on the problem. Can this be the universal and absolute objective function that some analysts seek? But how can this be in the face of the risks and uncertainties discussed earlier in our report? Surely, errors add to our entropy. Furthermore, how is the idea of negentropy reconciled with Heisenberg's uncertainty principle as applied in a social context? As mentioned previously in our report, social goals are bound to change because goals are a function of the social process itself.

Domokos presents what is described as a general theory on water restriction indices. These indices are defined in terms of indicator functions that take on the value of zero if supply equals or exceeds demand and other values if supply is less than demand. Three indices are defined: (a) Proportion of time that water deficiency exists with respect to a water deficiency tolerance index $K \geqslant 1$ that is set by prior water needs, (b) average relative water deficiency or proportional deficiency in required volume to satisfy demand, (c) average square relative water deficiency. However, the best water use in each case is not clear. In each case the demand is fixed and not stochastic; the social implications of this practice deserve careful analysis (Loucks, 1969; paper 4). Index (a) simply focuses on duration of water deficiency as a criterion (and is simply the area under the flow duration curve as used in American practice) and not on volume deficiency as do indices (b) and (c). In relation to this report, index (b) is a one-sided linear loss function and index (c) is a quadratic loss function. However, all three criteria are compared independent of any social or economic losses that surely are suffered with any water deficiency. Not all water deficiencies produce the same proportional economic loss. The implications of this important assumption require clarification. Domokos also assumes that the distribution function $F(X)$ for the streamflows X is truly known. In the context of the earlier-presented material on Bayesian decision theory [see equation (12)], Bayes risk considers both the loss function (if known!) and uncertainties in the parameters of $F(X)$.

In addition to the uncertainties in $F(X)$, other stochastic aspects of the streamflow record $R(t)$ deserve some amplification. For example, were statistical tests performed to justify the hypotheses about the ergodicity and thus the stationarity of the flow? $R(t)$ typically is oscillatory over the sequence of days and months within a year; hence, the author's judgment that the periodical component is zero needs justification. The proposed indices appear to come under the scope of the theory on crossings (runs) and the range.

Duckstein and Kisiel apply the theory on collective utility as a basis for comparing two pricing structures by an economic decision-maker. No specific goal need be defined and in this sense collective utility seems to have merit at the strategic level of the human action model. The concept may be deceptive in that the dollar return to the decision-maker is used as a proxy metric for the amorphous notion of utility. This point deserves special and frank attention, as do the social implications of some of the model assumptions. The introduction of uncertainty about the future states into the evaluation of the change in collective utility requires the determination of a probability, $P(u)$. This determination would pose a major problem of estimation for variables of secondary uncertainty, but $P(u)$ would appear to be meaningless with respect to inflation and technological change. Equation (11) in paper 4 (vol.2, p.885) has a form similar to the equation for Bayes risk and thus invokes the expected value criterion as a basis for decisions under uncertainty. Overall, in view of the explicit theoretical foundations of collective utility, its introduction into water resources systems analysis places in better focus the relative merits of operations research, cost–effectiveness analyses and cost–benefit analyses (see Figs.1 and 2 for comparison of these). The earlier material on the human action model has given some additional comparison of these methods.

Because of space limitations, explicit detailing of the above comparisons is not possible but may be found in Dupnick (1971). Not covered are the recent theoretical developments in handling multi-dimensional (vector) objective functions, satisfactums, and environmental values.

GENERAL TECHNIQUES

	STRUCTURAL TRANSFORMATIONS	MARGINAL TRANSFORMATIONS	MATHEMATICAL FRAMEWORK	INCLUDES EXTERNALITIES	HANDLES PRICE CHANGES	ADAPTABILITY TO DIVERSE PROJECTS	SOCIAL SCIENCE APPLICATION	INCLUDES UNCERTAINTY	RESOURCE ALLOCATION
COST-BENEFIT	✓								
COST-EFFECTIVENESS	✓	✓		✓		✓	✓	✓	
COLLECTIVE UTILITY	✓	✓	✓	✓	✓	✓	✓	✓	✓

FIGURE 1. Comparison of general techniques of cost–benefit, cost–effectiveness, and collective utility (Dupnick, 1971).

PITFALLS AND FALLACIES

	OMISSION DANGER	DOUBLE-COUNTING DANGER	RATIO	RANKING WEIGHTS	SOLE CRITERION	QUANTIFICATION	INTER-RELATION	DERIVITIVE	DEFINITION
COST-BENEFIT	✗	✗	✗	✗	✗	✗	✗	✗	✗
COST-EFFECTIVENESS	✗	✗		✗	✗			✗	✗
COLLECTIVE UTILITY					✗	✗	✗		

FIGURE 2. Comparison of pitfalls and fallacies of cost–benefit, cost–effectiveness, and collective utility, (Dupnick, 1971).

REFERENCES

Beard, L. R. (1971) Status of water resources systems analysis. *Meeting preprint 1308*: Amer. Soc. civ. Engrs National Water Resources Engineering Meeting, Phoenix, Arizona.

Bredehoeft, J. D. and Young, R. A. (1970) The temporal allocation of ground water – A simulation approach. *Wat. Resour. Res.* **6**, 3–21.

Buras, N. (1974) Mathematical modelling of water resources systems – a review. Committee on Mathematical Models in Hydrology, International Association of Scientific Hydrology, 15th General Assembly, IUGG, Moscow, USSR. *Hydrol. Sci. Bull.* **19**, No.4, 393–400.

Davis, D. R. (1971) Decision making under uncertainty in systems hydrology. *Technical Report No.2*: Hydrology and Water Resources, University of Arizona, Tucson, Arizona, Tucson, Arizona.

Davis, D. R. and Dvoranchik, W. (1971) Evaluation of the worth of additional data. *Wat. Resour. Bull.* **7**, No.6.

Dror, Y. (1970) Prolegomena to policy sciences. *Policy Sciences* **1** No.1, 135–150.

Duckstein, L. and Kisiel, C. C. (1968) General systems approach to groundwater problems. *Proceedings, National Symposium on the Analysis of Water Resource Systems*, pp.100–115: American Water Resources Association, Urbana, Illinois.

Dunn, E. S., Jr. (1971) *Economic and Social Development: A Process of Social Learning*: John Hopkins Press; Baltimore, Maryland, USA.

Dupnick, E. (1971) Collective utility in the management of natural resource use: A systems approach. *Technical Report No.5*: Hydrology and Water Resources, University of Arizona, Tucson, Arizona.

English, J. M. (editor) (1968) *Cost-Effectiveness: The Economic Evaluation of Engineered Systems, New York*: John Wiley.

Franksen, O. (1969) Mathematical programming in economics by physical analogies, Part III: System equilibrium and mathematical programming. *Simulation* **13** No.2, 63–87.

Jantsch, E. (1970) From forecasting to policy sciences. *Policy Sciences* **1** No.1, 31–47.

Kisiel, C. C. (1971) Efficiency of parameter and state estimation methods in models of lumped and distributed hydrologic systems. *Proceedings, US–Japan Bilateral Seminar in Systems Hydrology, Honolulu, Hawaii,* (edited by V. M. Yevjevich): Water Resources Publications, Fort Collins, Colorado, USA.

Loucks, D. P. (1969) Stochastic methods for analyzing river basin systems. *Cornell University Water Resources and Marine Sciences Center,* Ithaca, New York.

Matalas, N. C. and Langbein, W. B. (1962) Information content of the mean. *J. geophys. Res.* **67**, No.9, 3441–3448.

Simpson, E. S., Kisiel, C. C. and Duckstein, L. (1975) Space–time sampling of pollutants in aquifers. *Proceedings, Symposium on Groundwater Pollution* (Moscow, 1971). IAHS Publ. No. 104, pp.45–57.

Wilde, D. J. and Beightler, C. S. (1967) *Foundations of Optimization*: Prentice-Hall, Englewood Cliffs, New Jersey, USA.

DISCUSSION

M. Domokos:

In my paper, I deal with the general theory of water restriction indices. What is the use of them? According to one of the conceptions of water resources systems dimensioning, it is the following:

The question whether an existing (or planned) water resources system is (or will be) working efficiently or not, can be essentially answered by calculating the real or expected values of the water restriction indices of the water uses existing in the system (through probability theory reasoning or by means of the simulation of the system's working) and then comparing these values with their limit values, the so-called water deficiency tolerance indices, prescribed in advance on the basis of economic considerations.

The water restriction can have many different interpretations (measures), so from the point of view of economy it is important to use the index economically most characteristic in the given case. My paper tries to make this choice easier, presenting the general theory of water restriction indices and some concrete examples thereof.

The presented theory explains the water restriction index as an expected value of the water restriction indicator function, defined for the examination period. It is supposed that the indicator function

(1) depends on the time functions of water resources and water demands.
(2) its ordinates are dimensionless numbers between 0 and 1,
(3) its value increases for poorer water supplies.

The water restriction index must comply with the following requirements:

(1) Water restriction of the examination period must be characterized by a single dimensionless number (in order to compare it with the limit value of the water deficiency tolerance).
(2) The economic loss caused by water restriction must be a monotone increasing (if possible, linear) function of the index.

Included in this theory, to give some concrete examples, three water restriction indices are defined:

(1) proportion of time that exists during water deficiency surpassing a certain degree,
(2) average of the relative water deficiency,
(3) average square of the relative water deficiency.

The first of these indices is generally used in Hungarian water resources development.

If certain conditions are satisfied, an easily utilizable formula can be deduced for the numerical calculation of water restriction indices. Such conditions are:

(1) The water demand time function is a periodical one, with a period of 1 year.
(2) The water resources time funciton is an ergodic stochastic process.
(3) The year can be divided into part (not too short) periods, in which the process is stationary.

If these conditions are satisfied, such a division of the examination period can be found, that in each of the part periods, water demand can be given by a constant value and water resources may be characterized with a single probability distribution function, with a practically negligible error. In the paper, simple utilizable formulae are deduced by calculating the three mentioned indices for the conditions described above.

It must be emphasized that these simplifying suppositions are *not* necessary, the enumerated, and other, water restriction indices can be calculated by different, more general methods, too.

Professor Kisiel in his report criticized all these simplifying suppositions. To answer him, I would like to mention the following:

(1) The periodicity or constancy of water demand is, of course, a disputable. but not necessary supposition. If, for example, the common distribution function of demand and resource is known, the restriction index can also be calculated on the basis of the presented theory, certainly it is more time consumptive than the former case. Similarly, the calculation of the index is also simple, if the demand is a deterministic function of the resources; for example, if it is proportional to its reciprocal.
(2) In my opinion, the stationarity of water resources is admissible for each of the part periods given by a suitable division of the examination period. Otherwise, this supposition can be verified in a given case by means of a simple significance test.
(3) The ergodicity of water resources means that its average value must converge to its expected value. I believe that this supposition is practically admissible. Otherwise, I am sorry, but I ignore statistical tests serving for the exact control of this supposition.
(4) Theoretically, I agree with the necessity of taking into consideration the uncertainties of the water resources distribution function. I am inclined to think that until they are negligible in respect of those of water demands, the result of our calcu-

lations will not be damaged if the water resources distribution function is considered as truly known.

(5) Finally, Professor Kisiel makes the objection that all three indices are compared independently of any social or economic losses. Indeed, this proves true. As I mentioned in the last part of this paper, in the future we would like to categorize water uses according to the indices and to give concrete limit values on the basis of the theory presented. Apparently, one of the tasks of water resources development researches is the elaboration of these concrete empirical water deficiency tolerance limit values.

L. Duckstein:

It is, I believe, the first session in this meeting in which people have begun to ask economic questions. For example, there is not only the question of satisfying the demand for water, but also of asking at what price this demand could be satisfied. I would like to speak more specifically about the tool of collective utility to compare two economic states involving water. An economic state is defined as a flow of goods and services during a given period of time. I take a single period example to simplify the presentation. In the case which concerns us, the goods and services that are flowing are the various kinds of water taken from the hydrological cycle from the time when man starts utilizing it until the time when man stops utilizing it. It includes, in particular the time when people drink water out of a polluted river, whether voluntarily or not. Hence this tool of collective utility is more general than optimization since, strictly speaking, we do not have an objective function as the general reporter has pointed out. People that have an objective function are happy. They have a least squares criterion, they have an economic function of some kind which they want to maximize or to minimize. Decision-makers do not have this luck. They have to make decisions where objective functions are vector valued.

Collective utility compares two states in a rigorous mathematical manner. We start out by making hypotheses about our economy. One may accept or reject these hypotheses; however, if we accept the hypothesis then the mathematical derivation that follows is strict. I will show a little later the equation which is derived from this theory, which is a part of welfare economics; specifically we are using the derivation of Jacques Lesourne from France. The specific problem examined here is that of trying to determine what the water consumption will be as a function of the price structure of water.

In the western United States, at least, water is metered, and the price of water depends on the quantity consumed. However, the distinguished economists that have fixed the price structure of water, did so in a marginal manner, that is, the more water you consume, the less you pay; because the more water you produce, the cheaper it is to produce it; this is called economy of scale. When a resource becomes scarce we have to ask the question whether this was the proper way to look at things. This type of price structure may cause waste, because when you go beyond a certain quantity, it is very cheap to consume a little more. We have asked the question, whether or not having a progressive price structure (i.e. the more you consume the more it costs you) would not be more appropriate.

M. Roche:

La série de conférences qui nous a été si brillamment résumée par notre rapporteur général montre bien où se situe réelement le problème de l'objectivité. En matière de simulation, méthode qui doit servir à résoudre des problèmes d'aménagement des eaux, on remonte une cascade d'opérations qui doivent finalement aboutir à la solution d'un problème économique. Mais l'objectivité des jugements qu'on peut faire en se basant sur des critères économiques n'est pas toujours une chose aussi simple et absolue qu'il pourrait le paraître à priori.

D'abord, tous les usages de l'eau n'ont pas la même valeur et n'admettent pas les mêmes critères. Je vais prendre un exemple en simplifiant beaucoup. Supposons un aménagement, intégré à l'échelle d'un bassin, dans lequel on a à satisfaire une demande industrielle, une demande ménagère et une demande en irrigation. La rentabilité de l'eau ménagère ne se discute pratiquement jamais, on ne voit pas du reste sur quel critère économique on pourrait se baser; on doit fournir l'eau et on la fournit. Je ne connais pas d'exemple d'aménagement où on mette en doute la nécessité de fournir de l'eau potable quel qu'en soit le prix, tout en cherchant évidemment à ce que ce prix soit le plus faible possible. Pour l'eau industrielle, il est rare que se pose vraiment le problème de son prix; cela peut arriver pour certaines industries grosses consommatrices et produisant des produits relativement à bon marché, mais en général ce n'est pas un facteur telle-ment capital sur le plan de la production industrielle. Restent les irrigations. Le prob-lème économique de l'eau pour l'agriculture est une question bien difficile. Si on se base sur un calcul économique strict aboutissant à une rentabilité donnée, il se trouve peu de pays au monde où ce calcul économique, qu'on prenne les choses d'une manière ou d'une autre, aboutisse à un résultat satisfaisant, compte tenu des cours mondiaux des produits agricoles. On peut alors se demander si ce n'est pas justement la base économique qui est fausse au départ: par exemple des cours agricoles artificiels, qui sont presque toujours trop bas, pour de nombreuses raisons, soit que dans certains pays moins développés, plus défavorisés, on arrive à fournir des produits agricoles à des prix très bas parce qu'on paye très mal la main d'oeuvre, soit que dans les pays évo-lués, l'agriculture soit finalement soutenue par les autres secteurs économiques, notam-ment les secteurs industriels. Si donc on se basait uniquement sur ces critères écono-miques en matière d'agriculture, eh bien, Messieurs, vous seriez condamnés à manger de l'acier et à boire du pétrole; jusqu'à présent, je ne pense pas que cela soit une solution extrêmement satisfaisante pour l'humanité.

Que reste-t-il finalement de l'objectivité de ces critères sur lesquels sont basés les calculs économiques? Il en reste la possibilité, pour un ensemble à priori de réalisations, de comparer les solutions entre elles, et ça c'est déjà quelque chose. Que fait-on pratiquement quand on veut développer une région? Je prends un cas très simple pour ne pas embrouiller les choses, par exemple celui d'une région dans laquelle il n'y aurait rien ou presque rien et où on a décidé d'implanter quelque chose, une industrie consom-matrice d'eau, de l'agriculture irriguée, etc. On commence par faire la prospection des possibilités physiques du développement, ou l'inventaire des ressources. Il est bien évi-dent que, pour l'agriculture par exemple, il faut d'abord avoir des terres qui correspon-dent à ce qu'on veut cultiver, qui aient une vocation agricole déterminée, ou choisir une répartition des cultures qui soit conforme à cette vocation agricole. Cette pros-pection conduit à un ensemble de solutions possibles, duquel on va éliminer d'emblée un certain nombre de solutions, dont on sait pertinemment que ce n'est même pas la peine de les étudier pour un certain nombre de raisons: éloignement trop grand des sources en eau disponibles, conditions d'accès trop défavorables, moins bonne valeur des terres, densité de peuplement insuffisant etc. C'est une affaire de spécialiste et il n'existe pas de méthode mathématique pour réaliser ce premier choix.

Qu'en résulte-t-il? Il en résulte quelque chose qu'on appelle la prospective du déve-loppement de la région. Cette prospective, on pourrait la traduire en gros par une espèce de courbe, comme ceci, qui représente l'évolution de la fourniture d'eau dans le temps, donc les volumes annuels à fournir aux différentes étapes du développement. On part d'ici de l'instant initial zéro qui est souvent l'instant actuel, et on suppute d'une maière ou d'une autre quelle sera l'évolution des différents besoins. Si on se place en un point de cette courbe, dont l'abscisse correspond à un stade du développe-ment, l'incertitude sur l'estimation de la prospective se traduit par une distribution de

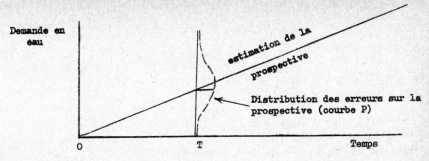

FIGURE 3

probabilité de la demand en eau prévue, dont le point situé sur la courbe est supposé représenter la valeur centrale (Fig.3).

Quand on choisit le stade de développement (temps T) pas trop éloigné de l'instant initial, on peut déterminer la prospective avec une assez grande précision; c'est-à-dire que la courbe P sera très pointue. A mesure qu'on s'éloigne dans le temps, l'estimation de la prospective devient plus incertaine et la courbe P s'aplatit, et si on va suffisam· ment loin on ne sait en fait plus rien de la prospective. En réalité, les choses ne se passent pas ainsi; en effet, les aménagements se font obligatoirement par paliers, en accord avec leurs différentes possibilités physiques, de sorte que le choix définitif au moment du passage à l'exécution porte finalement sur une prospective à court ou à moyen terme, la prospective à long terme n'indiquant qu'une tendance générale dont les conclusions peuvent être corrigées au fur et à mesure de la réalisation de l'aménagement.

Ceci montre que la sensibilité du calcul économique au choix des critères d'optimisation est étroitement liée à cette étude de la prospective qui permet en définitive de déterminer les données à introduire dans le calcul. Or cette prospective devient très difficile à saisir dès que l'on s'éloigne beaucoup du temps présent; son estimation fait appel à des hypothèses d'extrapolation dont il est parfaitement justifié de mettre en doute l'objectivité.

A. Becker:

I would like to comment on the paper by Tiemer and the questions raised by the general reporter on this paper. Tiemer's paper is, of course, only a very brief review of the extensive work which was carried out recently. Therefore, I shall recommend anybody who wants to know something more about the problem to look at earlier paper of Tiemer (1967). Using terms of the general report, Tiemer's paper is concerned with the identification, calibration and validation of systems modelling. The step 'calibration' is discussed only briefly without technical details. It seems to be necessary to give some additional information.

In the first stage of the investigations, the geohydraulic parameters T and S have been estimated in the usual way using the results of granulometric analysis, laboratory tests and aquifer tests. This information was available only for a small part of the aquifer under consideration. Therefore, an attempt was made to supplement the numerical values of T and S by simulating characteristic groundwater flow patterns.

The model predictions and actual observations were systematically adjusted using, as loss function, the mean value of the absolute deviation between calcualted and observed groundwater levels. This loss function is in close relation to the loss function G_{iae} in the general report [equation (9c)]. The information about the transmissibility, T, obtained from the simulation experiments was partly contradictory. Therefore, it seemed to be not justified to take into account the local dependency of the parameter

T. All results were used to estimate the representative mean value of *T.* The estimation of the storage coefficient, *S*, was based on the simulation of the test period in 1963. The improvement of the model, possible by varying the parameter *T*, is not very significant. At present there is no economic criterion for estimating the importance of these errors. The improvement of the model will be tested some years later if new observational data will be available. The homogeneous aquifer model has the advantage than an analytical solution of the groundwater flow problem may be studied. A model explicitly solvable may have great importance for optimization calculations.

The last remark corresponds to the question of the general reporter about the objective function. I think that the quoting of this question is quite correct. I want to confirm what the general reporter said about the subject, and what was said by Professor O'Donnell some sessions before. There is really a great need for dealing with this subject.

REFERENCE

Tiemer, K. (1967) Sur le problème de l'emmagasinement souterrain des eaux superficielles par infiltrations artificielles dans les zones de rabattement. *Recharge des Nappes Aquifères* (Proceedings of the Haifa Symposium): IAHS Publ. No.72.

U. H. Mudallal:

I would like to add a few comments on the application of mathematical models in groundwater hydrology. It is quite clear from the papers so far submitted that all, but two, or three, concentrated and stressed on the application of mathematical models on surface hydrology. The two or three papers that discussed the possibility of applying mathematical models on groundwater hydrology actually took an initial step in doing so. Although they did not come up with a certain conclusion about the success of applying such models, this should encourage the hydrologist to try some other aspects of application. It is not clear why the vast majority of the papers deal with the modelling of surface water hydrology, possibly it is because we do not like to use some of the parameters on groundwater hydrology because of the inhomogeneity of the aquifer, or because of some other factors that still have not been solved. Tiemer's paper this morning, and Elgizawi's paper the day before yesterday, were trying successfully to find the interrelationship between surface water and groundwater. Elgizawi in his paper tried to find the interrelationship between the river Nile fluctuation and the water level in the delta aquifer. Tiemer tried to find the simulation of the groundwater flow. Engineers, also, are trying to apply mathematical models to the study of the water balance. Unfortunately, we could not contribute with papers because we do not yet have deterministic figures, which might lead to definitive answers. The main problem in groundwater basins is to know the safe yield, especially in basins where the rainfall is so little and there is no direct recharge to the aquifer basin. The main recharge comes to that basin from distant regions as subsequent flow. By the help of pumping tests and flow analysis we have determined the storage coefficient (*S*) and transmissibility (*T*) of such a basin. We are now trying to apply the models in order to find out the safe yield. If we succeed, we would be able to know the amount of recharge annually to this basin. I wanted to mention this because I would like to suggest strongly that we should take the initiative to exchange experiences and experiments done so far on this aspect of the application of mathematical models in hydrology in order to promote and develop our knowledge and our results.

Closing remarks by the general reporter

I wish to thank the discussers for their most constructive remarks, particularly Domokos whose remarks are in the spirit, I believe, of the general reporter system and are most constructive for progress at this Symposium. The comments by Roche are well taken. The issues as emphasized by Roche will certainly be with us for a long time to come. It is so because there is so much we cannot quantify. It has been remarked many times that solutions to problems will lead to additional problems. Sometimes smaller, sometimes much larger. There exists in the world today a pessimistic view being pushed forward by some modellers of environmental systems; for example, there are some computer simulation results on world dynamics by Forrester from MIT. He projects rather substantial disaster conditions arising in another thirty to fifty years. This, of course, is prognosticating; hence a tremendous number of assumptions are involved and there are those who disagree with this approach. For example, Eiseley (1971) in a recent book, notes that the focal position of optimization, initiated more or less by Leibnitz, translates into the attitude that there is only one way into the future; that is the technological way, and that the approaches undertaken by computer simulation, as prognosticated by Forrester (1971), although they are pessimistic, do follow in the spirit of optimization procedures. Eiseley warns that computers have made us forget the old wisdom of fairy tales. There is a frontier to man's kingdom where predictability ceases, and the unimaginable begins. I do believe that, when many of us started to do some work in operations research, we were rather hopeful and very optimistic as suggested by Eiseley. And perhaps we should maintain that hope, and not believe too seriously in the computer simulation results of Forrester and others. I must admit that these are very challenging subjects, and the issues, as I said before, will be with us for a very long time to come. One final point, I wish to apolcgize to Tiemer and Becker for not having referred to the paper given at Haifa.

Session XVI   30 July 1971
General Reporter   Dr. Ven Te Chow, Professor of Hydraulic Engineering, University of Illinois, Urbana, Illinois, USA
Chairman   Dr. Arne Forsman, Swedish Meteorological and Hydrological Institute, Stockholm, Sweden

Optimal operation of water resources systems — I

Ven Te Chow

The following papers are reviewed and discussed in this general report. (All are published in volume 2 of the Symposium Proceedings.)

1. *J. A. Cole*   Optimal seasonal and short-term operation of a reservoir used for flood control and water supply
2. *Luis Lopez Garcia*   Optimization of a three-reservoir system by dynamic programming
3. *U. Maniak and W. Trau*   Optimal operations of reservoirs in the Harz Mountains
4. *A. J. Askew, W. W-G. Yeh and W. A. Hall*   Optimal design and operation of reservoir systems
5. *Tomasz Biernacki and Teofil Piwecki*   Optimization model of a system of two open-channel hydroplants

Modern water resources systems are often very complex, as they are of, for instance, multi-purpose, multi-constraint, multi-unit, multi-state, and multi-stage. Whereas assurance of the optimal planning and development of such a system becomes practically impossible by traditional methods empirically based largely on judgement and experience, an efficient design of the tangible (quantifiable) portion of the system may be achieved by modelling and can thus serve as a rational guide for planning and development of the system. For the optimization of a tangible water resources system, two categories of modelling techniques may be distinguished; namely the analytical technique and the simulation technique.

The analytical technique is to convert the optimization problem to a mathematical model which can be solved by an analytical optimization method. Structuring of the model, or modelling, begins with the selection of the physical variables for which an optimum quantity is desired, such as reservoir size, flood channel capacity, and hydropower plant scale. Mathematical expressions consisting of objective functions and constraints are then formulated to express the physical relationships among the variables and to limit the range of values that given variables may assume. The mathematical model so formulated should describe the functioning of the prototype system as closely as possible within the limitations of available analytical optimization methods. The analytical optimization method involves a procedure of specific mathematical manipulation to examine the model for the most desirable values for the physical variables according to the objective function. The optimum set of variables is determined systematically by the least computational effort within the range of constraints. Examples of analytical optimization methods commonly used for various types of problems are calculus of variation, linear programming, dynamic programming, and geometric programming (Chow and Meredith, 1969).

The simulation technique involves in the first step, representing the input, the physical functioning, and the output of the prototype system by a computerized algorithm. The input generally consists of hydrological data, economic events and structural measures. The hydrological data, or the economic events, may be of stochastic nature as it can be generated randomly by Monte Carlo methods from models based on historical information. The physical functioning involves essentially the relationship between the input and the output. The simulation technique is not just the modelling procedure but also involves the search for an optimal solution. A complicated water resources system can easily involve ten decision variables. If only three state variables are selected, such as high, low and average, the total number of different combinations is 3^{10}. The time taken for the calculations for each combination on a typical computer is about 30 s, hence the search for the optimal solution among all the combinations would require about 21 days. This calculation is too long for practical purposes. Therefore, a sampling technique is often used to select a portion of all the combinations and thus reduce the computation effort. Commonly used sampling techniques for the search of the optimum include the stratified and creeping random methods, the uniform-grid method, the steepest-ascent method, and the single-factor method (Chow, 1964).

The analytical technique, tailored for handling by mathematical optimization methods, has a number of advantages. One is that the formulation of a model required in this technique compels the analyst to abstract the essence of the system and thus he has to comprehend the problem as deeply and completely as possible. Also, the analytical technique permits a direct quest for optimality, often resulting in a great saving of computation time. However, such techniques have fundamental limitations. One is that the feedback in the system is very difficult, if not impossible, to deal with because the model is not so flexible. In general, the system must be relatively small, and its inputs fairly simple, such as in a deterministic form. Accordingly, the analytical technique may be applied with advantage to the optimization of system components or to many levels of sub-optimization which must precede the optimization of the overall system. For example, a two-level optimization model has been developed successfully for farm irrigation systems, utilizing linear programming and dynamic programming respectively for each level of sub-optimization (Windsor and Chow, 1971).

The simulation technique is a direct simulation of the system and its functioning and therefore offers great freedom and flexibility in modelling. Systems sensitivity to a particular input, transfer function, or policy feature is readily ascertained. The simulation technique thus avoids the difficulty of the analytical technique in condensing mathematical formulations describing the functioning of a prototype system into a form solvable by available analytical optimization methods. For complicated systems, for which no single production function can be written, direct simulation produces the output corresponding to a given array of input elements. The best of all possible outcomes can seldom be reached with certainty, and virtually all responses must be expressed in statistical terms. In other words, the technique does not yield direct answers. It may require a very large number of trials, the modelling cost can be enormous and the computation time may be substantial. Also, it is obviously influenced by the aptitude of the analyst for mathematical simulation which does not have a routine procedure. In general, the simulation technique is the most practical way for evaluating water resources sytems too complicated for direct optimization by analytical techniques. It is very justified for comprehensive and long-range planning effects. When approximate answers are needed quickly, however, analytical techniques applied to simplified models must be tried first.

Of the five papers being reviewed in this general report, two papers, by Cole and Lopez Garcia respectively, deal with the application of the analytical technique for optimization i.e. dynamic programming, and the other three papers by Maniak and Trau, Askew *et al.* and Biernacki and Piwecki respectively, deal with the application of the simulation technique for optimization. All five papers treat the operation of either a reservoir or a reservoir system. The paper by Askew *et al.* also considers the design of reservoir systems. A comparison of the essential features of these papers is given in Table 1.

Cole uses a hybrid dynamic programming technique to minimize the combined cost of deficient water supply and of downstream flood damage in the operation of a single reservoir serving two cities. The major findings of his analysis are: (1) the difference between a 10-division and a 37-division of the state variable, i.e. the reservoir content, does not distort the pattern of the operating rule obtained by the analysis, (2) the grouping of reservoir contents classified according to whether LQ is in the high, middle or low range produces a notable difference in the operating rule especially between the middle and low range groups, (3) the use of different cost coefficients makes the revealing of the balancing of release possible, (4) computer results converge relatively rapidly, in general, at the third iteration, (5) the VAL function at convergence behaves smoothly and monotonically, and (6) forecasts for flood-producing storms can be incorporated in the operating rule but the application of this theory is not demonstrated.

Usually, the flood control aspect is taken as a constraint within the system model. It is interesting to note that this aspect is included in the proposed optimization rather explicitly in the model in terms of the channel surplus index which, in turn, is used to express the penalty function. According to various selected penalty functions a sensitivity analysis can thus be made to show the variation in the operating rule. As Cole gives no details of the analysis, several questions and comments may be raised. Why is there a need to omit June and July and to create an artificial 10-season year? How does the channel surplus index, which is defined on monthly values, reflect the magnitude of flood damage, which depends significantly on momentary peak flood discharges? In order to introduce the stochastic characteristics of input hydrology, the paper suggests one of two recommended courses to introduce steady-state probabilities and to yield an expected cost estimate for the system. This approach is apparently objectionable, as pointed out in the paper by Askew *et al.*

Lopez Garcia applies conventional dynamic programming to optimize the net benefit for the operation of a typical three-reservoir system for the multiple purposes of irrigation, water supply, flood control, hydropower and low flow regulation. While constraints are used to meet the requirements of municipal and industrial water supply and low flow, only power and irrigation benefits are included in the economic function. The deficiency of the irrigation supply is introduced in the model as a penalty or 'loss' function. The input to the system can be stochastically generated by the Monte Carlo method. The proposed procedure can be applied to any system that can be reduced to the configuration of the assumed Y-pattern system.

In Lopez Garcia's paper, the consideration for flood control is mentioned, but it is not clear in what way the flood control effect was actually introduced into the analysis so that the difficulty of adequately including the momentary peak flood discharges in the monthly computation intervals can be avoided. The given example is simplified by considering water supply and low flow regulation as constraints. If these purposes are made variable, it is not clear whether the dynamic programming technique may become too involved for computer solution. The paper concludes that the need for large computation time by the Monte Carlo method is a 'shortcut' (or rather 'shortcoming'?) and suggests that some other methods could be used instead, but no specific methods are mentioned.

Maniak and Trau use the simulation technique to optimize the operation of a four-reservoir system in the Harz Mountains. The simulation technique is used because the

TABLE 1

Author	Optimization	System	Purpose	Function	Hydrology
Cole	DP	1 res.	Flood control, water supply	Operation	Stochastic
Lopez Garcia	DP	3 res.	Irrigation, water supply, flood control, hydropower, low flow regulation		
Maniak and Trau	Simulation	4 res.	Water supply, flood control, low flow regulation	Operation	Stochastic
Askew et al.	Simulation	1 res.	Hydropower, water supply	Design and operation	Stochastic
Biernacki and Piwecki	Simulation	2 res.	Hydropower	Operation	Deterministic

system is too complicated to have numerous design variables. The input synthetic hydrological data are generated by the Monte Carlo method on the basis of a logarithmic Pearson type III distribution without memory. Queuing theory is used to determine the probabilistic storages and monthly outflows of the reservoirs for the synthetic monthly inflow together with a certain release rule of the system. The release rule is determined from the season and the storage volume. Unfortunately, the paper does not describe in detail the procedure of analysis. It is not clear how the benefits of the operation plans are actually expressed, how the operation plans were exactly arrived at, and whether these operation plans are truly optimal or nearly optimal.

Askew et al. point out that analytical optimization techniques have a definite limitation when the input is stochastic in nature. They explain that if the expected value of the future return is used in the optimization, the output is a single value for design and hence does not convey any probability distribution of failure associated with this value. Also, if the Monte Carlo method is used to generate stochastic input for optimization, it is still not possible to establish the probability of failure associated with any particular combination of contract levels unless Monte Carlo analysis is applied again to check the probability of failure, thus requiring an additional effort.

Consequently, they advocate the use of simulation technique and discredit the analytical technique —especially stochastic dynamic programming. In the proposed simulation technique, the Monte Carlo method is used in deriving the optimal value for any particular parameter of a reservoir system. The constraints are imposed by setting maximum permissible probabilities of failure. An example is given for the analysis of a multi-purpose single reservoir. The operation of the system is simulated to supply the contract levels of hydropower and water while other parameters of the system are fixed. The input to the reservoir is a set of streamflow values generated by the Monte Carlo method. The length of the generated set is not specified, although it would apparently affect the accuracy of the result. Several initial conditions are used. Whether or not the system fails to meet its contract demands at any point in the generated record is noted. The percentage of generated records during which the system fails to meet the contract defines the probability of failure for the given conditions. These percentages are shown by a graphical representation or the probability table. From various probability tables the success or failure decisions and the search for optimal conditions are carried out manually although a computer procedure is also possible.

The major findings of the study by Askew et al. are: (1) the Monte Carlo analysis needs no definite form of mathematical model but requires the preservation of historical hydrological characteristics; (2) the simulation study can easily accommodate constraints and handle a variety of release rules, its concept is simple, and its computer programme is speedy and efficient; (3) it is desirable to design a water resources system to meet a given probability of failure which should be included in the design criteria, (4) the accuracy of the simulation results is related directly to the sample size and hence to the computation cost and also to the errors in the Monte Carlo method; (5) a complete probability table showing the results of simulation analysis can be used as a guide to system operations or as probabilistic predictions of failure to operators; and (6) hydrological forecasting can be incorporated in both the design and operation of simulation analysis and thus its practical and economic benefit can be ascertained. However, the paper acknowledges the amount of computer time as a limiting factor in the analysis and the errors that may occur in the Monte Carlo method.

Biernacki and Piwecki formulated a mathematical simulation model for a system of two reservoirs connected by an open channel for the production of hydropower. This model can be extended to more than two reservoirs connected in series, or perhaps also in branches, by open channels. The power output from a reservoir is a function of the head and the release from the reservoir. The flow profile and the discharge in the open channel are computed by the hydrodynamic equation of motion and the equation of continuity using the finite difference method. The total cost of power produced by the hydropower plants is then maximized by repeated computation for fitting the total hydropower into various portions of a given load diagram. In the model, the cost function in terms of the power and the head in terms of the reservoir storage and inflow and the channel discharge are provided. The releases from the two reservoirs are assumed equal although it seems that different releases can also be incorporated in the model without difficulties. It was found that the computer solution of the full mathematical model in this analysis increases the power output by some per cent as compared to the result obtained by hand computation.

The five papers being reviewed indicate the preference of a certain technique of optimization. However, it must be accepted that a fair comparison between the analytical technique and the simulation technique should be judged from the advantages and disadvantages with respect to its application to a particular problem as described at the beginning of this report. A blank statement such as 'that many of the seemingly sophisticated analytical methods are not easily comprehended, are very difficult to implement, and many even fail to provide the information' seems to be rather misleading and overstating. Experience shows that it is advantageous to use the analytical techniques whenever possible since it produces quick and direct answers if the system is not too complicated and can be modelled adequately. The Shasta Reservoir used in the Askew et al. paper is a single unit and the contract levels are only for hydropower and water supply. The dynamic programming used by Cole and Lopez Garcia, for example, may be applied just as well and perhaps be more time saving. When the simulation technique is used with Moante Carlo input, it would still be desirable to check the probability distribution of failure associated with the output so that the adequacy of the simulation model could be assured. Furthermore, the additional effort to determine the probability of failure in the stochastic dynamic programming might overcompensate the lengthy computation time usually needed in a simulation analysis.

Recently a new optimization technique known as the discrete differential dynamic programming (DDDP) has been developed (Heidari et al., 1971). This technique can not only greatly save the memory and computing time requirements of the computer being used, but can also handle more state variables than does the conventional dynamic programming. The use of DDDP in lieu of the common dynamic programming technique should make the analytical technique more attractive.

It is recognized that the ability of the Monte Carlo generating technique to simulate representative future streamflow records has yet to be improved. Experience has shown that the most efficient and accurate techniques for generating streamflow often produce generated data having quite different statistical characteristics from those of the historical data, mainly due to arbitrary and artificial procedures of adjustment that are built into such techniques. Future research on this subject is indeed needed.

REFERENCES

Chow, V. T. (1964) Water resources – Part II. Systems design by operations research. In *Handbook of Applied Hydrology – A Compendium of Water Resources Technology*, (edited by Ven Te Chow) pp.26–36 to 26–38: McGraw-Hill, New York.

Chow, V. T. and Meredith, D. D. (1969) Water resources systems analysis – Part IV. Review of programming techniques. *Civil Engineering Studies, Hydraulic Engineering Series No.22:* Department of Civil Engineering, University of Illinois, Urbana, Illinois.

Heidari, M., Chow, V. T., Kokotovic, P. V. and Meredith, D. D. (1971) Discrete differential dynamic programming approach to water resources systems optimization. *Water Resour. Res.* 7, No.2, 273–282.

Windsor, J. S. and Chow, V. T. (1971) Model for farm irrigation in humid areas. *J. Irrig. Drainage Div., Proc. Amer. Soc. civ. Engrs.* 97, No.IR3.

DISCUSSION

J. A. Cole:

I would like to explain that a 10-month year was purely a computational convenience and that in a real case a much finer time division could readily be introduced. By so doing, the problem of relating short-term damages to mean discharges was substantially overcome: even without a very fine time scale, a fair correlation would exist between mean flows and flood damage.

My paper offered a far more objective balance between water supply and flood control than had been possible by design flood methods. The debt to the latter, especially to the cited work of Beard, was acknowledged and the rapprochement of the two methods lies in simulation. I would like to introduce some recent results of simulating the system, these show the extent to which the various control rules (found by dynamic programming) differ in practical effect. Figures 1 and 2 show the deficits and

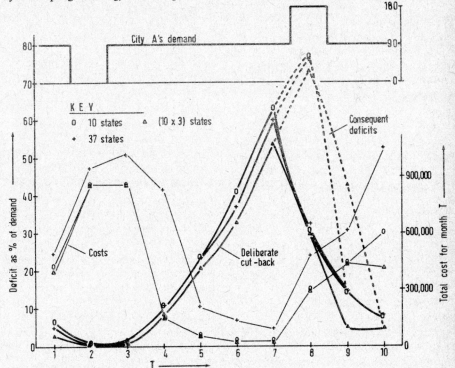

FIGURE 1. Mean deficits over 45 year simulation and total costs over 45 year simulation.

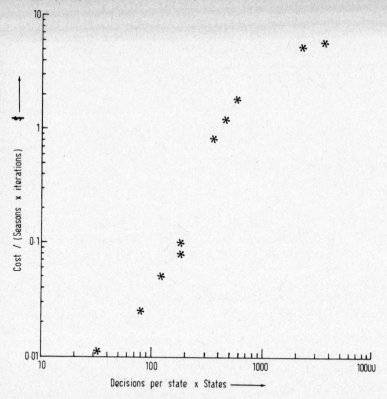

FIGURE 2. Costs of stochastic dynamic programming.

costs arrived at via a 45 year run of monthly data, which were read to 4 significant figures in contrast to the discretized values taken in by the dynamic programming (DP) calculation. Note was taken of how the DP created a strategy of anticipating shortfall, as seen where the peak of the deliberate outback occurs before the month no.8 of maximum demand by City A. The results in Fig.1 also show negligible cost difference between the 10 state and 10×3 state control rules, despite a slight disparity of deficit percentage between them. The fractionally higher costs for a 37 state simulation were attributed to the better detail given to high flow events.

Finally I would like to quote computational statistics of DP runs, made by myself and colleagues on both sides of the Atlantic (Table 2).

Costs are then compared as a function of problem dimensions as in Fig.2. Here the ordinates are the costs per stage of the calculation (taken from the bottom line of the table) and the abscissae are the dimensions of the state and decision variables at each stage. It was noteworthy that the type of computer made rather slight differences to the cost but that one's efficiency at reducing state dimensionality made an enormous difference.

A. J. Askew:

In our paper we have described one phase of a project which we undertook at the University of California.

In an earlier phase of the project we did some work on the comparison of generating techniques, but this was not its main purpose and I will not discuss this question here because it has already been dealt with at great length earlier in the Symposium.

Our association with the State of California both financially and practically through

TABLE 2

Program	LFAR		TWRS		FCØR		AUFGEP		STOUR	
Language	Fortran IV		Fortran IV		Fortran V		Fortran V		Fortran V	
Computer	IBM 360/75 Triangle Univ. Comp Center N. C., USA		IBM 360/85 Geological Survey Washington, D. C., USA		Atlas 1 London University Computing Centre, England		Atlas 1 London University Computing Centre England		Atlas 1 London University Computing Centre, England	
DP type	P. Imp†		P. Imp		V. Itn‡		P. Imp		V. Itn	
(Inflow x Content) states	30	30	90	144	30	37	11	51	79	31
Decisions/state	6	6	24	24	12	12	11	11	–	
Season/year (or equivalent)	10	4	4	1	10	10	6	6	12	12
Average iterations required	3	4	2	3	3	3	3	3	3.3	2.9
Compile store*	100K		94K		160K		112K		29K	
Execution store*	96K		188K	386K	57K		32K	56K	10K	
Compile time (min: s)	0:17		0:12		0:6		0:4		0:6	
Execution time (min: s)	0:33	0:26	7:36	2:36	1:18	3:20	0:2	2:23	0:84	0:33
Approx. cost of exec. time	$3	$1.25	$43	$18	$25	$36	$0.85	$33	$1	$0.4
Execution cost/(Season x Duration)	$0.10	$0.08	$5.40	$6.00	$0.85	$1.20	$0.05	$1.85	$0.025	$0.11

* K = thousand words (Atlas words of 48 bits or IBM words of 32 bytes).
† P. Imp = policy improvement DP.
‡ V. Itn = value iteration DP.

the State Department of Water Resources meant that our work was at all times practically orientated. For this reason we analysed real systems and used real design data, such as a fixed month by month distribution of the water demand. I accept that in practice such a demand would be stochastic; it was not within the terms of reference of our project to use other than a fixed time distribution of demand, but I feel that an extension of our work to cover such cases would be of great interest.

The general reporter has referred to an omission in our paper and I would like to correct this by stating that the design period for the system was taken to be 50 years, and the streamflow was therefore generated in sequences of 600 months.

Let me at this stage reiterate my statement that if you want to optimize a system for a given input then you should use dynamic programming or some similar analytical technique, and if you wish to maximize the expected value of the return, use some form of stochastic dynamic programme. In such cases I would wholeheartedly support the use of models such as those discussed at this session. But how often in reservoir design or operation are we in a position to confidently base our design on the optimization over a sequence of flows that is very unlikely to occur again? Or how often can we risk designing or operating in accordance with an expected value of return when this leaves us with a significant probability of failure?

Therefore I see dynamic programming and simulation as solutions to two different problems and what we have tried to do is to point out where we feel dynamic programming should not be used. In a number of places in this discussion, the general reporter has spoken as if the two are alternatives, for example by suggesting the use of simplified analytical techniques to give a first approximate answer. I am of the opinion that the application of deterministic dynamic programmes could give very misleading results in those cases where I have indicated that simulation should be used.

I feel that too much emphasis has been placed on the cost of simulation techniques in terms of computer time. By using such techniques, one can obtain far more information about the response of the system than is contained in the values given by dynamic programming. If we try to overcome the lack of any probability associated with the result by adding a stochastic element to the dynamic programme, then computation time is greatly increased and is very likely to exceed that required for an equivalent analysis using simulation. We had a very fast dynamic programme at the University of California, Los Angeles, but speed and efficiency are compensation for the lack of usable information provided by the analysis, and the accuracy is certainly no greater. The limiting factor in simulation is often the cost of computer time, so it is with most forms of analysis, but this does not mean that it is prohibitively expensive.

Finally, let me say that I cannot agree with the general reporter that 'experience shows that it is advantageous to use the analytical techniques whenever possible'. It is my experience that analytical techniques should be used for some problems and simulation for others. The question is not so much one of speed or accuracy, but of appropriate application. It was our purpose in writing this paper to show where and how simulation should be used in preference to analytical techniques.

L. Lopez-Garcia:

There is one point I would like to emphasize about the use of the three reservoir, dynamic programming model. As I have discussed in the paper, the model can be applied to optimize the operation either of simple systems fitting into the assumed Y configuration, or of complex systems which can be reduced to this pattern. In both cases the model can yield the final solution to the problem. However, due to the stochastic nature of the streamflow sequence, that solution may not be very reliable, even using some kind of Monte Carlo analysis. However, the optimal policy derived may be useful for setting the release rules of a subsequent simulation study. In this way the model would work as a secondary tool in the process of defining the final optimal solution.

Moreover, I agree with the general reporter in that it could even be used as the primary tool in finding the final solution of some particular problems. It seems that the model includes too many uses for a dynamic programming formulation. In fact, only four decision variables account for hydropower releases in all three reservoirs and the diversion between two of the reservoirs. The other decisions are either constrained or penalysed if not fully met.

So much for general remarks. I would like to try to answer the points raised by Dr. Chow in his general report. The first point is the flood control consideration. The method used is to constrain the reservoirs to reserve an empty storage space every month in order to damp the peak flow to a desired level. This storage is defined by taking into consideration the probability of occurrence of different floods and the permissible level of damage. I agree that this assumption may not be optimal since the flood damages are constrained. I think the model could be improved without further increase in computational difficulty by using the technique proposed by Cole in the paper discussed in this general report.

The second point is the consideration of the low flow regulation and municipal water supply as constraints. I believe that considering low flow as a constraint would be close to the optimum although this has to be proven. The water supply demand was based upon the fact that, in Spain, it has priority over other users. Therefore, in a multipurpose reservoir in which the irrigation demand is not negligible compared to the water supply demand the latter would have to be met; if not, in case a shortage of water supply was produced, the irrigation demand would not be met at all, so the reservoir storage would be obviously underestimated. Anyway, as I have discussed in the thesis in which this work was presented, a loss function similar to that of irrigation could be included in order to consider those cases in which the irrigation demand is irrelevant. The computational requirements would increase very slightly.

The last point is that of the Monte Carlo method as a 'short-cut'. This is a clerical mistake, since it should say 'short-coming'. The other methods discussed could be more properly defined as simplifications to be used when the Monte Carlo method is economically unfeasible according to the problem being solved.

W. Trau:

I wish to thank the general reporter for the comment on the paper we submitted. I would like to give some explanations concerning the paper and make a few remarks on questions raised in the general report.

For the reservoirs we investigated in the Harz Mountains, approximately two-thirds of the total annual runoff is observed in the winter season and only one-third in the summer season. The degree of low flow regulation and the extent of flood protection corresponding to a flood with a recurrence interval of approximately thirty years depend on requirements of water quality and the capacity of the downstream reaches.

The seasonal demand of water for municipal and industrial purposes is known. The maximum reservoir capacity, generally, is limited by the maximum storage level depending on the altitude of built-up areas in the region of the reservoir. These limitations often have to be taken into account. Thus, limited storage space and requirements of water supply determine the constraints for the operation of the reservoirs which have been investigated. The storage model, which provides information of a certain storage capacity with a certain probability at a certain season of year as a result of a certain operation plan, is used to select that operation plan which minimizes the failures of the reservoir. The failures are defined as states in which either emptying or spilling of the reservoir occurs.

In the procedure of computing, the operation plan is modified in a stepwise fashion, as long as a nearly optimal operation plan is found. This procedure takes on average, at least ten to fifteen runs. The computer time for one run is approximately 150 s on an

ICL 1907 computer and the period of reservoir operation covers 500 years or synthetic monthly flow. This means that the total computer time necessary to find a nearly optimal operation plan is approximately 30 min.

It is easy to assess benefits of supply of water for municipal and industrial purposes. Benefits of flood protection and low flow regulation, however, are difficult to eliminate in densely populated areas, especially when danger to human lives is involved. For this reason any objective function expressing the benefits of reservoir operation has been exlcuded from the investigations.

The procedure of selecting a nearly optimal operation plan and the method analysis is illustrated in Fig.3. Optimum criteria are achieved by introducing limitations of probabilities. These are contained in the assumed density functions of outflow or storage. Those operation plans are considered to be nearly optimal, which are in good agreement with the limitations of probabilities.

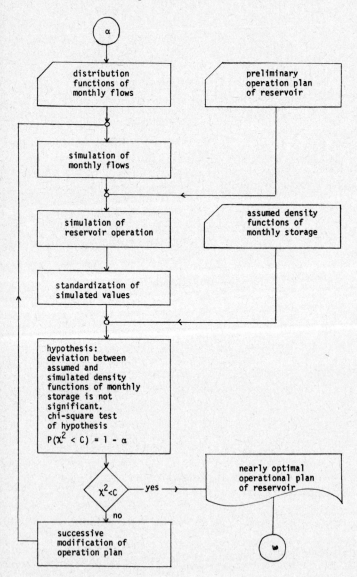

FIGURE 3. Approach of optimization of a storage system.

The system investigated is referred to in the general report to be a four reservoir system. This has not been interpreted quite correctly, for the system consists of four single reservoirs which have been individually investigated. However, the operation plans have been adapted to a certain extent. The storage model has been extended and includes now the variables: inflow, outflow, storage, storage level, and hydro-electric power generation, for each of which density and distribution functions can be easily determined.

V. Yevjevich

Two alternative methods are given to stochastic hydrology and to systems analysis, namely the analytical and the Monte Carlo or data generation methods. The first method developed for designing reservoirs, the so-called mass-curve, uses only historical data. It was designed by Rippl in England ninety years ago. This approach, called here the empirical method, is still widely followed at present. The question is whether one should neglect this empirical engineering method, and reduce the methods to only the above two alternatives. If the long-range regulation time interval is involved, say ten years or so, the two alternatives are either the analytical or the Monte Carlo (experimental) method, this latter method representing experimentation with computers. However, if one looks on the short-range problems, say what happens every year, then the historical sample and the empirical method give as many events or results as there are years. For thirty or forty years of data, there are 30 or 40 events or results, and it is reliable to use the current empirical method of the engineering profession. It is, therefore, necessary to define under what conditions one should use the empirical, analytical and experimental methods.

The point of controversy is whether the analytical method has a good future in practice, and how far one can push the application of the analytical method before being forced to use the experimental or Monte Carlo method. In general, the more complex a situation, the less feasible are the mathematical problems to tackle. Experience shows that the analytical method has been finding its way through, at least as the approximations, limitations, and boundaries of practical solutions. It is, therefore, safe to assume that the analytical method will be more and more developed with time. The limit potential of the analytical method is difficult to predict. In the meantime, one is forced to use the experimental method whenever the analytical method cannot be used reliably.

Z. Kaczmarek:

With reference to the general report, and in particular to the part of the report which deals wth the methods of optimization of the operation of reservoirs, I would like to describe a method which, jointly with Krajewski and other colleagues, we have applied in operational conditions of a storage reservoir, and which at the same time, may be treated as a kind of connection between the methods which were called analytical at this Symposium, and simulation methods. Let us first consider the time axis and a certain moment in time marked t_0, assuming that at this moment we know the volume of water stored in the reservoir, as well as the processes $q(t), x(t), y(t)$ describing the reservoir inflow in $t < t_0$, as well as other factors, such as, for example, the hydrometeorological factors. At the moment t_0 we have to make our decision about the operation rules of the storage reservoir. In order to be able to proceed further, I am going to divide the time axis into a certain number of discrete sections, t_1, t_2. As I said earlier, we used the method of deterministic dynamic programming for the operation of the reservoir, assuming as inflow either mean conditional values, or mean conditional model values. On the basis of thus determined inflow quantities, we have drawn up operational rules for the reservoir for a certain period in advance assuming, however, that these rules will apply to the initial period only. After a lapse of the initial period of, say, one month. we have repeated the whole procedure. Assuming that we are now at t_0, we have repeated the calculation in using the operational conditions of the reservoir. This made it possible to forecast the rules of operation of the reservoir. Applied to one reservoir in Poland, the method, in my opinion, produced satisfactory results. I think that it could be also used, as I have already mentioned, as a kind of connection between the method of dynamic programming, and the method of simulation when planning a reservoir, or drawing up the rules of operation for a reservoir. In that case we would pass through a certain historical or generated sequence of inflows having at our disposal actual, or generated, values of x, y, etc. It should be possible then to control the course of operation of the reservoir and evaluate the effects for a longer period of time. There is still the question of the necessary number of steps that should be taken account of in this process. As a basis for reasoning in our case, we have assumed such a number of steps which proved indispensable for the influence of the initial decision to be fading away. In the case of not very large reservoirs (we have evaluated reservoirs whose capacity constituted about 10 per cent of the average annual flow) and in such hydrological conditions which we have reviewed, that period could be defined to be a half-year period.

J. P. J. O'Kane:
In their paper Askew *et al.* state 'The stochastic dynamic programmes that have been developed optimize the expected value of the future return... The probability of failure associated with this value is quite substantial and yet is not controlled.' Penalty functions have long been used to control precisely this 'probability of failure' in stochastic dynamic programming. The mathematical basis for this was first pointed out by White (1969) and the necessary theorems may be found in Everett (1963). The theory is simple. Suppose we are dealing with an infinite planning horizon and we choose to use Howard's (1960) Policy Iteration Algorithm to optimize the gain, g, the average return per period from the use of the reservoir. The above problem may then be formulated as follows when the storage of the reservoir has been divided into discrete states:

$$g = \sum_i \pi_i q_i \quad \begin{matrix} \text{opt!} \\ \{k\} \end{matrix} \tag{1}$$

subject to the stochastic constraint

$$\sum_i a_i \pi_i = \eta \tag{2}$$

where π_i is the unconditional probability of being in state i; q_i is the expected immediate return from state i; a_i is the deficit in state i and η is the allowable average deficit. All variables except η are functions of k, the control rule, which may be found by Policy Iteration as follows.

First form the Langrangian function

$$g^* = \sum_i \pi_i q_i + \lambda \sum_i a_i \pi_i = \sum_i \pi_i \{q_i + \lambda a_i\} \tag{3}$$

Everett proves that if a value of λ can be found, such that the constraint (2) is satisfied, then the control rule $\{k\}$ which optimizes g^* without a constraint also optimizes g. The factorization in (3) clearly shows the Lagrange multiplier acting as a penalty. λ has all the usual economic interpretations (Everett, 1963). The resulting two-step algorithm is as follows:

(1) For a value of λ optimize g^* using Policy Interation.

(2) For the control rule found in (1), calculate π_i in the usual way and find the corresponding η, the expected deficit.

(3) If the value of η found in (2) is within ϵ (given) of the allowable value, stop with success; else search dichotomously on λ (or more efficiently if the properties of the deficit λ function are appropriate) and return to (1).

The Value Determination section of Policy Iteration is very expensive even when efficient decomposition techniques are used. Hence the obvious questions: why not (a) replace Policy Iteration in (1) by Value Iteration and (b) replace the direct calculation of η in (2) by a Monte Carlo experiment? This approach is of course less accurage but its practical use has been demonstrated in recent papers (Mawer et al., 1970). The main point of this contribution is the antithesis to the view of Askew et al. and contrasts somewhat to that of the general report namely, that stochastic dynamic programming (DP) and simulation in fact complement one another and are not in competition when used correctly. While the accurage probabilistic description of reservoir inflows in DP requires many expensive state variables, this is not so in simulation. On the other hand, simulation requires a control rule to operate and stochastic DP is eminently suited to provide it. They become self-correcting in the modified two-stage algorithm above. When DP has elucidated the functional form of the control rule then it may be replaced on occasion by a hill-climbing optimization routine. The above remarks also apply to the papers by Cole (this session) and Wyatt et al. (session XVII) where two constraints are given.

The remarks of Askew et al. are quite in order when applied to deterministic dynamic programming or its variant Monte Carlo DP (Young, 1967) when either is used in a stochastic environment. (Recall that the transition function in stochastic DP is a conditional probability distribution which depends on the chosen release of water and the probability distribution of reservoir inflows. Transitions from state to state in deterministic DP depend on the reservoir inflow *which is given exactly* and on the chosen release.) Monte Carlo DP is based on the misconception that the averaging and minimum operators are commutative. In simpler terms, assuming total knowledge of any number of equally likely future inflow sequences leads inevitably to super-optimal solutions. Clearly a clairvoyant can produce a far better response from a reservoir than a mortal blinded by the 'arrow of time'. No amount to averaging over a set of clairvoyants will alter this bias. Remarks similar to these apply to the papers by Garcia (this session), Mobasheri (session XVII) and Buras (session XVII).

REFERENCES

Everett, H. (1963) Generalized Lagrange multiplier method for solving problems of optimal allocation of resources. *Operations Research* 11, 399–417.
Howards, R. A. (1960) *Dynamic Programming and Markov Processes*: MIT Press.
Mawer, P. A., O'Kane, J. P. J. et al. (1970) Some conjuctive desalting studies in the United Kingdom. Paper presented at the *Sixth American Water Resources Conference*, Las Vegas, Nevada, USA.
White, D. J. (1969) *Dynamic Programming*: Oliver and Boyd, Edinburgh.
Young, G. K. (1967) Finding reservoir operating rules. *Proc. Amer. Soc. civ. Engrs, Hydraul. Div.* HY6, November.

Closing remarks by the general reporter

I am very grateful to the enthusiastic responses from the authors of the papers and other discussers, which prove that my report has served its purpose, that is to stimulate such an interesting discussion. Cole has mentioned that all of us arrived at this area of study rather recently, say, at most, in the last ten or fifteen years. So we have a great deal to learn in the future, particularly when we apply these theoretical methods to actual problems. Because our experience with actual problems is very limited, and what we are talking here is mostly academic, we still have yet to prove that these methods are workable. There is a great need for field experience in order to actually apply these methods. In application, it is necessary to develop a kind of dialogue or communication between the theoretical hydrologists and the practicing professionals. Also, in applications, there is a need to translate many intangible factors, such as the social and legal aspects of water resources problems into numerical quantities so that the theoretical computer analysis can be performed. I am sure that future improvements and developments will still change our way of thinking. Therefore, we should not hold our views so firmly that may hinder our progress. So this is the time, I think, at which all of us should be encouraged to do more investigations on the subject, regardless of the difficulties we have encountered.

General Reporter Mr. John A. Cole, The Water Research Association, Ferry Lane, Medmenham, Marlow, Buckinghamshire, UK

Chairman Professor Dr. Tomasz Biernacki, Professor of Water Resources Management, Gdańsk Polytechnic University, Gdańsk, Poland

Optimal operation of water resources systems — II

John A. Cole

The list of papers reviewed and discussed in this general report is given below. All are published in volume 2 of the Symposium Proceedings.

1. *Yu. A. Arkhanguelsky and L. V. Dounin-Barkovsky* Mathematical model of water resources utilization in a river basin
2. *Nathan Buras* Integration of aquifers in flood control projects
3. *J. Cuena and E. Custodio* Construction and adjustment of a two-layer mathematical model of the Llobregat Delta
4. *F. Mobasheri and V. Budhraja* The conjunctive use of a multi-reservoir system and a dual-purpose desalting plant
5. *A. Sh. Reznikovszky* Methods for control of the regimes for water resources systems
6. *B. S. Verkhovskii* Optimal complex use of controlled water resources of a basin
7. *T. Wyatt, D. G. Thorn and P. A. Mawer* Streamflow regulation by artificial recharge fed from upstream surface storage: derivation of control rules

My task here is complicated by the fact that the seven authors differ greatly in the importance they attach to presenting

    (1) the general philosophy of their subject, expressed in prose,
    (2) the mathematical framework of modelling and optimizing a chosen type of system, expressed in algebra and in flow diagrams,
    (3) numerical results from a particular example, expressed graphically or as tables,
    (4) the computational efficiency of their methods.

Here our emphasis is on mathematical models, so words must rapidly give place to algebra and numerical output. I am therefore presenting tabular versions of the papers (Figs.1—7), in which they are set into the framework:

System (diagram)
Mathematical modelling

    (a) Input data
    (b) Objective
    (c) Methodology
    (d) Output data

Computational details
Example(s)

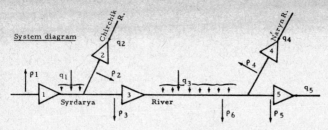

System diagram

P_i are supplies to irrigators
q_3 and q_1 are return flows from irrigation

Mathematical modelling

a)   Input data:-

updated each) month)	forecast inflows for monthly, quarterly and longer segments of a year; also limitation rule for water consumption when stored water happens to be inadequate.
at start) of year)	forecast of monthly evaporation and crop demand for irrigation season.

b)   Objective:

To maximize the expected value of irrigation water consumption, in face of uncertain demand.

c)   Methodology:

Both simulated and operational regimes are described, in which the optimization depends crucially on the authors' Algorithm 4. The block diagrams (Figs. 3 and 4 of the paper) show flow of data into and out of this algorithm, but the paper does not state how it works. (Presumably some dynamic programming scheme is used, either of the deterministic type operating on a simulation - or of a stochastic type where hydrology and water consumption lead to sub-divisions of storage states).

d)   Output data:

The simulation permits alternatives to be tested in respect of:

1.   the optimizing procedure itself

2.   how near to optimum are typical release patterns

3.   how water allocation to irrigation may be extended

4.   how reservoirs should have contents balanced

Given this knowledge one gets as the operational output:

a)   water releases from each reservoir ($U_{\alpha i}$)

b)   pre-allocation of water for irrigation ($P_{\alpha i}$)

c)   probabilities ($a_{\alpha i}$) of the water being required

Computational details   -

Example(s)   -

FIGURE 1. Tabular version of the paper by Arkhanguelsky and Dounin-Barkovsky.

Whilst hoping that this presentation will help you locate items of interest in each paper, I am duty bound to point out the absence of many relevant details from this set of papers. To present scientific and engineering work of this sort without numerical results, or without a system diagram, is almost unforgivable. Only one author saw fit to mention computer storage. The other side of the coin is seen in the care with which several authors spelt out the algebra of their method.

I was presented with such a diverse portfolio of papers to report that it did not at first seem profitable to compare them to one another. The majority (Arkhanguelsky and Dounin-Barkovsky, Buras, Mobasheri and Budhraja, Resnikovszky, Wyatt *et al.*) dealt with multi-storage systems; but Verkhovskii deals with a single storage and Cuena and Custodio deal with a distributed storage underground system.

As I have already said, the papers are not closely related in themselves and so it would be interesting to relate them to others in the literature. Luckily for me, the Secretary of our IASH Working Group on Water Resources Systems, Dr. Buras, has already done a useful general review (Buras, 1974) and it falls to me simply to refer each paper to its generic category. First note that of the seven topics here, five are addressed via hydrological simulation viz:

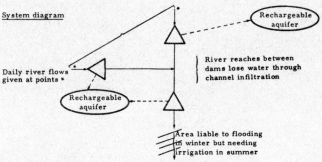

Mathematical modelling

    a)   Input data

                Sizes of dams and capacities of aquifers
                Initial storage of water in dams and aquifers
                63 days of river flow for flood season
                Maximum intake rates of aquifers
                Maximum safe flow in flood risk area
                Coefficients of channel infiltration losses

    b)   Objective

           The basic objective is to <u>minimize total winter flood volume</u>, in excess of the maximum safe flow, yet achieve full storage in the dams for summer irrigation. As it is evident that <u>maximizing infiltration losses</u> in the river reduces the downstream flood, the losses are included within the objective function. Obviously, if flood mitigation were the sole purpose, the system would be operated so as to release from all dams so as to satisfy infiltration losses and also to recharge the aquifers at their maximum feasible rate. But, to conserve water, the early releases from any dam can be inhibited (even at cost of later flooding). The paper illustrates this with release from dam 1 in the first time period included in the objective function, with an arbitrary weighting in favour of flood mitigation:

$$\text{Minimize}\left[\left\{2\sum_{t=1}^{T}\left(\text{flood volume}_t - \text{channel infiltration losses}_t\right)\right\} \pm \left(\begin{array}{c}\text{Dam 1}\\ \text{release}\end{array}\right)\right]$$

           The $\pm$ sign covers the alternative possibilities of making the release as large as possible (when −) or of only just satisfying channel losses in the uppermost reach (when +).

    c)   Methodology

           Linear programming subject to $(8T + 1)$ mass balance constraints, equations (5) to (14), and $8T$ bounds, equations (15) to (20).

    d)   Outputs

           Graphs of storage in dams, releases downstream and aquifer recharge.

Computational details

Example(s)   Figures 3 and 4 of paper.

FIGURE 2. Tabular version of the paper by Buras.

System diagram

The Llobregat delta's aquifer is in two main layers, separated by a semi-confining bed of lower conductivity (Fig. 2). The system is idealized as polygonal prismatic elements, generally asymmetrical in plan (Fig. 5). Transmissivity is taken as uniform within each segment of a prism.

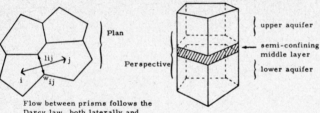

Flow between prisms follows the Darcy law, both laterally and vertically.

Boundary conditions are imposed on pumping rates to keep water levels within upper aquifer

Mathematical modelling

a)   Input data

From pumping tests and geological investigations, transmissivities and storage coefficients of upper and lower aquifers have been mapped. Correction for water level in upper aquifer can be assessed. Permeability of semi-confining layer is given. Recharge and pumping rates.

b)   Objective

To give a numerical simulation of the aquifer performance, under past and future pumping regimes. Test of fit is by comparison of observed and predicted well hydrographs.

c)   Methodology

Relaxation solution, for finite time interval, of new water level distribution given a starting level. The authors' equation (3) may be rewritten as:

$$H_{new} - H_{old} = (\text{Coefficient} \times \text{residual term of Eq. (1)})$$

i.e. adjustment of H at this iteration

depending on polygon size and conductance and storage coefficient

i.e. H_{old} values put for H_i and H_i and residual is difference arrived at between l.h.s. and r.h.s.

The particular interest here is that H_{new} affects the upper aquifer's transmissivity, which is adjusted if need be and repassed through the calculation.

Also if H_{new} encounters the upper or lower bound of the upper aquifer, pumping rates are increased or decreased locally to compensate.

d)   Output data

Hydrographs for water levels and piezometric head

Computational details

Model had 151 nodes and 394 connecting branches
Time step of 0.125 years was satisfactory.
Run on 32K word (16 bit) computer memory.

Example(s)   Graphs of observed and computed results (Fig. 6 and 7)

FIGURE 3.   Tabular version of the paper by Cuena and Custodio.

Arkhanguelsky and Dounin-Barkovsky show, in their block diagram (vol.2, p.928), a branched chain of reservoirs, with intervening water consumption areas (simulation of monthly hydrology, with return flows from irrigation expressed as a function of crop response to climate). Buras shows three reservoirs with intervening channel losses (daily simulation over 60 day flood risk period). Cuena and Custodio are simulating flows in an aquifer (1.5 month time increments, with the continuous aquifer discretized blockwise). Mobasheri and Budhraja have a reservoir chain [operation with a desalination plant is derived via deterministic DP, coupled to a weekly (?) simulation]. Wyatt *et al.* have a reservoir plus aquifer system; monthly simulations embedded within optimization via steepest ascents on DP for monthly control rules.

The other two papers, by Reznikovszky and Verkovskii deal with inputs that are

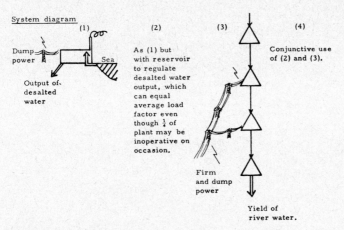

System diagram

(1)

Dump power

Output of desalted water

Sea

(2)

As (1) but with reservoir to regulate desalted water output, which can equal average load factor even though $\frac{1}{4}$ of plant may be inoperative on occasion.

(3)

Firm and dump power

(4)

Conjunctive use of (2) and (3).

Yield of river water.

Mathematical modelling

a)   Input data

Demand patterns for water and electrical power.
Desalting plant capacities, capital and running costs.
Reservoir capacities, power generation capabilities and monthly hydrological record.

b)   Objective

To obtain optimal operation of Systems (3) and (4), knowing the costs of operation of Systems (1) and (2). Firm yields of power and/or water are maximized, by postulating one and varying the other.

c)   Methodology

Use of Larson's state increment dynamic programming approach (cf. 3rd reference in paper) the basis of finding control rules. The algorithm is not stated in present paper, however. Author's 4th ref. reveals that it was applied to deterministic data, which assumes foreknowledge. One doubts whether control rules so derived are truly optimal.

d)   Output

Graphs of power and water yield of System (4), for various desalting plant size:

Computational details   35 iterations per computer run

Example(s)   Tables 1 and 2 of costs for Systems (1) and (2) respectively.
Graphs of output and costs in Fig. 1 'no desalting' curve is System (3), others relate to System (4).

Graphs in Fig. 2 show how cost per volume supplied additional to System (3) yield goes through minimum where plant load factor is low, then rises abruptly to limit at right hand side at 100% desalting load. Parameter on graphs is desalting plant capacity.

FIGURE 4. Tabular version of the paper by Mobasheri and Budhraja.

<u>System</u>

Described very generally, as comprising storages and channels
in series or in parallel with hydropower as main purpose.

<u>Mathematical modelling</u>

    a)   Input data

         Hydrological information

         Power demand forecasts

    b)   Objective

         <u>To seek optimal power or water supply</u>, with penalty for deficits
therein; the rule should cater for uncertainties of input and output
needs to include short term rules within a more general strategy
of operation.

    c)   Methodology

         Suggested by author's equation (1) which relates L_{ij}
(a release rule vector ?) to a polynomial containing:

             current contents in all reservoirs

             antecedent flows for various lag times,
into all reservoirs

         it is stated that the coefficients of the polynomial can be obtained
in two ways. These are not explained in the paper, other than by
passing reference to regression procedures.

         Where hydropower plants have very large regulating capability,
these act as compensators, with a release rule depending on the
power already assured from the smaller fixed rule plants: cf
author's equation (3).

         The "successive correction for the regimes", i.e. a progressive
adjustment of release rule, is mentioned as being appropriate,
particularly for large systems with over-year storage and with
multipurpose use.

    d)   Output   -

<u>Computational details</u>   -

<u>Example(s)</u>   -

FIGURE 5. Tabular version of the paper by Reznikovszky.

defined probabilistically and are not input via historical or synthetic sequences.

To draw some general threads of ideas together, let me point to the following well
known facts:

(a) complex storage systems, including aquifers, are nearly always treated by discrete element algebra, in both time and space.

(b) the more complex the system, the higher will be its state dimensionality, so
sacrifices must be made in order to retain computation feasibility.

The corollary of (a) and (b) is that fine detail of system analysis must be addressed to
small segments of a system (e.g. we optimize the operation of reservoir pairs) then
such sub-systems are aggregated into the whole. A necessary third step is to apply
simulation on a much finer scale of resolved detail to a system so analysed.

Let me furnish you with a visual example of what I am describing now. Figure 8
shows a set of four storages feeding three demand points. The optimal allocation of
water within a time frame is dependent, of course, on our objective function. For any
given state of known contents and probability of inflow, one may use LP or related
techniques to distribute water to the demand points within a single time frame, as
indicated by the lines joining the nodal demands. But we are left with the overriding
problem of how much carryover storage to allocate, namely to cope with transfers
from one time frame to another, as the broken lines on Fig.8 show. The question is

System diagram

In the calculation
one would normally
make $\Delta Q = \Delta W$

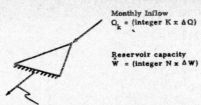

Monthly Inflow
$Q_k = (integer\ K \times \Delta Q)$

Reservoir capacity
$W = (integer\ N \times \Delta W)$

Release u_i in time period i goes to various
uses and causes effects (i. e. net monetary
benefits or other non-monetary returns)
which are functions d_i (u_i, uses to which the
u_i are put).

Mathematical modelling

a) Input data

Histograms showing probability of $(Q_k \pm \Delta Q/_2)$ over range of k
and as function of prior inflows.

Benefit functions d_i.

Reservoir capacity $\hat{W}$.

Revaluation coefficient β, which discounts net benefits within
a year to value at start of year.

b) Objective

To maximize the expected present value ('average summary effect')
of net benefits of operation, given an infinite time horizon of the
system.

c) Methodology

The basic mathematics detailed in Section 3 reduces to:

Step 1   Dynamic programming, using value iteration
(i. e. Bellman's algorithm, backward-stepping in time)
as in eq. (2) of paper, finds the u_i appropriate to various
reservoir contents (integer $r \times \Delta W$) for r = 0, 1...N
and a provisional probabilities of state to state transitions.

Step 2   Howard's algorithm for determining state values, given
as eq. (*) in the paper. An iterative method for solving
equation (*) is mentioned as a useful alternative to others
(e. g. matrix inversion).

Steps 1 & 2 are then repeated until such time as the control rule
converges.
Section 4 of the paper adds a rider showing how expected
net value of a particular starting state may be derived at
time of commissioning the reservoir, allowing for capital
costs of construction.

Section 5 remarks that this net value is itself a design target
to be optimized.

Finally, Section 6 of the paper expounds the chance constraints
and other bounds applicable to maximizing the net income achievable
by irrigation of a crop. The problem is reduced to one of selecting
an appropriate irrigable area, via a limit-balancing algorithm
referenced to one of the author's own publications.

d) Output data

Releases from a reservoir as function (k, prior inflow).

Irrigable area as function (available water, crop).

Computational details   -

Example(s)   -

FIGURE 6. Tabular version of the paper by Verkhovskii.

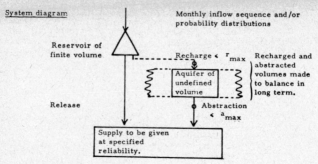

System diagram

Reservoir of finite volume

Monthly inflow sequence and/or probability distributions

Recharge $<$ r_{max}

Aquifer of undefined volume

Recharged and abstracted volumes made to balance in long term.

Release

Abstraction
$<$ a_{max}

Supply to be given at specified reliability.

Mathematical modelling

Alternative procedures are compared as shown below under two columns:

Hill climbing procedure	Dynamic programming with embedded simulation

a) Input Data

| ~30 years of historical monthly inflows to system. | Frequency histograms of historical flows (unconditional) in segments that match integer states of reservoir contents. The historical inflows are also required as a sequence. |
| Reservoir capacity.
Starting position of control lines. | Reservoir capacity.
Starting values of cost parameters λ and $\emptyset$. |

b) Objective — To meet a defined demand with minimum use of the aquifer subject to a tolerable value of deficit over the simulation ('target deficit').

c) Methodology — The optimal control strategy is pursued via the positioning of two control lines of an 'on/off' character determining when recharge to or abstraction from aquifer is required.

| 'Hill-climber' makes test changes to control lines and alters these for all 12 months in proportion to improvements achieved. Flow-chart (Fig. 2) shows step size varied as problem reaches convergence. | Value-iteration d. p. (Fig. 5) finds control rules, whose effect is tested by subsequent simulation (Fig. 4). Cost parameters λ (balance of aquifer) and $\emptyset$ (target deficit) are adjusted accordingly and d. p. rerun until constraints met. |

d) Output

Control rules.
Monthly, annual and long-term volumes of abstraction, recharge and deficit.

Computational details

	Convergence necessitated 20 iterations on only data tested.

Example(s)

Fig. 6 shows the differing form of control rule derived the two ways; end result in terms of recharge and abstraction very similar however.

Fig. 7 shows great variation of use of the aquifer according to reservoir storage provided. Also choice of one-line or two line control rule makes major difference.	

FIGURE 7. Tabular version of the paper by Wyatt *et al.*

△ source with storage
○ demand points

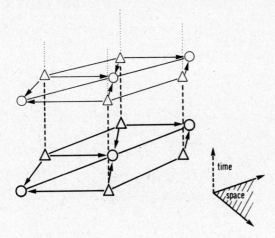

FIGURE 8. Allocation of stored water.

now whether to address the complete time and space problem, say via simulation and search techniques, or whether to decompose the problem either into vertical slices (i.e. reservoirs treated separately from each other, keeping inter-reservoir transfer rules constant) or as horizontal slices (i.e. treat each time frame separately, whilst maintaining a fixed carryover rule from time to time). I leave it to subsequent discussers to pursue this question in more detail.

Although we only have three authors of session XVII papers present we have no need to fear a shortage of discussion on this huge topic of multi-element water resources. Our Polish hosts have kindly consented to bring forward their papers from the end of the Symposium, so that we may learn of the planning and development of the Vistula project during this session. The three papers from topic 9 to be discussed now are the following.

1. *Maciej Jedrysik, Aleksander Łaski and Jan Zieliński*   General description of the Vistula River Project and basic planning data
2. *Z. Kaczmarek, K. Krajewski, T. Kornatowski, A. Filipkowski, J. Kindler and D. F. Kibler*   The multi-step method for simulation and optimization of Vistula River planning alternatives
3. *Ian P. King, Janusz Filimowski and Janusz Kindler*   The out-of-kilter algorithm of a single-step method for simulation and optimization of Vistula River planning alternatives

REFERENCES

Buras, N. (1974) Mathematical modelling of water resources systems (state-of-the-art report prepared on behalf of the IAHS Committee on Mathematical Models for the XVth General Assembly of IUGG, Moscow 1971). *Hydrol. Sci. Bull.* **19**, No.4, 393–400.
Chow, Ven Te (1969) Water resources systems analysis — part IV. *Review of Programming Techniques:* Report HES No.22, University of Illinois, Dept. of Civil Engineering.
Roefs, T. G. and Bodin, L. D. (1970) Multi-reservoir operation studies. *Wat. Resour. Res.* **6**, No.2 410–420.
Stephenson, D. (1970) Optimum design of complex water resource projects. *Proc. Amer. Soc. civ Engrs, J. Hydraul. Div.* **96**, 1229–1246.
Young, G. K., Moseley, J. C. and Evenson, D. E. (1970) Time sequencing of element construction in a multi-reservoir system. *Wat. Resour. Bull.* **6**, No.4, 528–541.

DISCUSSION

J. Kindler:

Three papers on the Vistula River Project have been presented today. The first of them contains an introduction to the problem, and a description of the situation in the Vistula basin. The next two discuss some of the tools which have been used in our search for an optimum solution in the Vistula basin.

I think it would be superflous and unnecessary for me to present a detailed description of the water situation in the Vistula basin because these problems were discussed, rather extensively, in the first paper presented to you earlier today. I think, however, that it would be worthwhile to mention three problems which we considered of paramount importance in regard to water economy in the Vistula basin to which, in a sense, we have subordinated our studies. These are, above all, problems of water supply for the population in urban areas, for industry and agriculture, protection against floods, and water quality control. As regards the hydrological basis of our project, we have confined ourselves to historical hydrology and the mean monthly flows for a 15-year period (only in the reservoir studies were stochastic properties of river streamflow taken into account). Thus the surface water resources are characterized by 180 variable mean values of monthly flows. Serious studies have also been devoted to the evaluation of groundwater resources. As regards water use demands, we have also tried to introduce, to the widest possible extent, the concept of substitute measures. Thus, trying to define the water demands of various users, we have also examined the possibilities of changing the technology of certain industrial or agricultural processes. In the final analysis of costs and benefits, the effectiveness of these and other solutions was compared.

I would like to discuss briefly the papers prepared by a Polish group and the specialists from Water Resources Engineers Inc., as the whole modelling work was completed by the Polish—American team acting within the framework of a UN project. In view of the fact that the Vistula basin covers a large area, it was decided to divide the basin into twelve subsystems. Simultaneously, in view of the complexity of the project, a problem decomposition was carried out.

Two methods were proposed in the course of work on the project. One is a single-step method, which is discussed in the last of the presented papers, the other is a multi-step method which, strictly speaking, contains three steps, which were instrumental in solving that problem. The ultimate objective of that part of the project was to carry out an optimal allocation of water resources to the different types of users in each of the river basin subsystems. The three-step procedure consists of the following three programmes: a so-called target-output programme which makes it possible to define the target releases from storage reservoirs in the given subsystem; the next step was to establish optimum policy rules for each reservoir, in which case we have applied stochastic dynamic programming. The final step was, after the transformation of natural water resources in control profiles on the basis of the already defined optimum policies for the storage reservoirs, to carry out the allocation of water to individual users. The first of the programmes, initially, was to use the linear programming algorithm. However, it turned out that, in view of a vast number of users represented in each subsystem, we were faced with matrices of very large dimensions, and it proved difficult to solve that problem with the computers available in Poland. That is why we used the out-of-kilter algorithm, which is a specialized algorithm of linear programming. With the aid of this algorithm, we have also solved the allocation problem and, in principle, the same algorithm was used in the first and the third programmes.

We have also used the out-of-kilter algorithm in the single-step method. This method is, in fact, very close to the last one of the programmes in the three-step method. Moreover, this method allows the allocation of the water resources to be effected simultaneously with the optimization of the reservoir operation policies.

This, of course, falls short of covering all the modelling work implemented within the framework of the Vistula project. Taken together, they make up a certain entity, but, unfortunately, we have to do with many elements which we have been unable to quantify. The final analysis of costs and benefits takes into account both the results achieved owing to the mathematical modelling which I have already described, and the results of many more traditional calculations which ran parallel to it.

Finally, I would like to underline an important thing, namely, that apart from our decision to base our project on the maximum possible use of modern planning techniques, we have agreed that a so-called post-optimization analysis, into which we have incorporated, as far as possible, all those factors which we were unable to quantify and for which we could not find a suitable model, is going to be the final step. We realize that much still remains to be done in relation to the Vistula project. We are fully aware of certain imperfections and some simplifications which are due, in considerable measure, to the extent of the problem we have to deal with. We hope that, as our studies advance, we shall be in the position to define more accurately, if not an optimum, then, at any rate, a near optimum investment programme for the development of the Vistula river and her main tributaries.

K. Krajewski:

I would like to make a few remarks about some problems which we were obliged to solve when working on the task which was previously mentioned by Mr. Cole, and which was discussed by Kindler. I should like to emphasize that the solution of such a big task could be based solely on modelling. We were obliged to choose the best of the possible investment variants for a very large area. We had the choice of a considerable number of hydrotechnical projects, above all, reservoirs. I would like to draw your attention to the fact that it is precisely modelling that can best help us in difficult and complicated situations. Modelling, in our case, was based on the utilization of the observation records of water resources. But we could have, equally well based our modelling on the sequences generated, for instance, by the Monte Carlo method. This would not have made the least difference to our proceedings. We felt, however, that our observation period was too short for generated flows to base on a large area. We have evaluated the problems of water supply for agriculture, industry, as well as problems of protection against floods. In a similar manner, simulating the operation of reservoirs given the assumption that we have at our disposal flow sequences for a given period, we were faced with the following alternatives: either to treat in more detail the operation policies of the reservoirs considering each of them separately, or to deal more closely with the operation of a system of reservoirs. Unfortunately, we had at our disposal a method defining a policy for one reservoir. It appeared to us, however, that failure to take into account the cooperation of the reservoirs within a system could result in many errors. Eventually, as Kindler has already mentioned, we adopted a method which takes into account the cooperation of reservoirs in which, by the nature of things, the reservoir policies are defined with a fair amount of approximation. I would like to underline that these rules result from the solution, at each stage, of a certain strictly analytical problem. Namely, we have tried to carry out the allocation of water resources in each of the 180 months using the out-of-kilter algorithm, or, (what makes for the same result) solving a large problem of linear programming. Does this suggest that our model is a deterministic model? I do not think so. Neither is it a purely stochastic model as it was understood by many discussers here. I think that we could try to look at this problem in the same manner as we look at a photograph or a film. We do not know that in a film we get the illusion of movement owing to a series of completely static and motionless shots. I am not going to take your time by describing the methodological details involved in the solution of that problem, but I would like to mention that the method is still in the course of development. The preliminary calculation results are being analysed, and, therefore, what is presented in the papers should not be regarded as a final proposal of a completed method. I think that, drawing on our experience so far, as well as experience gained at this Symposium, we should be able, at some time in the future, to define, in more detail, the recommendations concerning similar problems for large areas and problems which are connected with the operation of large water resources systems.

J. Cuena:

Je veux souligner, d'abord, que l'objet de notre papier est la déscription d'un modèle realisé pour étudier les problèmes concrets du delta du Llobregat en Espagne. Ce n'est pas, alors, un cas théorique, mais plutôt une compte rendue d'un cas pratique résolu avec les moyens théoriques dont on disposait. La justification mathématique demandée par M. le rapporteur générale, du procédé de maille asymmétrique s'appuie sur les travaux de Tyson & Weber (California Water Resources Development) et notamment dans l'aspect mathématique dans les travaux de MacNeal publiés en 1953.

Le type de maille choisie assymmétrique venait imposée par le besoin de s'adapter en conditions acceptables du point de vue précision de calcul et exploitation sur ordinateur, aux conditions de contour et aux courbatures atteindues dans la surface piezomètrique.

En effet, l'aquifère a modèle (Fig.5 de notre papier: tome 2. p.958) à la forme trangulaire avec un tronçon étroit en tête qui corréspond à la partie haute dans laquelle l'aquifère est alimenté par le fleuve Llobregat.

Aussi l'exploitation est très concentré en quelques points ce que produit dans ces zones localisées fortes courbatures à la surface piezomètrique. En cas d'utilisation de maille uniforme rectangulaire sa dimension devrait venir imposée par la corréspondante à la courbature dans ces zones, ce qui amenerait à une maille avec beaucoup d'éléments et par conséquence à une démande trop fort de mémoire ou à un temps de calcul très grand dû au nombre d'inconnues du système a resoudre chaque intervale du période a simuler (quoique ça peut être diminué par la forme de la matrice du système dans le cas de maille uniforme).

Avec la méthode de maille assymètrique on arrive s'adapter suffisament aux contours des aquifères et à la courbature de la surface piezomètrique avec un reseau de 151 éléments prismatiques de base polygonale avec une relation maximale de surface entre le plus grand polygone et le plus petit de la environ.

Le plus forte inconvenance de la maille asymmètrique est qu'elle faut faire la construction de la maille et le calcul de caractéristiques géomètriques et hydrogéologiques mannuellement et en cas de changement de maille il faut. refaire les calculs. dans le cas du delta du Llobregat on n'envisage pas le changement de situation des zones d'extraction et pour ça il est possible de definir une maille asymmètrique sur la mésure.

Dans le cas de maille uniforme il est possible pour bâtir un process automatique de calcul des caracteristiques de la maille, et même sa construction avec ordinateur, ce qui donne une plus grand flexibilité au changement de dimension de la maille. Cependant dans notre cas dû aux questions procèdentes on a preferé la maille asymmètrique, tenant compte qu'il n'était pas envisagé un changement en la distribution spatiale de l'exploitation.

Une autre remarque interessante qui n'a pas inclus dans notre rapport est la façon de definir dans l'étape de calibration la relation entre l'aquifère et le fleuve. Le fleuve Llobregat alimente l'aquifère dans la partie haute sans communication et perméable par la loi de Darcy. pour definir la quantité d'eau d'alimentation en fonction du débit superficiel du fleuve on a considéré les polygones de cet zone aient des niveaux imposes variables dans le temps égales a ceux dont on ait des données mesurées par piezomètre dans le période de calibration. de cette façon apres les differentes simulations du période de calibration il a été possible definir pour les différentes périodes de l'année une relation entre le débit infiltré et celui du fleuve, cette fonction a été utilisée dans la campagne de calculs opérationnels avec la modèle calibrée. Cette campagne qui est actuellement en train de réalization permettra definir la qualité des aquifères dane l'aspect conducteur et regulateur, fournissant des critères pour l'introduction de cet aquifère dans le système du Llobregat.

A. Łaski:

I would like to refer to the development of the Vistula basin, concentrating on a small segment of the problem. I would like to draw your attention to the problem of one of the groups of the initial data which is of great relevance to the project, and of basic importance to both the calculations and the results. I am speaking about a group concerning data on technical projects which are designed to ensure the solution of the problem of water economy in the basin. This applies, of course, to such basic technical projects as reservoirs and water transfer canals joining the individual parts of the basin. This group of data is of particular importance as it is linked with the basic task incorporated in the Vistula project. The task of preparing initial data concerning technical projects proved to be a very difficult, complicated and large-scale undertaking. It was necessary to take into account not only the existing projects, and not only those which have already been analysed or evaluated in the course of studies conducted in Poland for a number of years, but also to consider all the prospective possibilities of creating, in conditions prevailing in the Vistula basin, a network of storage reservoirs and water transfers. This was important, because the idea of the whole Vistula project was to estimate the possibilities of water supply for the national economy in the years 1985 and 2000. This required far-reaching projections, concerning not only water demands, but also of the technical possibilities which are necessary to expand the water projects. In the circumstances, we have decided to incorporate in the project certain assumptions which have a decisive impact on the model and course of calculations. Faced with the problem of several dozen storage reservoirs, more than a dozen water transfers, and many variants of solutions for each of the projects in regard of its size and technical parameters, we have agreed that a project with definite parameters constitutes a variant of the model. Any change of parameters of the given project entails the automatic introduction of a new variant of solution to the Vistula project. We had a total of about 150 variants of technical solutions, and, taking into account the possibility of various combinations, we have come to the conclusion that a step by step procedure would be the most appropriate one. The division of the Vistula basin into subsystems, already mentioned here, was one element of that method. Another was the introduction of partial analyses which preceded the introduction of the data on technical projects into the model. There were a number of constraints making it possible for those variants to be introduced in the calculation. What we have obtained were certain groups, sets of technical variants, as certain projects excluded one another, in some cases partial analyses showed that a simultaneous implementation of projects was unrealistic, and in other cases, consideration for the present construction potential in Poland spoke in favour of a gradual development of certain parts of the basin. For each of the twelve subsystems of the basin, we have worked out sets of groups of variants of technical solutions characterized by constant technical parameters typical for that group in an effort to limit the number of variants in each subsystem up to, at most, about a dozen by means of partial analysis of variants. Here, of course, we have encountered another problem, that of linking together the individual subsystems and variants occurring in those subsystems. This problem has been also solved by means of traditional partial analysis linked with economic analysis. The results obtained from one subsystem were accepted as initial data for the next subsystem. Our aim was not to multiply combinations of variants in all subsystems. In our view, this kind of procedure which envisages parallel partial analyses, based both on economic and non-economic criteria carried out for the purpose of the gradual elimination of variants, makes it possible not only to limit the time used for the calculation, but, to a certain degree, contributes to the clarity of the picture which is already obtained in the course of the procedure. Using that system, it is not necessary to complete the calculation in order to obtain a clear picture of the situation. Already in the transitional phases of the project, certain solutions prove either unrealistic, or do not bear comparison with other solutions. As a final point, I would like to emphasize that the technical solutions that were evaluated should not be treated as alternatives devised especially for that project. This is the result of many years of work by numerous teams of Polish practitioners and scientists. As a side-effect of the Vistula project, we are now in the possession of an almost full gamut of possible technical solutions for the entire basin, or for two-thirds of this country's territory. We realize that a proposal of the investment programme to be formulated at the end of the project will still be short of an optimum proposal. Most likely, there will be two or three solutions oscillating around a certain optimum. We hope that the years to come, and further work envisaged for the years 1971–1975, will enable us to specify our proposals by means of still more accurate calculation methods. At any rate, we plan, in the near future, to carry out another calculation based, not as hitherto on the division of the basin into subsystems, but to the results obtained so far for a limited number of variants and integrated groups of initial data in respect of demands and water resources. The entire basin will be analysed as one system, and the results compared with those obtained for the basin divided into subsystems. We think that the comparison will be interesting and may result in certain corrections to be made to the previously formulated programme.

J. P. J. O'Kane

In the paper by Wyatt et al., one may raise the question of whether the constrained use of the aquifer has in fact been minimized. The authors are obviously satisfied that they have found good control rules, but what mathematical guarantees are there that the constrained optimum will be reached by the arbitrary methods adopted in this paper?

A guarantee can be provided by formulating the problem as follows:

$$U = U(\vec{x}) \qquad \min_{\vec{x}}! \tag{1}$$

subject to

$$D(\vec{x}) = D_0 \tag{2}$$

$$R(\vec{x}) - A(\vec{x}) = 0 \tag{3}$$

where U, the aquifer use, is a function of the control vector $\vec{x}$. D_0 is the allowable deficit and $D(\vec{x})$ is the deficit resulting from the particular control vector $\vec{x}$. $R(\vec{x})$ and $A(\vec{x})$ are the corresponding quantities which are recharged and abstracted. Note that none of these functions in n (the dimension of $\vec{x}$:24) space are differentiable and that one pass of the simulation generates a single point. Now form the Langrangian function:

$$U^* = U(\vec{x}) + \lambda_1 D(\vec{x}) + \lambda_2 \{R(\vec{x}) - A(\vec{x})\} \qquad \min_{\vec{x}}! \tag{4}$$

By Everett's theorem, if a λ_1, λ_2 pair can be found such that constraints (2) and (3) are satisfied then the vector $\vec{x}$ which minimizes the unconstrained U^* also minimizes the constrained U. Hence the suggested algorithm for the first half of this paper is a hill-climber within a hill-climber. The inner one minimizes U^* for fixed values of λ_1 and λ_2. The outer one minimizes the sum of squares (or absolute values) of the slack in the constraints over λ_1, λ_2, until the sum reaches zero. The constraints are then satisfied and that $\tilde{x}$ which minimizes U^* also minimizes U. Everett guarantees that the constrained optimum has then been found. (The problem of local optima in U^* of course remains.) This approach, in contrast to the paper, does not require 'the development of specialist routines' which are based on the authors' insights into the nature of constrained surfaces in n-space. The authors might like to express their insights mathematically.

Precisely the same approach can be adopted when using stochastic dynamic programming. The setting up of the Lagrange multipliers as penalties follows the same lines as in the earlier discussion on topic 8, session XVI by O'Kane. The penalties in such a formulation should apply to the deficit and the difference between recharge and abstraction. However, in the paper, the penalties are applied to the deficit and the spill which corresponds to the flood control/water supply problem considered in the paper by Cole (session XVI). Hence there is no guarantee that these arbitrary penalties lead to the optimal solution. The onus lies with the authors to show that penalizing spills and penalizing recharge minus abstraction are equivalent. As outlined in the previous paragraph, a hill-climbing routine may also be used to find the pair of penalties for which the constraints are satisfied.

The most important aspect of the approach adopted in this paper has not been mentioned at all, namely the reduction in the dimensionality of the problem. Traditionally, the conjunctive control of an aquifer and a surface reservoir has necessitated the use of two state variables. By regarding the aquifer as a store, subject only to the balance constraint, one state variable less is required. However, considerable detail is lost and the question naturally arises of when it may be used. Where the Markovian transition probabilities are stated (vol.2, p.984), it would appear that the first condition, $j = m$ and $l = m$ is redundant since it is covered by the second one. (It is only necessary when a desalination plant feeds the reservoir.) The argument q in the functions $K_n(q)$ and $F_w(q)$ might be replaced by $j - l$ to accord with the arguments $P_n(j/l)$.

T. Wyatt:

In response to the suggestions made by the general reporter, I shall be pleased to give representative costs associated with the techniques described in our paper. I will also outline some of the modifications and extensions made to this work since our paper was submitted. As some of these improvements, specifically in the dynamic programming formulation, have resulted in substantial reductions in computing costs, I will begin with their description.

Modifications to the dynamic programming formulation

(1) Deficit states on the surface storage have been incorporated in order to cope with an optimal long-term strategy where demand is met initially from the reservoir with little or no abstraction from the aquifer. This may leave an immediate deficit, but inflows during the period under consideration can be used to reduce or wipe out such short-falls in supply. The formulation of expected deficits follows closely that adopted for expected spills given in the paper.

(2) A Newton–Raphson search routine on the natural logarithms of the λ and ϕ steering parameters has been introduced, in preference to the original dichotomous search. As a result, a more rapid convergence to the solution is achieved (Fig.9).

(3) The number of iterations required has been further decreased by using the final V values obtained from the DP routine to initiate computation of the new set of

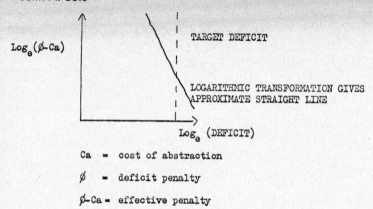

Ca   =   cost of abstraction

∅   =   deficit penalty

∅-Ca =   effective penalty

FIGURE 9

rules for a second λ/ϕ combination. This innovation was previously reported by Mawer *et al.* (1971).

Representative costs of the DP programme are given in Table 1.

Extensions to the hill-climbing technique
The type of response surface met with in conjunctive systems, and the inexplicit nature of the constraints imposed, render general search procedures such as the Rosenbrock algorithm, discussed by O'Donnell, unsuitable. The routines developed rely on the concept of steepest descent and are designed specifically to cope with the irregular constraints defined, while promoting rapid convergence.

Representative computational costs are given in Table 2.

Comparison of Tables 1 and 2 shows that similar order costs are involved for each technique when providing identical information. The system has now been successfully studied with a finite aquifer and preliminary runs with a simple aquifer loss function incorporated have proved successful.

With reference to the general reporter's question on the single and dual line rules used. it would appear that these give sufficient definition for the problem considered. The application of hill-climbing techniques to deal with more complex rules is currently being investigated at the Water Research Association (UK).

Comments on the paper by Mobasheri and Budhraja
There are two points which I consider need clarification in this paper.

(1) Fig.1. indicates the achievement of incremental firm water yield in excess of the installed desalting capacity.

(2) Confirmation on the stationarity of the control rules generated by the DP routine.

Reply to points raised by J. P. J. O'Kane
I would like to make the following remarks on the comments made by my former colleague, O'Kane.

Firstly, the search objectives used in the hill-climbing formulation are the aquifer balance and the overall deficit, as defined. In the DP formulation where a reduction in spill is assumed to improve aquifer balance, a subsequent revision to the spill penalty along the lines suggested produced no change result.

Secondly, the alternative formulation proposed for the hill-climbing approach using a penalty function relies on Everett's (1963) proof for the generalized Lagrange multi-

TABLE 1. Representative costs of dynamic programming. Costs dependent on: (a) overall degree of regulation and system constraints. (b) inflow record length. (c) variance of inflows, (d) no. of reservoir states. (e) search routine on steering parameters. Programme written in Fortran V and run on ICT Atlas 1 computer. All examples refer to infinite, non-leaky aquifer

| Record length (years) | No. of reservoir stages | Max. pumping rates as fraction of ADF | | Degree of regulation as fraction of ADF | | Abstraction load factor given by rule | Search type on steering parameters | Offline cost (£) | Computer storage | | Comments |
		Recharge	Abstraction	Overall	Reservoir only				Compile	Execution	
37	29	0.18	0.34	0.8	0.71	11.15	Dichotomous	5.50	29K	10K	–
37	79	0.11	0.11	0.86	0.76	13.5	Newton–Raphson	2.80	29K	10K	Loosely constrained
							Dichotomous	5.50	29K	10K	
34	31	0.5	0.5	0.80	0.45	40.3	Dichotomous	5.50	29K	10K	Highly constrained Large variance
34	39	0.35	0.35	0.7	0.6	5.4	Newton–Raphson	2.50	29K	10K	Large variance

ADF = Average Daily Flow.

TABLE 2. Representative costs of hill-climbing programme. Cost dependent on: (a) overall degree of regulation and system constraint, (b) record length, (c) simulation detail i.e. aquifer modelling, (d) single line evaluation requirement. Programme written in Algol 60 and run on ICT Atlas 1 computer

Record length (years)	Aquifer model	Max. pumping rates as fraction of ADF		Degree of regulation as fraction of ADF		Abstraction load factor given by rule		Offline Cost (£)	Computer storage		Comments
		Recharge	Abstraction	Overall	Afforded by surface storage	Single	Dual		Compile	Execution	
37	Infinite Non-leaky	0.18	0.34	0.8	0.71	-	10.35	3.75	39K	10K	-
34	Infinite Non-leaky	0.5	0.5	0.8	0.7	-	3.19	1.65	39K	10K	Loosely constrained system
34	Finite Non-leaky	0.3	0.3	0.8	0.65	13.00	4.41	5.25	39K	10K	Aquifer storage = surface storage
34	Infinite Non-leaky	0.5	0.5	0.8	0.5	41.9	15.4	5.95	39K	10K	-
34	Infinite Non-leaky	0.5	0.5	0.8	0.45	-	26.47	7.75	39K	10K	-
34	Infinite Non-leaky	0.3	0.3	0.8	0.56	44.61	33.33	13.75	39K	10K	Highly constrained system

plier method. This work shows that if a solution is obtained using penalty functions, then it will be optimal. The dominant condition is that the technique cannot produce solutions within a region bounded by convex constraints. The hill-climbing procedures described in our paper have been developed in order to cope with this problem while searching directly for convergence to the desired constraints.

The third point, again dealing with the optimality of results, is quite valid. In the absence of a detailed statistical analysis using generated inflow sequences, I would refer to the similar results obtained by the two techniques on the example given in the paper.

The final point was, I feel, adequately covered in my previous comments.

REFERENCES

Mawer, P. A., O'Kane, J. P. J., Sherriff, J. D. F., Wyatt, T. and Barrett, J. D. (1971) Some conjunctive desalting studies in the U.K. *Wat. Resour, Bull. (J. Amer. Wat. Resourc. Assoc.)* 7, No.3, August

Everett, H. (1963) Generalized Lagrange multiplier method for solving problems of optimum allocation of resource. *Operations Research* 11, 339–417

Closing remarks by the general reporter

I think this session has done a useful job in comparing approaches to the modelling and optimal operation of complex water resource systems. It was an occasion enriched by our Polish host's thorough experience with a variety of methods applied to the Vistula project, thanks to your agreement, Mr. Chairman, to my transferring three papers from tomorrow's programme to join session XVII.

The first most obvious point is that this is an open-ended discussion; we have certainly not concluded it. What I hope is that those of us, who are predominantly hydrologists, will not forget the fact, and therefore will not be totally trading our specialization for techniques of system analysis. Hydrologists have the duty to insist on representative data getting into the system analysis.

I thought there was a useful approach in the Vistula work, first employing historical data for testing segments of the system, then assembling those segments into a whole and simultaneously comparing results with a simpler whole system model. No doubt in future one will see simulated data fed in as well as the historical sequence; 'design droughts' come into this category too. We all know that there are many alternative approaches in the analysis and the coordination required in such studies is obvious.

Given the hydrologists' role in checking the validity of data fed into a system, and assigning limits thereto, one trusts that system analysts will in turn ensure that we do not become unreasonably finicky — there is not always a need to be exacting a fine accuracy from all inputs to a system, if certain components can be shown by sensitivity analysis to exert a minor effect on the outcome. In fact, in many water resources system models it is the economics that matter more than the hydrology; more dominant than either hydrology or economics may be the projection of water demand, which depends so much on technological change in the home and in industry and on population movements and trends.

I look forward to incisive developments in water resources systems operation being presented at a future symposium of IAHS.

General Reporter   Dr. Nathan Buras, The Lowdermilk Faculty of Agricultural
Engineering, Israel Institute of Technology. Due to the absence of Dr. Buras, Mr. Leo
R. Beard introduced and closed the discussion

Chairman   Professor Janusz Laszewski, University of Maracaibo, Venezuela

Planning of systems for regional development in water resources

Nathan Buras

The following papers are reviewed and discussed in this general report. (All are published in volume 2 of the Symposium Proceedings.)

1. *Leo R. Beard*   Hydrological evaluation of changes in runoff characteristics
2. *J. Cuena Bartolomé*   Modèle mathématique de simulation du système des ressources hydrauliques superficielles du Llobregat
3. *A. L. Velikanov and D. N. Korobova*   The methods of distribution of water resources in river development systems
4. *S. L. S. Jacoby and J. S. Kowalik*   Optimal planning of flows in multi-reservoir hydro-power systems
5. *W. James Morris and Neil W. Morgan*   Hybrid computer analysis of a combined surface water—groundwater system
6. *A. Sahuquillo*   Conjunctive use of the Tajo—Segura Aqueduct surface system and the aquifers of the La Mancha area
7. *Peter Rogers*   A sector model for regional and national water resources planning
8. *Jona Bargur*   A dynamic multisector programming approach to regional water resource management
9. *Emanuele Guggino*   La Science des Systèmes dans la planification des ressources en eau

The papers pertaining to this topic refer to a broad range of problems, from the evaluation of changes in runoff characteristics to the derivation of sector models for regional and nationwide water resources planning. Within this range, there are papers discussing systems based solely on surface water resources, while other reports deal with the more intricate situations where surface and groundwater resources are conjunctively used.

The paper by Beard describes a model that is designed to carry out hydrological and economic evaluations for any type of river basin in which urban development, reservoir construction, channel improvement, and other changes have taken place. These changes, many of which are structural and some may be a non-structural (such as operating policies for reservoirs) affect appreciably the runoff characteristics in a river basin.

Among the changes noticed above, urbanization is becoming a factor of growing importance resulting in increased volumes of runoff. The increasing amount of impervious construction (roads, parking areas, roofs) and storm drains tend to enlarge the peak rates of runoff. Similar effects can be attributed to channel and levee improvements on streams and rivers, which may speed up the concentration of flood waters in a natural hydrological system.

Reservoirs, in particular flood control structures, ordinarily tend to reduce flood flows. However, there may be conflicting situations when a full flood control reservoir is associated with a high rate of flow downstream. In this case, may it be necessary to release some of the water from the reservoir to accommodate the next flood wave, thus increasing the discharge of the river downstream from the reservoir.

The effect of development projects such as reservoirs within a river basin cannot be assessed without reference to meteorological phenomena. The time and areal distributions of storm precipitation greatly influence the effect that any particular development has. In addition, ground conditions may constitute an important factor, especially when considering imperviousness due to urban development. Thus many combinations of conditions have to be examined in order to evaluate a full range of effects.

To do so, it would be desirable to simulate accurately the entire rainfall—runoff process at all locations of interest within a river basin over a long period of time. The model representing this process would relate rainfall and ground conditions for each elemental area within the basin so that they can be integrated to yield consistent runoff values at the points of interest. This distributed-parameter model would probably require an unmanageable amount of computation. Therefore a simpler approach is sought.

A lumped-parameter model is proposed in which storm characteristics and loss rates are somewhat generalized. A series of eight or ten floods of different magnitudes having representative areal and time patterns of precipitation may be practical to compute. The probability of occurrence of these floods at each desired location within the basin can be estimated for any given development status, on the basis of runoff frequency studies for that status. Also, the probability of their occurrence for any other condition (status) of basin development can be estimated. In this way, the effects of river basin development can be evaluated.

A computer programme handling the approach described above has been developed by the Hydrologic Engineering Center of the US Army Corps of Engineers, which enables the computation of floods for a number of river basin development conditions and the evaluation of flood damages at selected points in the basin. A specific example is shown, in which is evaluated the effect of extensive channel improvement for the relief of overbank flooding.

The paper by Bartolomé refers to a study of basin management in northeast Spain. The Llobregat hydrological system is considered to be composed of a main stream and two tributaries, the Cardoner and the Noya. The problem is to produce a development plan for the basin so as to satisfy expected future demands for water, especially in the lower part of the basin where the urban complex of Barcelona and its satellite towns are found. A model is developed which consists of three kinds of elements: (a) a set of sources of water; (b) a set of demand areas; (c) a number of possible reservoirs and other regulating structures. The model enables the simulation of a given operating policy for the water resources system such that it is possible to determine the behaviour of the various components in satisfying a set of given monthly demands for water. The results of the simulation are used in an analysis of the economic aspects of the operating policy under study.

Synthetic hydrological series of monthly flows were used as inputs to the simulation study. These series were generated by the method described by Hufschmidt and Fiering (1966). The outputs from the various simulation runs were used in two ways: (a) to gain an insight into the functioning of this rather complex hydrological system; (b) to be included in an economical evaluation of the system. All these outputs were summarized in the form of histograms.

In the first case, the following histograms were developed: (i) frequency of the various states of every reservoir during a period of one year; (ii) frequency of inter-reservoir diversions; (iii) frequency of monthly 'improved' discharges from every reservoir. An 'improved' monthly discharge was defined as the difference between the total monthly outflow from a reservoir and the natural monthly inflow into the same storage facility.

In the second case, mean values and standard deviations were computed in addition to the histograms. The following variables were defined: (i) monthly shortages of water, relative to the types of demand; (ii) flows in reaches of streams where hydroelectric generating plants exist; (iii) flows in various conduits and inter-reservoir diversions.

In order to analyse better the relationship between sources and demand areas, operating policies specified the order of preference in which a given demand area would be supplied by different sources. Similarly, priorities were specified for demand areas to be supplied by a given source. In this way, strategies for using the various water resources were devised for different demand conditions.

The various development programmes studied by means of the simulation model were evaluated on the basis of annual costs, which consisted of the following elements: (i) an annual payment to a sinking fund for the amortization of the system; (ii) annual operating costs; (iii) annual returns generated by the use of water (negative costs). The returns were derived from three sources: (i) irrigation, (ii) domestic; (iii) industrial supply.

An irrigation return function was established in the form of monetary units per unit area versus shortages. This is a monotonically decreasing function with a maximum at the point of zero shortages.

The urban supply return function was based on a reasonable tariff for domestic water. The losses generated by deficient supply were based on the same tariff up to a certain threshold. Beyond this threshold the losses were based on the cost of reclaiming waste water in the inland areas, and of desalinizing sea water along the coast.

For industrial supply, the return function was discontinuous at the point where enterprises had to invest in facilities to recirculate water.

The computer programme was written in Fortran IV and the computations performed on the 1130 IBM digital computer.

Velikanov and Korobova use dynamic programming formulations in their papers. The problem of the rational utilization of regional water resources is formulated in general terms, and a distinction is made between consumptive users of water (termed in the paper 'water consumers') and non-consumptive users ('water users'). This distinction is quantitatively expressed in the models developed. Let x_i be the amount of water diverted to the ith user. Then, if there is any return flow to the river system, Y_i, it is assumed that $Y_i = Y_i(x_i)$. Thus, for a consumptive user, $Y_i = 0$; for a strictly non-consumptive user, $Y_i = x_i$.

A general recurrence relationship is derived in order to maximize the objective function

$$D(x_1, x_2, \ldots, x_m) = \sum_{j=1}^{m} D_j(x_j)$$

where $D(x_1, x_2, \ldots, x_m)$ is the overall return from allocating the amount of water x_j to user j, and $D_j(x_j)$ is the specific return from that user. The recurrence relationships are

$$R_j(w_j) = \max_{a_j \leqslant x_j \leqslant b_j} \left\{ D_j(x_j) + R_{j-1}(w_j - x_j) \right\}$$

$$\sum_{j=1}^{m} x_j \leqslant W$$

$$R_1(w_1) = D_1(x_1)$$

where

$R_j(w_j)$ = total return derived from allocating the amount of water w_j to j users and following an optimal policy

$$w_j = \sum_{j=1}^{j} x_j;$$

a_j, b_j = minimum and maximum limits, respectively, specified for the allocation of water to the jth user;

W = total available flow in the river basin.

Three different cases are presented. Case 1 deals only with a main stream (without tributaries) which has an inflow W, and a given terminal flow (beyond the mth (last consumer) of W. The general recurrence relationships apply with the proviso that the upper limits on x_j may have to be adjusted (reduced) to satisfy all constraints. Thus

$$a_i \leqslant x_j \leqslant b_j$$

Case 2 considers only consumptive users (as does case 1), but includes the contribution of tributaries (w_k), to the main stream. Again, the general recurrence relationships apply. The limits on allocating water to the consumer $k + 1$ (past the kth tributary), are

$$A_{k+1} \leqslant x_{k+1} \leqslant \min \left\{ b_{k+1}, \sum_{k=1}^{k} w_k \right\}$$

Case 3 is the most general of all. It includes tributaries, consumptive users, as well as non-consumptive users. The recurrence relationship has the form

$$R_j(w_j) = \max_{a_j \leqslant x_j \leqslant b_j} \{ D_j(x_j) + R_{j-1}(w_j - x_j + y_j) \}$$

Two variations of this case are discussed briefly: (a) when $x_j = y_j$; (b) when $y_j = x_j - c$, i.e. when each user consumes a fixed amount of water c.

The models presented are deterministic and refer, presumably, to situations obtained in an average year.

The paper of Jacoby and Kowalik deals with the problem of the optimal planning of flows for a considerable number of relatively long time periods in a multi-structure hydropower system. Specifically, it is desired to generate a set of rule curves, one for each reservoir, determining water levels as a function of time. These rule curves refer to a hydropower system which may include up to 50 interconnected reservoirs, each curve covering a total period of about 4–5 years. Each year is subdivided into 14 time periods, ten of which are of monthly length and four covering two weeks each. Thus up to 60 (unequal) time periods may be considered.

A mathematical model is constructed which minimizes power deficit. 'Power deficit' is defined as the difference between power load (demand) and total power generated by all plants in the system (supply). This deficit is to be at a constant level at all time periods. An alternative formulation minimizes the maximum deficit in any time period, when this deficit is allowed to vary within prescribed lower and upper limits.

Defining D = total deficit; D_j = deficit in period j; L_j = total load in period j; P_j = total power produced in period j, the alternate objective functions can be written as

$$\min\{D \mid D_j - L_j + P_j = 0, \text{ for all } j = 1, 2, \ldots, J\}$$

or

$$\min\{\max_j D_j \mid D^{\min}_j \leqslant D_j \leqslant D^{\max}_j, \text{ for all } j = 1, 2, \ldots, J\}$$

The problem as stated seems to be a linear programming problem in $N = 50 \times 60 = 3000$ decision variables and a number of equality and inequality constraints several times N. In fact, it is a nonlinear optimization problem, because the power generation relationships are nonlinear. The size of this nonlinear optimization problem is very large, but certain features of it can be taken advantage of, in order to solve it in a reasonable time on a digital computer. In fact, it is proposed to solve this problem several times, subject to different systems configuration, varying generating plant characteristics, different initial and boundary conditions and different constraints.

It is considered that the optimization problem is, in general, nonlinear in J equality constraints, the remaining constraints being essentially linear. Three different solution methods are presented and discussed: (a) the Modified Feasible Direction method – MFD; (b) the Method of Approximate Programming – MAP; (c) the Generalized Reduced Gradient method – GRG.

MAP starts with an initial approximation, linearizes the nonlinear constraints, solves the resulting linear programming problem which yields a new approximation, and the process is continued until the change in the objective function is less than a prescribed tolerance.

MFD is similar to MAP, but it includes at each iteration a search and interpolation procedure which may yield a better approximation. The particular structure of this problem enables the application of MFD in conjunction with a *decomposition technique*.

GRG, as the other methods, is also an extension of the simplex method oriented to the solution of nonlinear problems where the objective function is subject to upper (or lower) bounds on the variables and to equality constraints. Nonlinear equations are solved by iterative techniques, such as the Raphson–Newton method which uses sequential linearization.

A fourth method is suggested, the use of *penalty functions* where the sum of squares of all nonlinear equations is minimized subject to linear constraints.

Flow charts of the possible approaches to this complex, large-scale problem are given, including the MFD and GRG methods. No numerical examples are offered and it seems that this conceptual approach had no computational experience.

An interesting contribution, showing the use of analogues in conjunction with digital computers for the analysis of hydrological systems is the paper by Morris and Morgan. The main advantage of hybrid computers is that the analogue component eanbles the study of the effects of parameter changes as displayed on an oscilloscope or a plotter, and, at the same time , the digital component is used for rapid and accurate arithmetical computations. In this way, integrations are carried out on the analogue and numerical calculations on the digital computer.

A computer model was developed which consisted of two components. The surface water component included the hydrological and hydraulic processes at ground level. The groundwater component included all water in the saturated zone. The model assumed isotropic-homogeneous conditions within an unconfined aquifer with negligible compressibility effects, and was based on Darcy's law. A second-order partial differential equation with nonlinear terms was developed. The solution on the hybrid computer was derived by dividing the aquifer into elemental cells, each cell having horizontal dimensions Δx and Δy and a vertical dimension h which measured the elevation of the groundwater table above an impervious barrier. The solution was considered acceptable when the successive drawdown—time curves at each node agreed within a specified tolerance.

The model was used to evaluate the effect of its linearization (by neglecting the nonlinear terms), when water levels were obtained for discharge from a single well. For an aquifer thickness of 100 ft (about 30 m), a storativity coefficient of 0.1, a hydraulic conductivity of 10 ft/day (about 3.5×10^{-3} cm/s), and a rate of pumping of 10^5 ft³/day (roughly 120 m³/h), the linearized form yielded a drawdown less than 10 cm higher than the nonlinear equation at the centre line of the well. Both results were compared with Theis's theoretical solution. At about 200 ft (60 m) from the well, the two versions of the model were hardly distinguishable, while the Theis solution was about 0.5 ft (15 cm), higher. This appeared to be the maximum difference between Theis's equation and the proposed model.

Hybrid computer analysis can be used in the optimization of the water balance parameters, so that it can be developed as a tool for planning the regional development of water resources.

A large scale regional plan for the development of surface and groundwater resources in Spain is presented in the paper by Sahuquillo. The Tajo—Segura aqueduct currently under construction will be capable of transferring large quantities of water from the Tajo into the Segura basin. The aqueduct crosses an area underlain by large aquifer formations which appear to have, in addition to a significant safe yield, groundwater reserves estimated to be of the order of thousands of millions of cubic metres. The planning process of this large water resource system involves at this stage, among others, the formulation of alternative operating rules based on a forecast of irrigation demand for an area of 380,000 ha.

The major storage structure in this system is the Alarcón reservoir on the River Júcar, midway between the Tajo and the Segura basin. The integration of the aquifers, which are mainly in the Júcar basin, would result in added flexibility to the operation of the system. Preliminary hydrogeological investigations indicated that wells drilled in a Jurassic limestone aquifer in the neighbourhood of the major aqueduct route yield more than 350 m³/h with small drawdowns. The La Mancha formation, to the west of the Tajo—Segura aqueduct, also has a large water-bearing capacity. It seems, therefore, that the integrated development and use of surface and groundwater resources can provide an ample supply of water to that part of Spain. When the demand will exceed this supply, it would be possible to effectuate a link with another large water resources system further to the northeast, that of the Ebro River.

The basic operating rule is shown in Fig.1. The pumpage from the aquifers, C, and the portion of the pumped groundwater allowed to flow into the Júcar River are, in

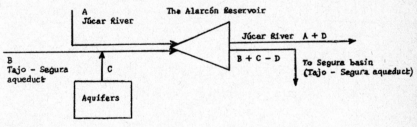

FIGURE 1

fact, decision variables which will have to be set at values which will optimize a given criterion.

Several alternatives to the basic operating rule are presented. For example, the possibility of D being negative, i.e. Júcar River water would contribute to the supply in the Segura basin. Another possibility is to use the aquifers for yearly or long-term storage. Furthermore, the economic depth of pumping from aquifers can be explored in the light of the investment necessary for additional power generating facilities.

The paper, without going into the details of constructing a mathematical model, presents a very valuable initial conceptual framework for planning a large system for regional development of water resources in Spain.

A comprehensive approach regarding the planning of water resources systems for regional development is offered by Rogers in his paper. The conceptual model presented in this paper was constructed in connection with the planning of water use for agricultural development in East Pakistan.

The overall approach was to construct, in fact, two complementary models: one dealing with variables controllable by government agencies engaged in investment in the agricultural sector (such as choice of projects and programmes), and another dealing with decision variables at the farm level (such as the selection of cropping patterns). The two models should answer the following question: which combination of projects and programmes should be selected for the implementation in the water sector out of a very large number of possible projects and programmes? Furthermore, where should these projects and programmes be located, and when (in the next fifteen years) should they be started, if at all? The model, thus, is aimed at regional intertemporal and interproject interactions and variations within the water resources sector.

The methodology for sector planning was developed in the last decade with the application of systems analysis techniques and operations research methods to public investment decision making. An important contribution to this methodology was that of Marglin (1963). Marglin's approach was based, among others, on three assumptions. Firstly, net present value functions must be differentiable concave. Although this assumption reflects the law of diminishing returns, it does not take into account the situation prevailing in a large number of multi-purpose water resources projects where substantial outlays are required before the project can be started (e.g. site clearing). Secondly, the benefits of any one project are independent of the existence (or nonexistence) of any other project. This, clearly, is seldom so in the field of water resources. Thirdly, all inputs to a project except capital (which is scarce) are available at given prices. This is unrealistic in developing countries.

In order to overcome these difficulties, it is proposed to consider the same project at different levels of development as a separate project, so that a project is either built or not built at a particular point in time. This is then the classical zero-one integer capital budgeting problem. Furthermore, in order to take into account economic interdependence between projects and regions, the proposed model includes an iterative routine as follows:

Step 1. Solve the government model, using world prices for all commodities traded on the world market and shadow price all inputs at assumed levels.

Step 2. Introduce projects and programmes derived from step 1 into former model.

Step 3. Iterate the former model until a set of equilibrium prices for the agricultural production is obtained.

Step 4. Return to government model and revise benefit coefficients based on results of step 3. If they imply radically different investment decision, go to step 2 and repeat.

The East Pakistan government model included 91 zero-one integer variables (projects), a 15-year development programme divided into four unequal periods, and 18 regions, yielding 840 continuous variables. The former model was a linear programming problem with 216 variables and 378 constraints. Because of the difficulties involved in data collecting and processing in the form required by the models, no computational results are reported.

REFERENCES

Hutschmidt, M. M. and Fiering, M. B. (1966) *Simulation Techniques for the Design of Water Resource Systems*: Harvard University Press, Cambridge, Mass., USA.
Marglin, S. A. (1963) *Approaches to Dynamic Investment Planning*: North-Holland, Amsterdam.

Introduction to the discussion by Leo R. Beard

As unfortunately Dr. Buras cannot be here, I will try, at short notice, to review very briefly the contents of the papers for this session and to lead some discussion that may be fruitful on the subject of this session. There were two papers that Dr. Buras was unable to obtain in time to review. One of these is a paper by Bargur which deals entirely with economic aspects, and does not involve a hydrological model. I will read a brief portion of the abstract. 'Market forces have failed in the area of development, allocation and management of natural resources in general, and of water resources in particular. The shaping of proper investment programmes for water resources development has long been dominated by partial equilibrium analysis, such as benefit—cost analysis. However, this approach is rather limited for an inter-regional, inter-temporal management programme of water resources. The present study offers a multi-sector planning and management approach for water resources based on a general equilibrium analysis, employing input—output models, and linear programming techniques.' The paper deals entirely with inputs to, outputs from, and the management of the economic system. It does not contain hydrological models, so I will not review it any more except than to state that it is important for the hydrological engineer to understand the manner in which the hydrological models will be used in economic analysis and in the general planning programmes generally.

Dr. Buras has not posed any questions to instigate discussion at this session. I have considered a few questions that might be of concern here. There were three subject areas that I thought would be important. One is the variety of hydrological models that have been considered at this Symposium, and that should be considered by this Committee and by the Association at future symposia. We have restricted ourselves largely to models of rainfall and runoff. Although we have had some discussion of groundwater models, of water temperature and reservoirs, there is a much broader scope of hydrological models that should be considered by the Association. A model of sedimentation for the entire river basin would be a very great challenge to hydrological engineers. I know of no comprehensive model that deals with the entire topography of a basin, the nature of the channels, and of the stream beds in relation to the transport of sediment, as it varies throughout the basin, and as it varies throughout the years. This is a model that needs developing in the future. Many other types of models, particularly water quality models, will be very important. The temperatures in reservoirs were discussed. Temperatures in rivers are also important. The three-dimensional variation of temperatures in reservoirs, of dissolved oxygen in the reservoirs, and of other quality parameters pose a tremendous challenge for formulating models by hydrological engineers. That is one thought that I would like to leave for the purpose of discussion, if we have the time.

There are two other aspects that I thought were important. One has been discussed, and that is the advisability, or the value of obtaining highly detailed, accurate models, if we do not have sufficient data for using such sophisticated models. There have been two thoughts given at this Symposium already. One has stated that, where data are scarce, more sophisticated models are needed in order to obtain more information from the data, because the information is very precious. The other has stated that it is foolish to work too hard on a model if the data are not worth such detailed analysis. This is a subject of discussion that I think will be very fruitful. The third subject that I think is worthy of discussion is the relationship between the complexity of a hydrological model, and the manner in which it must be employed for developing operation rules and plans of development. As we know, in the last 10 or 15 years, the water resources problems have become far more complex. We need to develop projects for many purposes. Some of these purposes complement each other, many are conflicting with each other. The operation of a project must be examined in detail with respect to all of the purposes. A project must now serve many locations. Years ago, one reservoir could generate power only for one load centre. Now the power is generated for many load centres. The other purposes, such as water supply, low flow regulation, are also designed for serving many locations. Many projects must also be operated together. This is a challenge, as it is more complicated than one project serving many locations. It is many projects serving many locations. People are now interested in many facets of water resources development. Not only the economic benefits, but benefits that cannot be measured in monetary terms, such as ecological effects, environmental effects of a project. These are challenging questions, and it is necessary to develop hydrological models that would demonstrate the output of proposed projects in great detail, in order to answer these questions. Also, whenever a project or a system is being developed for water resources, it is necessary to consider many alternative types of projects. Sometimes dozens or hundreds of alternative projects must all be examined in terms of the hydrological impacts. Some of the papers in this session concentrated on these problems. We must consider the complexities of the mathematical models as a limiting factor in employing them to examine so many complicated alternative sets of projects. Sometimes the computation is prohibitive, because the model is too complex. Another thing that we must consider in developing our mathematical models is that they must be used maybe thousands of times just for one project. If they require a great deal of computation, perhaps the model should be simplified.

DISCUSSION

Z. Haman:

I would like to make some comments in relation to the mathematical models of groundwater aquifer systems. I will not comment on any particular paper although every paper on this topic is, in some way, included here.

During this Symposium, a large number of papers about mathematical models of surface water systems have been presented, and this will be of great help to the professional people dealing with this part of hydrological cycle. However, the part of this cycle which is related to groundwater was not represented in such scale, and, as you know, there exist many problems which are of great importance to scientists as well as to engineers. My knowledge of groundwater is based on a deterministic approach, and I believe that the reasons for such an approach are justified. When representing groundwater systems by a model, we operate with a certain amount of data obtained from the investigations carried out on that particular field. Usually, we know the hydrogeological condition of our aquifer more or less accurately. In fact, we know the aquifer geometry, which is one of the important factors for building the model. The usual technique in interpretation here is the analysis of available well logs or data from geophysical investigations such as geoelectric or seismic surveys. The next step is to determine the hydraulic properties of our aquifer. There are always some hydraulic data, for example, the specific capacities of the wells in the investigated areas, on which to base the preliminary analysis of the T and S values of the aquifer. Furthermore, when some groundwater reservoirs are being developed among other investigation programmes, the pumping test is our most useful tool for the determination of aquifer hydraulic properties. However, tests require wells, and these are expensive in most cases so we do not get all we want. Therefore, I would like to stress that each available test should be performed as accurately as possible because, only in this way, can we obtain data on whose analysis we can base our determination of the aquifer hydraulic properties and choose a model. I would like to extend the words of Yuvjevich about using as many methods of analysis as possible in the case of groundwater. We have to look into the cone of depression not only from the hydraulic point of view but also as a tool by which we can extend our knowledge about aquifer geometry and composition beyond the points of observations. Due to the fact that the cone of depression during pumpage develops in the physical system, it must reflect the properties of this system; for example, the existing boundaries which probably have not been determined due to insufficient hydrogeological data. One should also recognize aquifer homogeneity or inhomogeneity by interpreting observation data from different wells, and all these data can serve to judge whether or not our model is representative of the aquifer. If future withdrawal will influence the aquifer within known extensions, there are great chances that our model will represent the aquifer with a high degree of accuracy. However, in most cases, the planned withdrawal exceeds several times the rate of the pumping test. Thus, the new area of influence is much larger, and we come to the question whether our investigation wells have been properly distributed so as to cover the new area of influence with a sufficient degree of accuracy for building a model. As I have said previously, the number of wells at our disposal is usually limited, therefore we must consider carefully the distribution of available points of investigation. The technique we employed in our survey was to make the distance between the points so that the created radii of influence slightly overlapped, thus covering the new area entirely. In such a way, we can be sure that our cones of depression will discover the hydraulic as well as the physical properties of the new part of the aquifer. Of course, on the basis of such data, it would be possible to build a model which will be representative of our aquifer. However, the degree of accuracy of the model should be governed by the scale of importance of the project and this is the social and economic issue coming into our analysis.

I could spend much more time in discussing these matters in detail, however, very important conclusions may already be drawn. Our approach is deterministic not stochastic. Perhaps I am wrong, but it does not appear to me that we can use the stochastic approach in the underground due to the complexity of the hydrological set up which determines the behaviour of all other variables.

Nevertheless, in our studies of groundwater, we usually come to the point where we are in doubt. As we all know, recharge is based on infiltration into the aquifer, which is again based, among other factors, on precipitation. This is the link with surface hydrology and I believe that the stochastic approach may be of great help to us.

Usually, there are available precipitation and evapotranspiration data, although I know that many of us dealing with undeveloped countries will disagree with this statement. There are also data about the piezometry of the aquifer. I would like to ask the question whether stochastic hydrology can give us operational models which can help us in solving the recharge problem which is very important in groundwater studies. I would also like to suggest that the Organizing Committee emphasize the problem of groundwater models so that during the next symposium there may be presented far more useful examples and solutions in this part of the hydrological cycle.

A. Sahuquillo:

I would like to comment on two aspects of my paper which are not explicitly stated. The first point is related to a characteristic of water resources systems which is not frequently mentioned. Most of the literature deals with systems having fixed elements which may vary *in situ*. It is the optimization of the design and/or operation of a system. However, in many cases, the inclusion of some additional element or of an unscheduled type of operation may improve our system. Some examples are the inclusion of a new reservoir or canal, the use of an aquifer, the development of artificial recharge facilities, and the installation of a pumped storage hydropower plant. This point is particularly true in the case of aquifers in arid and semiarid countries, which are frequently neglected as dependable sources of water in a first planning stage.

The second point is the use of the one time reserve of water from many aquifers. This may postpone the heavy financial loads incurred by large hydraulic structures for a substantial period of years.

As proposed by Mandel, in order to use the one time reserve of an aquifer, it is necessary to initially pump a volume larger than the so called 'safe yield', and this can be done, even if the aquifer characteristics are not well known.

It may be stated that even if the aquifer properties are not well known, it may be imbedded in a water resources system.

Moreover, these properties may be determined after a shorter observation period than that required by the surface hydrology. This is due to the damping effect produced by the usually large underground storage capacity.

Obviously, there is a problem in the planning and use of optimization techniques when the aquifers are in semi-natural conditions, since then their physical characteristics are not well known. However, reasonable assumptions can be made which will be continuously improved as new information is collected.

J. W. Delleur:

I would like to comment briefly on two aspects of the paper by Beard. The statement is made in the general reporters review that to simulate accurately the rainfall–runoff process at all locations of interest would be impossible. I would like to comment to the effect that Rao and myself have conducted a study of urbanization on runoff, and have compared several conceptual models for this purpose (Rao *et al.*, 1972). One of the conclusions was that, for basins less than approximately 5 mi^2, the single linear reservoir was very effective. It was found that the reservoir constant depended upon two types of characteristics, one being climatological, and the other one representing basin properties. The climatological quantities involved in multiple regression for the reservoir constant were the precipitation volume and duration. The two basin quantities were an urbanization factor and the basin area. The urbanization factor was defined as unity plus the fraction of impervious area. This made it possible to use this model for fully urbanized, partly urbanized, and the fringe zone around the city. A study was made on the peak discharge, the time to peak, and the probability distribution of the flows for different storm types and for different urbanization factors. It would then appear that it would be possible to use an extremely simple model for basins having reasonably uniform characteristics, and being probably less than 5 mi^2 in area.

The second part of this paper that I would like to comment upon and can agree with, is the statement that the meteorological phenomena are of importance. In fact, an interesting phenomenon occurs in the small town of Laporte, located east of Chicago. It happens to be down wind of an industrial centre which increases the number of nuclei of condensation due to the pollution production in the vicinity of the city of Chicago. I should like to add that, at present, a study is in progress by my colleague, Rao (1972), dealing with the sequences of wet and dry periods for urbanized and non-urbanized areas for the purpose of understanding the effect of urbanization on the time series sequences of rainfall.

REFERENCES

Rao, R. A. (1972) The 'Laporte anomaly' as analyzed by crossing theory, Abstract H50, *EOS,
Trans. Amer. geophys. Un.* **53**, No.4, p.376.
Rao, R. A., Delleur, J. W. and Sarma, P. B. S., (1972) Conceptual hydrologic models for urbanizing
basins. *J. Hydraul. Div., Amer Soc. civ. Engrs* **98**, No.HY7, 1205–1220.

Z. Kaczmarek:

I would like to raise two problems connected with water resources systems. The first problem is connected with the analysis of large systems. As you know, the theory of large systems has been developing within the framework of other disciplines, in particular within the framework of cybernetics and automatic control. In my opinion, the simplest definition of a large system is that it is a system which must be divided. We have to apply decomposition to be able to carry out an analysis. I would like to discuss precisely this particular problem of decomposition of large systems. We have come across that problem in the course of investigating the water management in the Vistula basin, which was discussed during yesterday's session. One of the papers presented today, namely that by Jacoby and Kowalik, also deals with such systems which contain 3000 decision variables and many thousand constraints. It, too, stresses the necessity of decomposition. Thus, having agreed on this necessity, we should look for ways of how best to carry out this decomposition. There exists the possibility of spacial decomposition which, quite simply, divides a large number of reservoirs or a large number of water users, into certain subsystems. Problem decomposition is another possibility. It could deal with problems of quantitative water supply independently of, say, water quality, or would deal separately with such problems as water supply and floods. I, personally, am in favour of the first conception. I am in favour of spatial decomposition and for evaluating, within the framework of subsystems, all possible aspects of water management. This, of course, raises the problem of the connections between subsystems, but, I think, this could be achieved more easily than in the case of problem decomposition. I should like to put forward a suggestion for consideration by our Committee on Mathematical Models. The fact is, that over recent years, studies on large systems have begun in various countries. Some of these efforts were described at this Symposium. I think it would be worthwhile to call a meeting at a convenient time with a limited number of participants. Gathered at the meeting would be people from various countries engaged precisely in studies of large systems. I would suggest that the issues of decomposition of a system and the connections between the individual water management tasks considered within the framework of subsystems be adopted as the main subject for such a meeting. Another thing which I would like to discuss concerns a problem raised earlier at this Symposium and also mentioned by Mr. Beard. It is the problem of data and methods. I, personally, share the opinion of those of my colleagues who hold the view that, when we have a limited amount of data at our disposal, we should apply methods which enable us to utilize best the information contained in these data. As a rule, they will be more complicated, sophisticated methods. On the other hand, we should remember, when applying statistical methods, that we are obliged first to test the compatibility of the method with the size of the sample which is at our disposal. This is necessary as it may happen that errors of estimation, in the case of small tests, could seriously distort the results of a calculation if the number of parameters to be defined is very large. This is the reason why the size of a sample must always fit the number of such parameters.

Closing remarks by Leo R. Beard

I certainly agree with Kaczmarek in his statement that areas of sparse data must be investigated with more sophisticated methods in order to utilize most effectively the information that is contained in the data. I was interested in hearing Bartolomé's remarks about simulation and the manner in which it is used in Spain, because this is very similar to the methods that we are now using in the Corps of Engineers (USA) in planning and developing water resources projects. We have found that the more analytical types of techniques, such as dynamic programming and linear programming, have so far had limited application in the planning of real projects, because of the great amount of detail that is involved in evaluating the alternative plans of development. It has been necessary then to simulate in great detail all of the different hydrological sequences that might occur in the various parts of the basin being investigated, and under various plans of development and operation, and to evaluate the great amount of detail in terms of economic values on which decisions must be based. I was interested in Delleur s comments relative to the models that he and Rao have developed, wherein they have been able to simulate the hydrology of areas of 5 mi^2. That was a relatively simple model, as i understand it. The statements in my paper on which he was commenting were with respect to rather complex systems regardless of their size, where the determinations of effects must be made at many locations. In some systems, we must evaluate the effects at 10 or 15 locations, and some times, as many as 100, or 150 locations. For example, in the Chicago metropolitan area we have about 150 evaluation points, and that is only a small drainage basin of about 100 or 150 mi^2. In this flood control study, it is necessary to consider storms of all sizes, because the developments that are proposed, such as storm drains. small storage facilities, sometimes affect small floods entirely differently than they affect large floods. They must be investigated for their effects on floods resulting from rain storms that concentrated in upper regions of a basin, or in lower regions of a basin, or at different locations within a basin, because their effects will be different on the floods. It is necessary then to evaluate the effects on various sizes of floods and various centrings of the floods, also floods of short duration and high intensity as well as floods of long duration and low intensity. So it becomes necessary to evaluate a great many combinations of meteorological and hydrological factors. In some cases we are finding that we should compute flood hydrographs for all the locations within a basin. That is why I was saying that it is virtually impossible to evaluate properly the alternative plans of development in a complex river basin problem. The subject of this particular session is of particular interest to me, because we are concerned, in the Corps of Engineers (USA), with utilizing the new methods that are being developed in the planning and operation of our projects. There is a long delay between the time when these new techniques and methods are developed and the time that it is practical to apply them to the real problems in all of the complexities that we must face in the planning process. In fact, as I was stating earlier, we are still simulating the operations of a system in great detail, and have been unable to make direct analytical solutions of the problem of selecting the best projects. These are trial and error successive approximations. That is the state of the art as we employ it in the Corps of Engineers. We are very interested in using the more powerful analytical techniques, but there is yet some time to go before they are developed to the level at which they can be employed directly in the work that we are doing in planning complex water resources systems and in operating these systems. I would certainly encourage all of you, not only to develop better mathematical models, but to keep in mind particularly the manner in which these models would be used by the engineers, who must plan, design, and operate the water resources projects and the systems that we need in order to supply the water and meet all the needs in the increasingly complex societies that we are in.

Closing meeting — Séance de clôture

31 July 1971 — 31 juillet 1971

Closing Speech by Professor Dr. Zdzisław Kaczmarek

Mr. Chairman, Ladies and Gentlemen,

This Symposium has been attended by more than 240 participants from thirty countries. In the course of eighteen sessions we had the opportunity to acquaint ourselves with over 120 papers. As Chairman of the Organizing Committee, I would like to express my conviction that the papers which have been presented at this Symposium constitute a valuable material for further work and inspiration to all of us. On behalf of the Organizing Committee, I would like to thank the authors of the papers, the general reporters, and chairmen of the individual sessions, who have helped us so much with the organization of this Symposium.

Making no attempt at summing up the discussion, I would like, nevertheless, to put forward a few questions and remarks. I think that in the course of the Symposium, we had the opportunity to understand better the necessity of links to be formed between various divisions of hydrology, between hydrology and mathematics, between physics and hydrology, and between hydraulics and hydrology. I think that the mathematical modelling of hydrological phenomena must be based on physical analysis of phenomena which are subject to modelling. This is why cooperation is necessary between all who work in the field of hydrology. I also think that it is extremely important, when promoting work on modelling of the hydrological phenomena, to secure proper initial data. Such data will make the practical application of the models much more concrete. I would like to emphasize that, in the course of this Symposium, we have often raised the problem of practical application. Indeed, I believe that, as scientists and engineers, we have the duty to direct our research, above all, towards these fields which will find application in water economy, protection of the environment, and other related problems. In this connection I would like to put forward some proposals for future consideration by the participants.

Firstly, I believe that symposia of this kind, convened at intervals of a few years, would be extremely useful because they make it possible for people interested in a specific branch of hydrology to come together. At the same time, however, I think that at all symposia of our Association, the methods we are dealing with should be discussed, explained, and efforts should be made to transfer them into various fields of hydrology. The purpose of such meetings like this one is not only to help in the solution of some scientific and technical problems. In my opinion, they help remarkably towards making better mutual acquaintance among the participants. I think that a lot of misunderstanding in the world today, not only in the field of science, is due, simply, to the lack of reciprocal information and lack of mutual acquaintance. I would like to express my particular pleasure at the fact that so many persons have come to this country to take part in the Symposium. As the Organizing Committee, we have been very pleased with such a large attendance.

I would like to add that taking part in this Symposium have been people linked with Poland either as the place of birth, or by family ties. Addressing these colleagues of ours, I would like to assure them that they are always cordially welcome in Poland. In closing my speech, I would like to thank again very cordially Professor Dooge and all the members of the Committee on Mathematical Models for the very valuable assistance extended to us throughout the period of preparations for this Symposium. Thank you very much.

Closing Speech by Mr. Frederik H. Verhoog

Mr. Chairman, Ladies and Gentlemen.

At the forthcoming Moscow General Assembly of IUGG which most of you will attend. UNESCO will, give a more detailed official statement on relations between UNESCO and IASH. I can therefore be very brief. This Symposium has been very useful and the Polish Organizing Committee with its Chairman, Professor Kaczmarek, IASH, the participants, the reporters, and the chairman, may be congratulated upon it. At the opening session, most of the representatives of WMO and UNESCO stressed the need for operational mathematical models. I will go somewhat further. Regarding the general idea of UNESCO's aims, namely education, culture and science, most people think that UNESCO is only interested in academic methods. This is not the case. In hydrology as well as in most other fields, UNESCO is promoting the development of the hydrosciences, the exchange of information, and the assistance to Member States. For us, this Symposium falls under exchange of information and, as such, it has been very successful. UNESCO and WMO statements on the need for more 'operational' models show us that the general feeling in the world is that the time has come for mathematical modelling to exchange know-how and to go to the developing countries. As an example, I may give the resolution adopted during the Mid-Decade Conference of the IHD whose objective is to establish a computer programme retrieval system. We here, in this room, know that this is enormously difficult.

This is an example of what the international community asks from the international organizations in the field of mathematical modelling.

Closing Speech by Dr. A. J. Askew

WMO's association with the preparation of this Symposium, the progress it has made in projects related to mathematical models in hydrology, and my experience this week, has more than convinced me of the importance of this subject to the international hydrological community. This Symposium has given me an opportunity of explaining WMO's work, both to individuals and to the participants as a whole, but this was certainly not the reason for the support we have given to this meeting. WMO is an intergovernmental organization, as is UNESCO, and it deals with operational meteorological and hydrological agencies of its Member States. It is our responsibility to provide a channel for international cooperation and coordination at this level. It goes without saying that we are primarily interested in the application of research results, such as those discussed here, to the design, and the operational use of these in the hydrological systems. I fully support Professor Kaczmarek's views on this method. IAHS and its attendant national committees addressed themselves to individual scientists and engineers, and there is, therfore, no conflict between our fields of work. Representatives of the international scientific community attend many of our meetings and, in return we support such symposia, such as that on world water balance in Reading, and this one, here, on mathematical models. Much of what I have heard here this week is of direct relevance to the work that WMO is undertaking, and it has been valuable to have had this opportunity of learning the views of the participants on various methods. We shall certainly make every endeavour to take account of these views in planning our future work. I am, personally, in the unusual and advantageous position of being able to view this Symposium both from the standpoint of its organization, and as an individual participant and author. From both respects, I have been most impressed by the organization of the meeting. The problems associated with such organizations are immense. However, there can be no doubt, I think, that this Symposium has been a great success, and I should like to echo the comments of my colleague from UNESCO in congratulating the Organizing Committee, and, in particular, Professor Kaczmarek and Dr. Kindler, for the magnificent job that they have done. It has been a pleasure to be associated with this Symposium and, on behalf of WMO, I should like to wish the Committee on Mathematical Models in Hydrology and IASH in general every success in their deliberations in Moscow next week. Thank you, Mr. Chairman.

Closing Speech by Professor J. C. I. Dooge

The final speech this morning falls to me, and the first thing that I would like to do is to read to you the text of the telegram that was sent on your behalf to the General Secretary of IASH, Professor Tison. The telegram was sent in French, but I will read the English text.

> Participants in the international symposium on mathematical models in hydrology, assembled in Warsaw, express deep regret at the absence, due to illness, of Professor Tison, their wise counsellor and kind friend, and express their hope that his recovery will be continuous, rapid and complete.

I hasten to assure all those of you who are already feeling the heat that I have no intention of attempting to summarize this Symposium. Even were I able to do so, I imagine that I would find myself addressing an audience which would show an exponental decline, or perhaps even a faster one than that.

We all have our own reactions, our own stimulation from this particular meeting. I think I speak for everyone when I say that this has been a remarkable gathering. We all have met old friends, sometimes discussed old problems with them once again, sometimes debated new problems. More important, we have met new friends, friends we have only known through correspondence or from their publications. In such informal talks, as much as in the actual presentation and discussion of the papers, this Symposium has been a notable success.

On the opening day I mentioned that this Symposium – due to the cooperation of the Polish Organizing Committee – became the culmination of the four-year work of the Committee on Mathematical Models of IASH. But it did much more than this and, in my opinion, it has marked a decisive stage in this particular approach to hydrology. I think that this Symposium will have an effect, far beyond the work of the present Committee.

As I said, we all have our reactions to this Symposium and I would like to spend a few minutes giving some of my own personal reactions to the Symposium. I was particularly interested with the extent, to which the ideas from the different groups had come to interpenetrate one another. The mixing of ideas between the various groups of research workers, which we always hoped for, is, I think, becoming more and more of a reality.

For example, it has always been the concern of those in the water resources group of our Committee and engaged in this particular field that objectives be clearly recognized. They cannot set up their particular problems without a clear recognition of objectives. It is interesting to see this same concern about objectives becoming more apparent in the papers and in the discussions involved in the stochastic and the parametric fields. We had in the discussion of the structure of the hydrological sequences, an increasing realization that everything cannot be preserved, and that we must be clear about the objective of our exercise before we choose our particular techniques. Similarly, we had in the discussions on deterministic hydrology the emergence, I think for the first time, of direct discussion of the effect of the structure and degree of complexity of a model on its performance and their relationship to the objectives of the study.

We have had an increasing concern with the problem of parameter optimization and a realization, I think throughout all the fields we discussed this week, that the determination of parameters which are not merely hypothetical, but the determination of parameters that would be useful for further reliable predictions is a far more difficult problem than we had thought some years ago. These, as I say, are impressions which come to me from some of our sessions.

Also we have had concern, it was mentioned several times and again during the last session this morning by the general reporter, with the fact that we have concentrated, perhaps a little too narrowly, on the classical models of rainfall and runoff. Of course we still have problems to solve in regard to these particular models, but many of the techniques that have been developed could be adapted without too much difficulty to the wider variety of models, those concerned with water quality, with sediment transport and with the environment generally. After this Symposium, perhaps some of our minds will turn in this direction.

Another question which emerged from discussion and tended to recur throughout the week was the problem of the proliferation of models, the inflation of models as my Vice-Chairman, Professor Kalinin, described it earlier in the week. We had the growing realization among the scientists discussing the problem here of the need for the objective comparison of models, and for the validation of techniques for the comparison of models.

These seem to me a number of themes which, as in a work of musical composition, tended to recur but were not always orchestrated in the same way and not always dominating in the same way, but, nevertheless, themes which tended to recur throughout the whole symphony of this Symposium. As I say, this is one participant's reaction. I am quite sure that others would react differently. Indeed, if they did not, there might be no room for further discussions.

Having given these, my personal reactions, all that remains for me is to give, on your behalf, the final word of thanks to those responsible for this Symposium. Our very heartfelt thanks are due to the Polish Organizing Committee, to their Chairman, Professor Kaczmarek, to their Secretary, Dr. Kindler, to all those others who worked so hard, and to the staff who assisted them throughout this week. I would ask the participants here to express their appreciation, not just of those mentioned but of everybody in the organization who has helped to make this Symposium, not only scientifically a success, but also a pleasurable occasion.

I would like to express thanks on your behalf to the UN agencies who have supported this particular symposium, to UNESCO and to WMO. I would also like to express thanks and make clear the indebtedness of this Symposium to the officers of IASH, in particular to Dr. Szestay, President of IASH, who fostered this Symposium from the beginning, and who was helpful at all its stages, to Professor Tison, who was equally helpful, until last December when he was stricken by illness, and to Dr Kovacs, who at short notice picked up the threads on behalf of IASH, and without whom things would not have gone as smoothly as they did.

Finally, I do not know on whose behalf I make these thanks, possibly on behalf of the Committee of which I am Chairman, but I want to thank the participants themselves. They indeed were the only people who could, during the week, have finally determined the success or failure of this Symposium. I would like to express appreciation at the way the authors presented their papers, the general reporters gave their views, and the discussers picked up the various points. We will, when we receive the proceedings, be able to go further along the lines that have been raised. In many instances, probably in all instances, what we have had this week is not a final discussion, not the reaching of conclusions, but merely the opening of a discussion which will carry on.

Some of us go to Moscow for the General Assembly of IUGG. One of the symposia at Moscow, of which Professor Kalinin and I are scientific convenors, is concerned with mathematical models in geophysics, in which we will be talking to oceanographers and to meteorologists concerning common problems. We go to this symposium very much strengthened by what we have heard this week. We go fresh with the views of more than 200 hydrologists as to our own particular standing at the moment in this regard. Others of us will meet soon, we hope, as working partners on the Committee on Mathematical Models. We will meet, perhaps, at other symposia of IASH, but whether we meet next week, or whether do do not meet for a few years, I think, we will still have something from these discussions to talk about when we meet again.

We will have the opportunity to make our personal goodbyes to one another before we leave this building. It only remains to me to make the formal goodbye, to declare the Symposium closed, and to say to you all 'goodbye, au revoir, do svidaniya'.

List of participants — Liste des participants

Adams, William M.: University of Hawaii, Honolulu, USA.
Alexeev, G. A.: State Hydrological Institute, Leningrad, USSR.
Amorocho, J.: Department of Water Science and Engineering, University of California, Davis, California 95616, USA.
Andersen, Lars J.: Geological Survey of Denmark, Thoravej 31, DK-2400, Copenhagen NV, Denmark.
Anderson, Henry W.: Forest Experimental Station, PO Box 245, Berkeley, California 94701, USA.
Andrew, Vivian A.: World Meteorological Organization, Geneva, Switzerland.
Appleby, Frederick V.: Gatehouse Farm, Stebbing, Dunmow, Essex, UK.
Arab, Chawkat: Abon Samra, Tripoli, Lebanon.
Arnold, K. H.: Wasserwitschaftsdirektion Dresden, Büro für Forschung und Entwicklung, Julian-Grimau-Allee 23, Dresden, GDR.
Askew, A. J.: Hydrometeorology Division, World Meteorological Organization, Geneva, Switzerland.
Barnes, Ivan K.: 345 Middlefield Road, Menlo Park, California, USA.
Beard, Leo R.: The Hydrologic Engineering Center, US Army Corps of Engineers, 609 Second Street, Davis, California 95616, USA.
Becker, A.: Institut für Wasserwitschaft, Schnellerstrasse 140, 119 Berlin, GDR.
Bernier, J.: Service des Etudes et Recherches Nucleaires Thermiques et Hydrauliques, 6 Quai Watier, 78400 Chatou, France.
Biedrzycki, Marek: Gdańsk Polytechnic University, ul. Majakowskiego 11, Gdańsk, Poland.
Biernacki, Tomasz: Gdańsk Polytechnic University, ul. Majakowskiego 11, Gdańsk, Poland.
Bissell, Vernon C.: Research Hydrologist, 8060-13th Street, W 232 Silver Spring, Maryland, USA.
Bloomer, R. J. G.: Water Resources Board, Reading Bridge House, Reading, Berkshire, UK.
Bobée, Bernard: CEQUEAU INRS, Université du Québec, 2050 Bird. St-Cyrille, Québec 10, Qué., Canada.
Bobiński, Eryk: National Institute for Hydrology and Meteorology (PIHM), ul. Podleśna 61, Warsaw, Poland.
Bogárdi, István: Alkotmány utca 29, Budapest V, Hungary.
Bogdan, Jon: Meteorological and Hydrological Institute, Sos Bucuresti Ploiesti 97, Bucharest,
Bonnet, Marc: Ingenieur ENSG, EIH BRGM, BP 6009, 45 Orleans, France.
Borovikova, Ludmila N.: 85/5 Observatorskaya St., Tashkent 52, USSR.
Bouwknegt, J.: Technical Engineer, c/o KNHM, Postbus 33, Arnheim, Holland.
Brajković, Milivoje: Ustanicka 178, Belgrade, Yugoslavia.
Byczkowski, Andrzej: Warsaw Agricultural University, Irrigation and Drainage Department ul Nowoursynowska 166, Warsaw, Poland.
Cabart, Jan: Výzkumný ústav Melioraci Zbraslav, Czechoslovakia.
Cavadias, G. S.: Faculty of Management, McGill University, Purvis Hall, 1020 Pine Ave. West, Montreal 2, Canada
Cetnarowicz, Maria: National Institute for Hydrology and Meteorology (PIHM), ul. Podleśna 61, Warsaw, Poland.
Chin, William Q.: Planning Section, Water Survey of Canada, Vancouver, BC, Canada.
Chow, V. T.: Hydrosystems Laboratory, University of Illinois, Urbana, Illinois 61801, USA.
Ciepielewski, Andrzej: Warsaw Agricural University, ul. Nowursynowska 166, Warsaw, Poland.
Clarke, R. T.: Institute of Hydrology, Howbery Park, Wallingford, Berkshire, UK.
Cluis, Daniel: CEQUEAU INRS, 2050 Bird. St-Cyrille, Québec 10, Qué., Canada
Cole, J. A.: The Water Research Association, Ferry Lane, Medmenham, Marlow, Buckinghamshire, UK.
Cuena, J.: Gabinete de Calculo Ministerio de Obras Publicas (Nuevos Ministerios), Madrid, Spain.
Cunnane, Conleth: Institute of Hydrology, Howbery Park, Wallingford, Berkshire. UK.
Curran, Charles D.: UNDP Vistula River Project, 20 Emilii Plater Street, Warsaw, Poland.
Czamara, Wlodzimierz: Agricultural University, ul. Bartla 6, Wroclaw, Poland.
Czetwertyński, Edward: Warsaw Polytechnic University, Pl. Jednosci Robotniczej 1, Warsaw, Poland.
Dahms, P.: Universität Rostock, Satower Str., 25 Rostock, GDR.
Dawdy, David R.: US Geological Survey, Engineering Research Center, Colorado State University, Fort Collins, Colorado 80521, USA.
Das, M. D.: UNDP Vistula River Project, 20 Emilii Plater Street Warsaw, Poland.

de Marsily, Ghislain: Ecole des Mines de Paris, 35 rue Saint-Honoré, 77 Fontainbleau, France.

Delleur, Jacques S.: School of Civil Engineering, Purdue University, Lafayette, Indiana 47907, USA.

Déri, József: Kerékgyártó u. 97, Budapest XVI, Hungary.

Derix, L.: Inland Waters Branch, Department of the Environment, Ottawa, Canada.

Diaz Arenas, A.: Grupo Hidraulico-DAP, Humboldt 196, Vedado-Habana, Cuba.

Domokos, Miklós: Herékgyártó u. 97, Budapest XIV, Hungary.

Dooge, J. C. I.: Department of Civil Engineering, University College, Merrion Street, Dublin 2, Ireland.

Druzhinin, I. P.: Siberian Power Institute, Lermontov Street 130, Irkutsk 33, USSR.

Duckstein, Lucien: 2940 E. Hawthorne, Tucson, Arizona 85716, USA.

Dürbaum, Hans-Jürgen; 3001 Neuwarmbüchen Waldstr. 8, GDR.

Dubicki, Alfred: National Institute for Hydrology and Meteorology (PIHM), ul Norwida 34, Wroclaw, Poland

Dyck, S.: Technical Universitat Dresden, Mommenstrasse 13, 8027 Dresden, GDR.

Ehlert, Kurt: Swedish Meteorological and Hydrological Institute, Box 12 108, 10223 Stockholm 12, Sweden.

Elgizawa, M. G.: Nile Delta Authority for Nile Drainage Projects, PO Box 1694, Cairo, UAR.

Fal, Barbara: National Institute for Hydrology and Meteorology (PIHM), ul Podlesna 61, Warsaw, Poland.

Ferral, R. Larry: National Weather Service, River Forecast Center, 1416 9th St., Rm 1641, Sacramento, California 95814, USA.

Filimowski, Janusz: Water Economy Research Institute (IGW), Cracow, Poland

Filipkowski, Andrzej: Water Economy Research Institute (IGW), ul. Partyzantów 6, Warsaw, Poland

Filipkowska, Alina: Water Economy Research Institute (IGW), ul. Partyzantow 6, Warsaw, Poland.

Finke, Werner: Wdringerstr. 24, 3018 Magdeburg, GDR.

Forsman, Arne: Swedish Meteorological and Hydrological Institute, Box 12 108, 10223 Stockholm 12, Sweden.

Fortin, Jean-Pierre: CEQUEAU INRS, 2050 Bird. St-Cyrille, Quebec 10, Qué., Canada.

Friedlaender, Michael: CNRS-GRI, 45 Orleans La Source, France.

Gabriel, Pavel: Katedra Hydrotechniky CVUT Smetanovo nábr. 6 Praha 1 – Staré Mesto, Czechoslovakia.

Garcia, Luis Lopez: Servicio Geologico de Obras Publicas (MOP), Avda. Portugal 81, Madrid 11, Spain.

Gazovic, Frantisek: VUUH Bratislavia, Karloveská 9, Bratislava, Czechoslovakia.

Geringer, Józef: Gdańsk Polytechnic University, ul. Majakowskiego 11, Gdansk, Poland.

Girard, Georges: ORSTOM, 19 rue Eugène Carrière, Paris XVIII, France.

Guiliano, Giuseppe: Institute for Water Research, National Research Council, 1 Via Reno, 00198 Rome, Italy.

Golek, Julian: National Institute for Hydrology and Meteorology (PIHM), ul. Podlesna 61, Warsaw, Poland.

Gonzales, Lillian: Escuela Ingenieria Civil CUJAG, Universidad de la Habana, Cuba.

Gottschalk, Lars: Inst. Vattenbyggwad, Lth Fack 725, 22007 Lund, Sweden.

Gras, Raymond: EDF, 6 quai Watier, 78 Chatou, France.

Gruszewskij, M. S.: c/o USSR Committee for IHD, 12 Pavlik Morozow Street, Moscow D-376, USSR.

Guergov, Guergui: Institute for Hydrology and Meteorology, Blv. Lenin 66, Sofia 13, Bulgaria.

Guggino, Emanuele: Instituto di Idraulica Agraria, Università Via Valdisavoia 5, Catania, Italy.

Guillot, Pierre: EDF, DTG, 37 rue Diderot, 38 Grenoble, France.

Hall, Michael John: Imperial College, Imperial Institute Road, London SW7, UK.

Hamlin, M. J.: Department of Civil Engineering, University of Birmingham, PO Box 363, Birmingham BI5 2TT, UK.

Hansen, Eggert: Hydraulic Laboratory IOF, Østervoldgade, 1350 Copenhagen, Denmark.

Hastinge, Warren William: Water Resources Division, US Geological Survey, 345 Middlefield Road, Menlo Park, California 94025, USA.

Hartwig Laszlo: Alkotmany ut. 27, Budapest V, Hungary.

Heitmann, Marie-Luise: Pankow Miltenberger Weg 11, 110 Berlin, GDR.

Heras, Rafael J.: Po Bajo de la Virgen del Puerte 3, Madrid 5, Spain.

Herbst, Marcin: National Institute for Hydrology and Meteorology (PIHM), ul Podlesna 61, Warsaw, Poland.

Herrmann, Reimner: 63 Giessen-Kl., Linden Schledhorn 12, GFR.

Hendriks, Martinus Hubertus: De Drijen 8, Wageningen, The Netherlands.

Hetager, Sven-Erik: Rektorhavgen 31, Oslo 8, Norway.

Hofius, K.: Secretariat for IHD, Bad Godesberg Kennedyallee 40, GFR.

Hoorn Ten, Willem H. C.: Brittenburg 32, Leiderdorp, The Netherlands.

Indelicato, Salvatore: Instituto di Idraulica Agraria, Università Via Valdisavoia 5, Catania, Italy.
Jacoby, Samuel: Mathematical Analysis Unit, Boeing Computer Services Inc., PO Box 24346, Seattle, Washington 98124, USA.
Jelen, V.: Vysoka skola zemedelska Suchdol u Prahy, Czechoslovakia.
Jensen, Helger: VAW Hydrologie, Voltastrasse 24, 8044 Zurich, Switzerland.
Johnson, Philip: Civil Engineering Department, University of Newcastle-upon-Tyne, Newcastle-upon-Tyne, UK.
Johnson, Paul Meredith: Department of Civil Engineering Imperial College, London SW7, UK.
Jouhet, Pierre: SOGREAH, 84–86 Avenue Léon Blum, 38 Grenoble Gare, France.
Jovanović, Slavoljub: Kursulina 29, Belgrade, Yugoslavia.
Jurak, Danuta: National Institute for Hydrology and Meteorology (PIHM), ul. Podlesna 61, Warsaw. Poland.
Kaijser, Thomas: Mathematical Department, Royal Institute of Technology, 1004 Stockholm 70, Sweden.
Kaczmarek, Zdzislaw: Warsaw Polytechnic University, Institute of Environmental Engineering, Pl. Jednosci Robotniczej 1, Warsaw, Poland.
Kalinin, G. P.: Moscow University, Moscow D-376, USSR.
Kanzow Dietz: Institut für Wasserbau, Technische Universität München. Arcisstrasse 21, München 2, GFR.
Khomeriki, J. R.: Power Research Institute. Lenin Str. 70, Tbilisi 15, USSR.
Kicinski, Tadeusz: Warsaw Agricultural University, ul. Nowoursynowska 166, Warsaw, Poland.
Kindler, Janusz A.: Bureau of the UNDP/UN Plan of Operation 'Vistula', ul. Emilii Plater 20, Warsaw, Poland.
Kisiel, Chester C.: Hydrology and Water Resources Department, University of Arizona, Tucson. Arizona 85721, USA.
Kloss. Adam: Hydroprojekt, ul. Swietokrzyska 12, Warsaw, Poland.
Kluge, Christian: Kunzstrasse 4, 8023 Dresden, GDR.
Knenicky, Stanislav: Vodni zdroje Praha Makarenkova 32, Praha 2, Czechoslovakia.
Kontur, Istvan: Kazinczy 28, Budapest VII, Hungary.
Konovalenko, Z. P.: Siberian Power Institute, Lermontova Street, 130, Irkutsk 33, USSR.
Korzoun, W. I.: Glavgimet 12, Pavlik Morozow Street, Moscow, D-376, USSR.
Kolev. Nicolai: Water Problems Research Institute, N.36, Str. Bl. IV, Sofia 13, Bulgaria.
Kolago, Cyryl: Geological Institute, ul. Rakowiecka 4, Warsaw, Poland.
Kordas, Boleslaw: Cracow Polytechnic University, Cracow, Poland.
Kowalik, Piotr: Gdansk Polytechnik University, Hydrotechnical Institute, ul. Majakowskiego 11, Gdansk, Poland.
Koczorowska, Romana: National Institute for Hydrology and Meteorology (PIHM), ul. Dabrowskiego 174/176, Poznan, Poland.
Kottegoda, N. T.: Civil Engineering Department University of Birmingham, Birmingham B15 2TT, UK.
Kovacs, Gyorgy: Rakoczi ut. 41, Budapest VIII, Hungary.
Kraijenhoff van de Leur, D. A.: State Agricultural University, Duivendaal 1A, Wageningen, The Netherlands.
Krausneker, Peter: Gregor Mendel Strasse 33, 1180 Vienna, Austria.
Krebs, Roland. Alkotmany Street 27, Budapest V, Hungary.
Krukowicz, Andrzej: Hydroprojekt, ul. Swietokrzyska 12, Warsaw, Poland.
Krzeminski, Wojciech: Polish Academy of Sciences, National Committee for IUGG, PKiN, Room 2315, Warsaw, Poland.
Krajewski. Karol: Warsaw Polytechnic University, Institute of Environmental Engineering, Pl. Jednosci Robotniczej 1, Warsaw, Poland.
Kupczyk, Elzbieta: Warsaw University, ul. Krakowskie Przedmiescie, Warsaw, Poland.
Kuchment, L. S.: Hydrometeorological Research Center, 13 Bolshevistskaya, Moscow D-376, USSR.
Laski, Aleksander: Hydroprojekt, ul. ul. Swietokrzyska 12, Warsaw, Poland.
Ladyzynska, Alicja: Water Economy Research Institute, ul. Partyzantow 6, Warsaw, Poland.
Laszewski, Janusz. Av. 2C No.71B-06, Maracaibo, Venezuela.
Lorenc, Alicja: National Institute for Hydrology and Meteorology (PIHM), ul. Podlesna 61, Warsaw, Poland.
Lørum, Leif-Egil: NVE, Box 5091, Majorstua, Oslo 3, Norway.
Maczuga, Tomasz: Cracow Polytechnic University, Institute of Water Engineering, ul. Warszawska 24, Cracow, Poland.
Majewski, Wojciech: Polish Academy of Sciences, Institute of Hydraulic Engineering, ul. Majakowskiego 11, Gdansk, Poland.
Maniak. Ulrich: Leichtweiss-Institut, Technische Universitat Braunschweig, GFR.
Matalas, Nicholas C.: US Geological Survey Washington DC, USA.

Maul, Christian: Wasserwitschaftsdirektion Mittlere Elbe-Sude-Elde, Domplatz 819, 301 Madgeburg, GDR.

Meier, Rupert C.: Institut fur Wasserbau, Technische Universitat München, Arcisstr. 21, München 2, GFR.

Mendel, Hermann Gregor: 54 Koblenz, GFR.

Meyer, Wlodzimierz: National Institute for Hydrology and Meteorology (PIHM), ul. Podlesna 61 Warsaw, Poland.

Michajlova, G.: c/o USSR Committee for IHD, 12 Pavlik Morozow Street, Moscow D-376, USSR.

Mikulski Zdzislaw: National Institute for Hydrology and Meteorology (PIHM), ul. Podlesna 61, Warsaw, Poland.

Mitosek, Henryk: Warsaw Polytechnic University, Institute of Environmental Engineering, Pl. Jednosci Robotniczej 1, Warsaw, Poland.

Mohler, I.: Water Resources Research Institute, Karloveska 9, Bratislava, Czechoslovakia.

Morin, Guy: CEQUEAU INRS, 2050 Blrd. St-Cyrille, Québec 10, Qué., Canada.

Morris, William James: Civil Engineering Department, The City University, St. John Street, London, UK.

Mudallal, Usama: Water Resources Division, Natural Resources Authority, PO Box (7) (39) (2220), Amman, Jordan.

Muzsnay, Géza: Alkotmany u.27, Budapest V, Hungary.

Myczka, Jacek: Agricultural University, ul. Miciewicza 24–28 Cracow, Poland.

Nachazel, Karel: Smetanovo na br. 6, Praha 1-Stare Mesto, Czechoslovakia.

Nawrocka, Elżbieta: National Institute for Hydrology and Meteorology (PIHM), ul. Podlesna 61, Warsaw, Poland.

Nerhevlidze, J.: c/o USSR Committee for IHD, 12 Pavlik Morozov Street, Moscow D-376, USSR.

van de Nes, Theodorus J.: Diuvendaal 1a, Wageningen, The Netherlands.

Nieczajeva, N.: Bolshevistskaja 13, Moscow D-376, USSR.

Nikolova, Nadejda: Water Problem Research Institute, No.36, Str.Bloc IV, Sofia 13, Bulgaria.

Nilsson, Lars Yngye: Kulturteknik Kth Drottning Kristinas Vag 30, Stockholm 70, Sweden.

O'Connell, Enda Patrick: Department of Civil Engineering, Imperial College, London SW7, UK.

O'Connor, Kiernan M.: Engineering Department, University College, Galway, Ireland.

O'Donnell, T.: Department of Civil Engineering, Imperial College, South Kensington, London SW7, UK.

O'Kane, J. P. J.: An Foras Forbartha, St. Martins House, Waterloo Road, Dublin 4, Ireland.

Olejnik, Konrad: National Institute for Hydrology and Meteorology (PIHM), ul. Dabrowskiego, Poznan, Poland.

van Ouwerkerk, J. H.: c/o KNHM, PO Box 33, Arnhem, The Netherlands.

Overton, Donald E.: USDA-ARS-SWC, Plant Industry Station, Beltsville, Maryland 20705, USA.

Panfilova, W.: c/o USSR Committee for IHD, 12 Pavlik Morozow Street, Moscow D-376, USSR.

Parczewski Wladyslaw: Warsaw Polytechnic University, Institute of Environmental Engineering, Pl. Jednosci Robotniczej 1, Warsaw, Poland.

Parzonka, Wlodzimierz: Agricultural University, ul. Bartla 6, Wroclaw, Poland.

Paszov, L.: c/o USSR Committee for IHD, 12 Pavlik Morozow Street, Moscow D-376 USSR.

Pal, Karkus: VIKOZ, Pf. 3 5 2, Budapest V, Hungary.

Papazov, Radoy: Institute of Water Problems, Gagarin 18 St. bl.37, Sofia 13, Bulgaria.

Patera, Adolf: Smetanovo Nabr, 6 Praha 1 – Stare Mesto, Czechoslovakia.

Pazdro, Zdzislaw: Warsaw University, Department of Geology, ul. Krakowskie Przedm., Warsaw, Poland.

Petrlik, Jaroslav: Výzkumný ústav vodohospodarský Podbabská 30, Praha 6 – Podbaba, Czechoslovakia.

Pfafstetter, Otto: Av. Presidente Vargas 62, Rio de Janeiro GB, Brazil.

Pinder, George Francis: 10200 Addison Ct., Fairfax, Virginia 22030, USA.

Pinkayan, Subin: Asian Institute of Tehcnology, Henri Dunant Street, Bangkok, Thailand.

Plate, Erich J.: Institut für Wasserbau III an der Universität Karlsruhe 75, Kaiserstr. 12, Karlsruhe, GFR.

Poitrinal, D.: Ecole des Mines, Service d'Hydrogeologie Mathematique, rue Saint-Honoré, 77 Fontainebleau, France.

Punzet, Jerzy: National Institute for Hydrology and Meteorology (PIHM), ul. Borowego 9, Cracow, Poland.

Quimpo, R. G.: Department of Civil Engineering, University of Pittsburgh, Pittsburgh, Pennsylvania 15213, USA.

Qureshi, A. Saleem: Suite 2, 1309 Henryfarm Drive, Ottawa 5, Ontario, Canada.

Rank, Dieter: Bundesversuchs und Forschungsanstalt Arsenal Objekt 210, 1030 Vienna, Austria.

Rao, A. Ramachandra: School of Civil Engineering, Purdue University, Lafayette, Indiana, USA.

Reinicke, Horst: VEB Rationalisierung Braunkohle, 7805 Grossräschen-Süd, GDR.

Reznikovszky, A. Sh.: Department of Hydropower of the Institute, 'Energoproject', 2nd Baumanskaya str. No.7, Moscow B-5, USSR.

Roche, M.: ORSTOM-EDF, Proiesseur à l'ENGREF, 10 rue Eugene Carriere, Paris XVIII, France.
Rodríguez-Iturbe, Ignacio: Civil Engineering Department, MIT, Cambridge, Massachusets, USA. (PO Box 1827, Caracas, Venezuela).
Romero, Marcello: Velasquez 132-8°, Madrid, Spain.
Rosemann, Hans-Jürgen: 81 Stolzingstr. 21/II, 8000 München, GFR.
Rossi, Giuseppe: Instituto di Idraulica Agraria, Universita, Via Valisavoia 5, 95100 Catania, Italy.
Rossi, Giorgio: ENEL-CRIS, Croso del popolo 245, 30100 Mestre, Italy.
Rutkowski, Kieczyslaw: Hydroprojekt, ul. Swietokrzyska 12, Warsaw, Poland.
Sasim, Marianna: National Institute for Hydrology and Meteorology, ul. Podleśna 61, Warsaw, Poland.
Sahuquillo, Andrés: Servicio Geológico de Obras Públicas, Avenide de Portugal 81, Madrid 11, Spain.
Schreiber, Helmut: Hydrographisches Zentralbüro im Bundesministerium für Land und Forstwirtschaft, Maxergasse 2, 1030 Vienna, Austria.
Schultz, Gert A.: Institut für Wasserbau III an der Universität Karlsruhe, Kaiserstrasse 12, 75 Karlsruhe, GFR.
Suena, Pertti: Nahkela, Finland.
Siekanska-Trylska, Danuta: National Instutute of Hydrology and Meteorology (PIHM), ul. Borowego 9, Cracow, Poland.
Skibiński, Jan: Warsaw Agricultural University, ul. Nowoursynowska 166, Warsaw, Poland.
Sochorec, Rostislav: Hydrometeorologický ustaw Ostrava 8 Kmyslivne 1, USSR.
Soczyńska, Urszula: National Institute for Hydrology and Meteorology (PIHM), ul. Podleśna 61, Warsaw, Poland.
Soloviova, A.: Siberian Power Institute, Lermontova Street 130, Irkutsk 33, USSR.
Spaziani, Fausto Maria: Institute for Water Research, 1 Via Reno, 00198 Rome, Italy.
Srinivasan, S. K.: Department of Mathematics, Indian Institute of Technology, Madras 36, India.
Strupczewski, Witold: Warsaw Polytechnic University, Institute of Environmental Engineering, Pl. Jedności Robotniczej, Warsaw, Poland.
Stachy, Juliusz: National Institute for Hydrology and Meteorology (PIHM), ul. Podleśna 61, Warsaw, Poland.
Stanciu, Petre: Meteorological and Hydrological Institute, Sos. Bucuresti Ploiesti 97, Bucharest, Romania.
Stolarska, Anna: Water Economy Research Institute, ul. Partyzantów 6, Warsaw, Poland.
Svanidze, G. G.: Tbilisi State University, I. Chavchavadze Av. 1, Tbilisi 28, USSR.
Svoboda, Ales: Institute of Hydrology and Hydraulics, Slovak Academy of Sciences, Trnarská 20, Bratislava, Czechoslovakia.
Szpindor. Adam: Agricultural University, ul. Bartla 6, Wroclaw, Poland.
Thunvık, Roger: Department of Land Improvement and Drainage, Royal Institute of Technology, Stockholm, Sweden.
Todsen, Marius: Institute of Geophysics, University of Oslo, Oslo 3, Norway.
Toebes, C.: National Water and Soil Conservation Organisation, Ministry of Works, PO Box 12-C41, Wellington North, New Zealand.
Tomik, Teresa: National Institute for Hydrology and Meteorology (PIHM), ul. Podleśna 61, Warsaw, Poland.
Trau, Wolfgang H. O.: Leichtweiss-Institut, Pockelsstr. 4, 33 Brunswick, GFR.
Treunert, Eckhart: Landesanstalt für Gew ässserkunde und Gewässerschutz, Steinstrasse 137, 514 Krefeld, GFR.
Trippler, Klaus: Schiffhornfeld 27, 3 Hannover, GFR.
Tietze, Klaus D.: Hamunstrasse 2, 3 Hannover-Bucholz, GFR.
Twardowska, Irena: Polish Academy of Sciences, Environmental Protection of Industrial Regions, Zabrze ul. Hagera 17, Poland.
Utrysko, Bohdan: ISTW, Warsaw Polytechnic University Pl. Jedności Robotniczej 1, Warsaw, Poland.
Vinogradov, J. B.: Kazakh Hydrometeorological Research Institute, Alma-Ata 72, USSR.
Verhoog, Frederik H.: Office of Hydrology UNESCO, Place de Fontenoy, Paris VII, France.
Virta, J.: Institute for Geophysics, Vironicato 7, Helsinki 17, Finland.
Visser, J. H.: c/o UNDP, PO Box 3216, Beirut, Lebanon.
Visser, Willem C.: Foulkesweg 85, Wageningen, The Netherlands.
Wallis, James R.: IBM Watson Research Center, Yorktown Heights, New York, USA.
Waszak-Haladyj, Marianna: National Institute for Hydrology and Meteorology (PIHM), ul. Podleśna 61, Warsaw, Poland.
Winter, Janos: KISZ Lokotelep C/5.IV.3, Budapest XVIII, Hungary.
Witkowska, Hanna: Cracow Polytechnic University, Cracow, Poland.
Woloszyn, Julian: Agricultural University, ul. Bartla 6, Wroclaw, Poland.
Wyatt, Timothy: The Water Research Association, Ferry Lane, Medmenham, Marlow, Buckinghamshire, UK.

Wicher, Malgorzata: Warsaw Agricultrual University, ul. Nowoursynowska 166, Warsaw, Poland
Yevjevich V. M.: Department of Civil Engineering, Colorado State University, Fort Collins, Colo-
    rado 80521, USA.
Zajbert, Mieczyslaw: Water Economy Research Institute. ul. Partyzantow 6, Warsaw, Poland.
Zawidzka, Joanna: Polish Academy of Sciences Organizing Committee, PKiN, Room 2315, Poland.
Zevetanov, I. P.: Bulgarian Academy of Sciences, Institute of Engineering, Cibernetics bl. IV, Sifia-
    13, Bulgaria.
Zdanowski, Roman: Hydroprojekt, ul. Swietokrzyska 12, Warsaw, Poland.
Zelaziński, Janusz: National Institute for Hydrology and Meteorology (PIHM), ul. Podleśna 61,
    Warsaw, Poland.
Zeleny, Václav: Vyzkumná stanice Hnojnik u Ces. Tesina, Czechoslovakia.
Zezulák, Jiri: Water Resources Department, Prague Agricultural College, Prague 6 – Suchdol,
    Czechoslovakia.
Zmijewa, E.: c/o USSR Committee for IHD, 12 Pavlik Morozov Street, Moscow D-376, USSR.
Zsuffa, Istvàn: Szechenyi utca 2, Baja, Hungary.
Zvonimir, Haman: Geological Survey of Denmark, Thoravej 31, 2400 Copenhan NV, Denmark.

Unesco series *Studies and reports in hydrology*
Collection Unesco *Études et rapports d'hydrologie*

* Quadrilingual publication: English–French–Spanish–Russian.
  Publication quadrilingue: anglais–français–espagnol–russe.